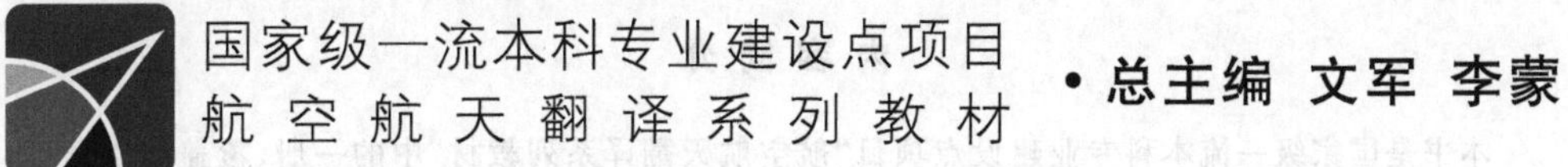

航空航天口译教程

Interpretation Course for Aeronautics and Astronautics

主 编 郑 薇

副主编 频 超 王 沛

北京航空航天大学出版社

内容简介

本书是国家级一流本科专业建设点项目“航空航天翻译系列教材”中的一册，覆盖航空管理、航空安全、航空减排、商用飞机、载人航天、对地观测、探月工程、北斗导航、航天科普等多领域知识。

本书适用于英语专业和翻译专业本科生以及英语水平相当的翻译爱好者学习交替传译技巧，了解航空航天领域的相关背景知识，提升相关技术领域的口译实践能力，为航空航天领域国际交流与合作培养人才。

图书在版编目(CIP)数据

航空航天口译教程 / 郑薇主编. -- 北京 : 北京航空航天大学出版社，2024.8

ISBN 978-7-5124-4408-9

Ⅰ. ①航… Ⅱ. ①郑… Ⅲ. ①航空工程—英语—口译—教材 ②航天工程—英语—口译—教材 Ⅳ. ①V

中国国家版本馆 CIP 数据核字(2024)第 094705 号

航空航天口译教程

主编 郑 薇

副主编 顿 超 王 沛

策划编辑 赵延永 蔡 喆

责任编辑 张 凌

*

北京航空航天大学出版社出版发行

北京市海淀区学院路 37 号(邮编 100191) http://www.buaapress.com.cn

发行部电话：(010)82317024 传真：(010)82328026

读者信箱：goodtextbook@126.com 邮购电话：(010)82316936

北京建宏印刷有限公司印装 各地书店经销

*

开本：710×1 000 1/16 印张：21 字数：448 千字

2024 年 11 月第 1 版 2024 年 11 月第 1 次印刷

ISBN 978-7-5124-4408-9 定价：69.00 元

若本书有倒页、脱页、缺页等印装质量问题，请与本社发行部联系调换。联系电话：(010)82317024

编 委 会

总 主 编 文 军 李 蒙

编 委 （按姓氏笔画排序）

孔 杰 朱殿勇 许明武 李 雪

李京廉 张化丽 张艳玲 范祥涛

赵雪琴 秦明利 梁茂成 彭 静

丛书策划 赵延永 蔡 喆

本书主编 郑 薇

本书副主编 顿 超 王 沛

本书编委 胡晓剑 李占伟 洪旻昊 邱颖茵

孙源泽 宗子昕 郭昕悦 张安琪

赫玮涵

总序 Foreword

科学技术的发展离不开交流与合作，航空航天的发展也不例外。在中国航空航天发展史上，这种交流与合作在很大程度上与翻译相关，概括起来，大致经历了两个阶段：早期的翻译引进，后期的翻译引进与翻译输出并举。最早与航空航天有关的翻译引进活动始于 1903 年到 1907 年期间中国掀起的“凡尔纳热”，其中的航空航天科幻小说翻译包括鲁迅的《月界旅行》、商务印书馆出版的《环游月球》以及谢祺的《飞行记》等。1910 年，高鲁翻译出版了《空中航行术》，这是中国航空航天科技书籍和资料汉译的开端。而在译出方面，随着我国航空航天事业的飞速发展，近些年的科技新闻、政府白皮书等都有大量航空航天方面的信息对外发布，及时而系统地向全世界展现了中国在此领域的发展现状和巨大成就。

总体而言，航空航天的领域宽广，翻译多种多样。从翻译的主题看，航空航天话语以科技语言为主，其一般特点有七个方面：无人称、语气正式、陈述客观准确、语言规范、文体质朴、逻辑性强和专业术语性强①，与之相关的科技论文等的翻译是航空航天翻译的主体。此外，航空航天话语中还包括与商务活动相关的商贸翻译（如合同、谈判等）、与航空航天新闻活动相关的新闻翻译（如新闻发布会、各种媒体的相关新闻报道等）、与航空航天文学相关的文学翻译（如航空类小说、航天类科幻小说等）、与航空航天影视活动相关的影视翻译（如纪录片、科幻电影）等。从翻译活动的方式看，航空航天翻译包括了笔译、视译、交替传译、同声传译、机器翻译＋译后编辑等几乎所有的翻译方式。

航空航天翻译主题和体裁的多样性及翻译方式的全面性，对翻译人才的培养提出了新的、更高的要求。为此，我们特设计和编写了这套“航空航天翻译系列教材”，其特色主要体现在以下几个方面：

① “入主流”与“显特色”并举。“入主流”主要指各种教材的设计都体现了翻译这一核心要素，其内容选择都以“怎么翻译”为焦点；“显特色”则体现在教材内容的选

---

① 冯志杰. 汉英科技翻译指要[M]. 北京：中国对外翻译出版公司，1998：6－7.

择上，无论是例句还是练习，都选择了与航空航天密切相关的语料，力求解决航空航天翻译中的实际问题。

② 理论与实践并重。在教材设计上，突显理论融于实践的理念，对理论不做大篇幅的阐释，而将翻译策略、翻译方法等融于对例句和语篇的讲解之中，而这些例句和语篇都选自真实的航空航天语料，以着力提升学生的翻译实践能力。

③ 阐释与练习并立。对各种翻译现象的解释与阐释在教材中必不可少，是教材的主干；与此同时，各教材采用按“节”的方式设置练习，其用意在于着力加强练习与教材正文的关联性，以方便学生的学习和操练。

本系列教材可以作为翻译专业、英语专业和大学英语相关课程的课堂教学材料，也可供对航空航天翻译感兴趣的读者使用。

迄今，本系列教材中已规划了英汉翻译、汉英翻译、口译、视听翻译等方面的教材。今后，我们还可增加与航空航天翻译相关的品种，如航空航天文学翻译、航空航天新闻翻译、商贸翻译、航空航天同声传译等方面的教材。

为使本系列教材的编写更具广泛性和权威性，我们组建了高水平的编委会。编委会委员有北京航空航天大学的文军、李蒙、梁茂成，北京理工大学的李京廉，重庆大学的彭静，大连理工大学的秦明利，哈尔滨工业大学的李雪，哈尔滨工程大学的朱殿勇，华中科技大学的许明武，南京航空航天大学的范祥涛，南京理工大学的赵雪琴，西北工业大学的孔杰，西安航空学院的张化丽和中国民航大学的张艳玲等专家学者。

本系列教材的编写是一种尝试，希望能得到业内专家学者、学生和其他读者的反馈和意见，以使教材更臻完善。

文军　李蒙

2023 年 3 月于北京

前言 Preface

本教材是国家级一流本科专业建设点项目“航空航天翻译系列教材”中的一册，体现了北京航空航天大学外国语学院多年的口译教学理念与教学方法。其宗旨是帮助英语专业和翻译专业本科生以及英语水平相当的翻译爱好者学习交替传译技巧，了解航空航天领域的相关背景知识，提升相关技术领域的口译实践能力，为航空航天领域国际交流与合作培养人才。

教材特点：

1. 话题全面性。本教材选择的语篇涉及领域广泛，例如：航空管理、航空安全、航空减排、商用飞机、载人航天、对地观测、探月工程、北斗导航、航天科普等。

2. 材料多样性。口译训练材料涉及多场景、多类别，既有新闻发布会的主持介绍，也有国际会议上的正式发言；既有专业研讨会上的学术报告，也有行业论坛上的前景展望；既有机构介绍，也有新闻动态。多样的训练材料能帮助学生适应口译实践中的诸多场景和源语类型。

3. 内容系统性。本教材系统讲授交替传译的技巧和训练方法，涉及记忆、笔记、数字口译、语言转换、跨文化交流、译前准备和现场应急情况处理等方面，不仅可夯实口译基础，也贴近口译实践，让学生由浅入深地了解口译学习和实践的全流程。

教材结构：

本教材共有八章，每一章有两节，第一节为口译技能介绍，通过分类别、举例子、作分析的方法帮助学生理解口译技巧。第二节为“篇章讲解与练习”，包括“重点解析篇”和“实战训练篇”。“重点解析篇”通过“词汇表”和“背景知识”帮助学生做好课前热身，课上老师重点训练、讲解这部分篇章，在“原文”之外还提供“要点讲解”和“参考译文”，以帮助学生了解译员是如何梳理源语逻辑、准确输出源语意思的。“实战训练篇”包含大量贴近实践的口译语篇，可作为学生课上补充训练和课后自学的材料。本教材中所有语篇原文都配有录音，读者扫描二维码即可收听。

郑薇负责全书的整体设计、个别章节的撰写和全书的统稿；顿超和王沛负责具体章节的撰写和校对。参与编写的同学有胡晓剑、李占伟、洪旻昊、邱颖茵、孙源泽、宗子昕、郭昕悦、张安琪和赫玮涵，在此由衷表示感谢。

本教材内容广泛，部分语篇和译例选自外交部、国家航天局、国际民航组织、国际宇航联合会、地球观测组织、美国国家航空航天局、美国联邦航空管理局、欧洲航天局、欧洲导航卫星系统管理局、中国日报、凤凰网等网站，在此特别表示感谢。

由于时间仓促，书中难免有错误和不当之处，欢迎批评指正。

编　者

2024 年 6 月

目　　录

第一章　口译绪论

随着人类社会的发展和全球化进程的加快，口译在国际交往中发挥着不可替代的作用。然而，口译活动并非新生事物。陈福康在《中国译学理论史稿》中指出，早在周朝，我国就已经有“重译”的活动[①]。古代丝绸之路、商旅往来、外事交流、宗教传教等活动，都需要口译人员进行语言转换和文化传递。这些早期的实践积累为口译行业的萌芽与发展奠定了基础。在现代社会，口译变得更加专业化和系统化，在国际政治、经济、科技和文化交流中发挥着重要的作用。国际会议、外交活动、商务洽谈、学术交流等，都需要译员通过实时的口译活动帮助参与者相互理解、有效沟通。

第一节　口译概述

首先，口译是一个语言转换的过程。译员需要倾听源语，理解其含义和语境，并运用目标语准确输出，传达信息。其次，口译是一种跨文化交流的工具。仲伟合指出，口译是人类交流思想所借助的重要手段之一[②]。译员需要了解源语文化和目标语文化之间的差异，以便在口译过程中规避文化误解和翻译偏差。因此，口译还涉及人的思维和心理过程。此外，口译的发展受到科技进步的影响。随着计算机辅助口译（computer-assisted interpreting）和技术辅助口译（technology-assisted interpreting）等口译技术的蓬勃发展，译员需要适应新兴的工具、平台和技术，以提高口译的质量和效率。总而言之，口译通过实现跨文化沟通促进全球交流与合作，为不同语言、文化和民族背景下的人们提供了沟通交流的桥梁。

一、口译的类型

根据场合和情境的不同，口译可有多种分类。从传送方式、场合和内容以及传译方向等角度来看，口译可以分为以下几类：

（一）按照传送方式分

① 同声传译（simultaneous interpreting）：同声传译是指在讲话进行的同时进行口译的形式。译员通过专业设备（例如耳机和话筒）实时收听源语，并立即将之转换

---

① 陈福康.中国译学理论史稿[M].上海：上海外语教育出版社，2000：2.

② 仲伟合.口译理论概谈[J].语言与翻译，1993(4)：20-23，32.

为目标语的口头陈述。同声传译常用于大型国际会议，具有节省时间、保证交流连贯的特点。

② 交替传译(consecutive interpreting)：交替传译是指在讲话完成后进行口译的形式。讲话者和译员交替进行：讲话者将一段内容表达完毕后停顿，然后由译员进行翻译。交替传译常用于小型会议、学术研讨和个别对话等场合。交替传译工作场景下，译员有充分的时间进行信息整理和记录，可以有条不紊地转译源语意思。由于交替传译在实际工作中使用最为广泛，本教材重点介绍其技巧和训练方法。

③ 耳语式传译(whispered interpreting)：耳语式传译是指在不借助同传设备的情况下，译员通过耳语将源语内容转译给少数几位听众的口译形式。这种形式通常用于目标语听众少的情况，不必搭建同传间，具有成本集约、移动灵活的特点，既适用于国际会议的固定工作场景，也适用于参观访问的移动工作场景；缺点是为了不干扰现场其他人员聆听源语，译员传译声音较小，有可能影响耳语服务对象接收信息。

（二）按照场合和内容分

① 会议口译(conference interpreting)：会议口译是指在国际会议、研讨会和座谈会等活动中进行的口译。译员需要在会议进行过程中实时转译与会人员的发言，以确保与会者能够相互理解，参与讨论。

② 陪同口译(escort interpreting)：陪同口译是指在个别或小型团体交流中的口译服务。译员陪同特定人员参加文化交流、参观访问和外出旅游等活动，并提供实时的语言转译服务，以确保跨文化场景中的顺畅交流。

③ 法庭口译(court interpreting)：法庭口译是在司法系统中进行的口译服务。译员在法庭审判、调解和律师辩论等过程中，将涉案方的口头陈述、证词和法律解释转译为另一种语言，以确保相关各方能够理解和参与司法程序。

④ 商务口译(business interpreting)：商务口译是在商务领域中进行的口译服务。译员在商务会谈、商业谈判和合作洽谈等场合中，实时转译与商务活动相关的内容，促进商业合作与交流。商务口译可能涉及商业机密，因此，译员还需确保此类信息不流入外界。

（三）按照传译方向分

① 单向口译(one-way interpreting)：单向口译是指将一种语言转换为另一种语言的口译形式。译员将源语言转译为目标语言，通常用于场合中只有一种源语的情况，或者与会各方译员负责各自代表团成员讲话的译出，例如双边会见中，中方译员只负责汉译英，对方译员只负责英译汉。

② 双向口译(two-way interpreting)：双向口译是指在口译过程中进行双向转译的形式，例如译员在某些场合既需要英译汉，又需要汉译英。

以上分类可以帮助我们理解口译的多样性和复杂性。不同类型的口译在不同的场合和需求中发挥着重要的作用，为跨文化交流和理解提供了宝贵的语言中介服务。

二、口译的特点

仲伟合认为，相较于笔译，口译的难度更大，因其受到时间的限制和讲话人的支配[①]。可以说，口译之所以难度较大，与其以下特点密不可分。

（一）即时性

不同于能够事先准备和事后润色修改的写作或笔译工作，口译是一种即时双语交流活动。译员需要在现场迅速反应并进行双语转译，因而要求译员具备高度的灵活性和适应能力，能够迅速理解和转译各类主题和语言风格。

（二）个体性

口译是一项个体性操作。虽然在口译的准备过程中，译员可以得到外界的帮助，但口译活动一旦开始，译员就需要独立地完成“听、记、译”的过程，不能时时处处指望他人的帮助。因此，译员责任重大，需要有过硬的心理素质和强大的抗压能力。

（三）综合性

口译要求译员综合运用各种语言技能，包括视觉能力（观察和解读非语言表达）、听觉能力（理解和捕捉语音信息）、口头表达能力（流利输出目标语）、书写能力（记录重要信息）以及阅读能力（阅读相关文档）。译员需要在短时间内充分发挥这些技能，以实现准确的信息传达。

（四）广泛性

口译涉及的主题和内容非常广泛，包罗万象。译员需要涉猎多个领域的知识，例如政治、经济、科技、社会、文化等。译员需要对不同领域的专业知识有所了解，以便准确理解和转译专业术语和概念。

以上特点使口译成为一项充满挑战性的语言交流活动，要求译员具备广博的知识和技能，并在复杂的环境中高效、准确地进行口译。

三、口译的标准

法国释意派口译理论研究者塞莱斯科维奇（Danica Seleskovitch）认为，口译的标准是“达意、通顺”[②]。仲伟合将口译标准简化为“忠实与及时”[③]。鲍刚认为，口译标

---

① 仲伟合. 口译理论概谈[J]. 语言与翻译，1993(4)：20-23，32.

② Seleskovitch D. Interpreting for International Conferences[M]. Washington D. C.：Pen and Booth，1978：5-6.

③ 同①。

准可以概括为“全面、准确、通畅”[①]。综合以上，评判口译好坏的标准大致可以归结为以下三点：

（一）准确性

口译的首要标准是准确性（accuracy）。译员需要正确理解源语信息，并将其转换为准确无误的目标语表达。准确性需要译员深入理解语言和文化背景，并掌握特定领域的知识。译员需要忠实地传达讲话者的意图，确保信息不失真、不遗漏，同时避免添加个人观点或解释。

（二）通顺性

译员应以连贯、顺畅的方式传达信息，使目标语言听众轻松理解口译的内容。为了保证通顺性（coherence），译员需要熟悉目标语言的语法规则、词汇选择等，使用恰当的语言结构和过渡词语，确保句子、段落、语篇之间衔接顺畅，避免造成信息断裂和理解障碍。

（三）流利性

流利性（fluency）是评估译员表达能力的重要标准。流利的口译意味着译员能够自如地运用语言，流畅地转译信息。流利的口译能够提高信息传递的效率，同时也能够促进听众的理解。为了提高流利性，译员应该不断练习口头表达，以产出自信和自然的语言表达。

四、口译工作和译员素质

口译工作需要大量的实践和经验积累。同学们应不断学习和练习，注重反思和改进，练就过硬的口译能力，从而高质量地完成口译工作。

首先，译员需要具备良好的听辨理解和记忆能力。通过大量的听力训练，提高对各种语速、口音和语调的适应能力。在记忆方面，可以借助复述法、综述法、成像法、信息组块法和信息联想法等辅助手段记忆源语言的信息。

其次，译员需要学会借助笔记和符号，高效地梳理和记忆源语信息。在交替传译中，译员可以采用缩进式和叠加式的方式记笔记，并通过反复训练，形成一套简洁易懂的符号体系，快速而准确地记录源语中的句子主干、逻辑关系词和容易遗忘的信息等。

---

① 鲍刚.口译理论概述[M].北京：中国对外翻译出版公司，2005：300-330.

最后,译员需要掌握重组法、拆句法、合并法、转换法、增译法、减译法、简化法等转换技巧,并学会控制语速、语调和停顿,使目标语言的传达更加流畅自然。

除了以上方法和技巧,译员还需要具备以下三点基本素质:

① 广博的知识储备:除了语言能力,译员还需具备广博的非语言知识,包括对不同领域专业知识的了解。此外,译员还应了解各国文化背景,以便更好地理解源语言的文化内涵,避免造成误解或转译失真。

② 强大的心理素质:口译工作常常面临巨大的现场压力,因此译员需要拥有过硬的心理素质,在高压环境下保持冷静和专注,随时处理各种突发情况。优秀的译员能够迅速调整自己的状态和表达方式,以应对各种口译场合的要求。

③ 良好的职业道德:译员应遵守职业准则和保密要求,维护客户和参与者的利益。译员应以诚实、诚信和尊重为原则,维护口译行业的声誉和信誉。

五、航空航天科技口译

20 世纪科学技术飞速进步、社会生产迅猛发展,催生了航空航天事业。直至今日,航空航天领域的成就已成为科技创新能力的集中体现,其影响也扩展到政治、经济、军事以及社会生活的方方面面,成为代表人类文明发展和国家科技水平的关键因素。

随着中国航空航天事业蓬勃发展,国际交往日益深入,航空航天科技口译的需求不断增加。典型的应用场景包括:航空市场开拓、经贸往来、商务合同谈判、技术交流、技术培训与讲座、售后保障等[①]。航空航天科技口译要求译员具备广博的航空航天知识,并能正确理解和运用专业术语,熟悉相关技术要求,从而准确传达信息。

科技口译的特点基本适用于航空航天科技口译。科技口译是指以口译(交传或同传)为媒介传播科学与技术的相关知识与信息的社会交际活动[②]。刘和平指出,科技口译逻辑性强、概念清楚、用词准确、表达简练且专业性较强[③]。董金道认为,科技口译的特点是:时间性强、技术性强、政策性强、精确度高、对译员的素质要求高[④]。航空航天科技口译聚焦细分专业领域,对译员提出了更高的要求,其特点大致可以归结为以下三点:

---

① 夏皓.基于释意理论的航空科技口译策略研究[J].译苑新谭,2020,1(1):120.

② 张威.科技口译质量评估:口译使用者视角[J].上海翻译,2010,104(3):43.

③ 刘和平.科技口译与质量评估[J].上海科技翻译,2002(1):33-37.

④ 董金道.科技口译特点和标准的探讨[J].上海科技翻译,1993(4):18-19.

（一）术语专业性强

航空航天科技领域的专业知识和术语较为复杂，涵盖了航空工程、宇航科学、导航技术、航天器制造等诸多专业领域，对译员的专业背景提出了较高要求。为了在口译过程中准确传达信息，译员需要具备深厚的专业知识，了解各种航空器的设计原理、发动机技术、飞行动力学、航空器设计等方面的术语和概念。另外，该领域的工程和科学概念通常十分复杂，因此航空航天口译与其他行业口译相比，技术含量更高，译员需要充分理解该领域的相关原理，以保证口译输出通顺流畅。

（二）句子逻辑性强

航空航天科技领域的表达往往结构清晰、信息有条理，发言人一般能够精准流畅地传达复杂的技术概念，这对译员的信息接收能力和逻辑组织能力有很大的挑战。在航空航天口译中，句子逻辑性的构建通常以时间序列、空间结构或逻辑推理为基础。译员必须结合丰富的专业知识，巧妙组织信息，确保信息在表达中不产生混淆或歧义，通过合理运用连接词和过渡词，提高句子之间的关联性和衔接度，从而实现听众对复杂技术内容能够理解到位。

（三）语篇政策性强

从事航空航天科技口译，译员除了熟悉专业知识，还需对国家和国际层面的政策、法规等有深刻理解。首先，译员需要深入了解国家航空航天领域的政策框架，包括技术发展战略、产业政策、资金支持以及与其他国家的合作框架。其次，由于航空航天领域常常涉及国际合作与标准制定，译员需要熟悉国际组织制定的相关政策和协议，包括安全标准、环境保护准则等，确保在不同文化、法制体系中能够保持准确的信息传达，促进国际间的合作与共识。

除了掌握上述口译技术层面的特点，译员在从事航空航天口译工作时还需注意，该领域的国际合作往往涉及多个国家和多方机构，且经常涉及敏感的科学技术和商业机密，需要严守保密要求，确保会议和交流过程中的信息安全。航空航天科技口译在科技领域的国际交流中扮演着重要的角色，通过不断的发展，将为促进我国航空航天科技领域的发展和国防交流与合作贡献更多力量。

第二节　篇章讲解与练习

一、重点解析篇

（一）地球观测组织

1. 词汇表

Group on Earth Observations (GEO)	地球观测组织
satellite imagery	卫星图像
remote sensing	遥感
in situ data	现场数据
Sustainable Development Goals (SDGs)	可持续发展目标
2030 Agenda for Sustainable Development	2030年可持续发展议程
global Indicator Framework	全球指标框架
geospatial data	地理空间数据
derived information	衍生信息
demographic	人口的，人口统计的
variable	变量
spatial, spectral and temporal resolutions	空间、光谱和时间分辨率
UN Framework Convention on Climate Change (UNFCCC)	《联合国气候变化框架公约》
Intergovernmental Panel on Climate Change (IPCC)	政府间气候变化专门委员会
World Meteorological Organization (WMO)	世界气象组织
United Nations Environment Programme (UNEP)	联合国环境规划署
Committee on Earth Observation Satellites (CEOS)	地球观测卫星委员会
2015 Paris Agreement	《巴黎协定(2015)》
global stocktake	全球盘点
carbon accounting	碳核算
Conference of the Parties (COP) of the *UNFCCC*	《联合国气候变化框架公约》缔约方会议
Subsidiary Body for Scientific and Technological Advice (SBSTA)	附属科学技术咨询机构
marine heatwave	海洋热浪
ice flow	冰流
vegetation greening/browning	植被绿化及褐化
Disaster Risk Reduction (DRR)	降低灾害风险
GEO Work Programme (GWP)	地球观测组织工作计划

2. 背景知识

(1) 地球观测组织

2005 年 5 月,地球观测组织第一届全会在瑞士日内瓦召开,标志地球观测组织正式成立。该组织的宗旨是以协调、全面、持续的地球观测,为决策和行动提供信息支持,提升人类福祉,现有成员国 116 个、参加组织 151 个、关联组织 19 个。

地球观测组织作为地球观测领域最大的政府间国际组织,将落实联合国 2030 年可持续发展议程、气候变化《巴黎协定》和仙台减灾框架作为合作优先事项,并将“韧性城市与人居环境”列为 GEO 第四大优先事项,通过协调、全面、持续的地球观测支持在生物多样性和生态系统管理、防灾减灾、能源和矿产资源管理、粮食安全与可持续农业、基础设施和交通系统管理、公共卫生监测、城镇可持续发展、水资源管理等八个领域开展工作①。

(2)《联合国气候变化框架公约》

20 世纪 80 年代以来,人类逐渐认识并日益重视气候变化问题。为应对气候变化,1992 年 5 月 9 日通过了《联合国气候变化框架公约》(以下简称《公约》)。《公约》于 1994 年 3 月 21 日生效。截至 2023 年 10 月,共有 198 个缔约方②。《公约》的最终目标是将温室气体浓度稳定在“防止气候系统受到危险的人为干扰的水平”上。它指出,这一水平应当在足以使生态系统能够自然地适应气候变化、确保粮食生产免受威胁并使经济发展能够可持续地进行的时间范围内实现③。

3. 原 文

⑤

Group on Earth Observations ④

The Group on Earth Observations (GEO) is an intergovernmental partnership working to improve the availability, access and use of open Earth observations, including satellite imagery, remote sensing and in situ data, to impact policy and

---

① 外交部. 地球观测组织[EB/OL]. (2023-04-01)[2023-08-14]. http://svideo.mfa.gov.cn/wjb_673085/zzjg_673183/gjjjs_674249/gjzzyhygk_674253/dqgczz_698107/gk_698109/.

② 外交部.《联合国气候变化框架公约》进程[EB/OL]. (2023-10-19)[2023-08-14]. https://www.mfa.gov.cn/web/ziliao_674904/tytj_674911/tyfg_674913/201410/t20141016_7949732.shtml.

③ 百度百科. 联合国气候变化框架公约[EB/OL]. (2023-12-16)[2023-08-14]. https://unfccc.int/zh/jinchenghehuiyi/shenmeshilianheguoqihoubianhuakuangjiagongyue.

④ Group on Earth Observations. Earth Observations for Impact[EB/OL]. (2023-07-13)[2023-08-14]. https://old.earthobservations.org/index.php.

⑤ 扫描二维码即可收听本语篇的录音。全书同。

decision making in a wide range of sectors.

GEO's global priorities include the Sustainable Development Goals, Climate Action, and Disaster Risk Reduction (DRR).

Sustainable Development Goals

Earth observations play a major role in achieving the Sustainable Development Goals (SDGs). GEO is instrumental in integrating Earth observation data into the methodology of measuring indicators and achieving the SDGs.

The 2030 Agenda for Sustainable Development provides a universal development agenda for all countries and stakeholders to use as a blueprint of action for people, the planet and prosperity. The agenda is driven by seventeen SDGs, associated Targets, and a global Indicator Framework. Collectively, these elements enable countries and the global community to measure, manage, and monitor progress on economic, social and environmental sustainability.

Earth observations, geospatial data, and derived information play insightful roles in monitoring targets, planning, tracking progress, and helping nations and stakeholders make informed decisions, plans, and on-going adjustments that will contribute toward achieving the SDGs. Combined with demographic and statistical data, these sources enable nations to analyze and model conditions, create maps and other visualizations, evaluate impacts across sectors and regions, monitor change over time in a consistent and standardized manner, and improve accountability.

Climate Action

Earth observations relevant to climate action are not limited to weather or climate, but are much broader and include terrestrial and socio-economic variables. GEO makes available Earth observations in support of effective policy responses for climate change adaptation, mitigation and other specific provisions, working with partners to enhance global observation systems for climate action.

Climate change is one of the biggest challenges facing the world. It is also recognized as one of the challenges where use of Earth Observations (EO) can make the most difference, as EO has the capability to capture environmental and socio-economic data over a range of spatial, spectral and temporal resolutions.

The impacts of climate change are faced by all, but poor and vulnerable communities and groups are the most affected. Supporting sustainable development agendas while tackling the effects of climate change illustrates the inter-linkages between GEO's Engagement Priorities.

GEO uses its unique convening power to connect Members and key partners such as the *UN Framework Convention on Climate Change* (*UNFCCC*), the Intergovernmental Panel on Climate Change (IPCC), the World Meteorological Organization (WMO), the United Nations Environment Programme (UNEP), and the Committee on Earth Observation Satellites (CEOS) to lead national, regional and global climate action efforts.

GEO and the *UN Framework Convention on Climate Change* (*UNFCCC*)

The Earth observations community can play a crucial role in global efforts to address climate change and implement the *UNFCCC* and the *2015 Paris Agreement*. The data and knowledge derived from Earth observations helps governments and other stakeholders at regional, national and sub-national levels to respond in many areas, including mitigation, adaptation and other specific provisions of the *Paris Agreement*, as well as provide input to the process including through the global stocktake.

Earth observations contribute near real-time data on Greenhouse Gas (GHG) concentrations and emissions for carbon accounting in relation to mitigation responses. When Earth observation data is combined with other critical socio-economic information at the local scale and over extended timescales, efforts to monitor progress on adaptation responses can all be enhanced in addition to impact, vulnerability and risk assessments and the development of measures to increase resilience.

GEO and the Intergovernmental Panel on Climate Change (IPCC)

Earth observations are important for the work of the IPCC. The IPCC provides scientific input to inform the Conference of the Parties (COP) of the *UNFCCC* and the Convention bodies, in particular the Subsidiary Body for Scientific and Technological Advice (SBSTA). The SBSTA has been increasingly emphasizing the value of systematic observations—a term that encompasses Earth observations in the *UNFCCC* context. Earth observations, and in particular satellite data, provide benchmark measurements on variables which contribute to the accuracy of climate models and projections that inform policy decisions.

The findings of the latest IPCC special reports have already benefited from the enhanced use of Earth observation data and there is scope for improvement. For instance, satellite observations were used to monitor the frequency of marine heatwaves over several decades, and other variables including ice flows. Satellite data was also used to monitor vegetation greening/browning. Increased availability of open Earth observation datasets can increase the quality of monitoring and help address the gaps identified by the IPCC.

Disaster Risk Reduction

Earth observations contribute to disaster preparedness and better mitigation and response. GEO supports disaster resilience by increasing coordination of Earth observations to forecast and prepare for disasters, to reduce damage and to better manage and recover from disasters.

GEO supports efforts to build disaster resilience and Disaster Risk Reduction. The GEO Work Programme (GWP) is currently implementing more than two dozen activities using Earth observation for disaster preparedness and prevention, mitigation of potential damage, and better management of and recovery from disasters. Significant reductions in fatalities and property damage can be achieved by strengthening cooperation and data sharing for satellite and surface data to manage risks posed by fires, floods, earthquakes, and other hazards. Better information, made widely accessible, leads to improved understanding of disaster risk.

4. 参考译文

地球观测组织[①]

地球观测组织，即GEO，是一个政府间伙伴关系组织，致力于更好地获取并使用开放的地球观测资源，具体包括卫星图像、遥感和现场数据等，从而影响各部门的决策。

GEO的全球工作重点包括可持续发展目标、气候行动和减灾。

可持续发展目标

地球观测在实现可持续发展目标方面发挥着重要作用。地球观测组织将观测数据纳入衡量各种指标和实现可持续发展目标的方法之中。

2030年可持续发展议程为各国及利益相关者提供了一份共同的发展议程，是人类、地球和繁荣发展的行动蓝图。该议程涵盖17个可持续发展目标、相关目标和全球指标框架，支持各国乃至全球衡量、管理并监测经济、社会以及环境可持续性方面的进展。

地球观测、地理空间数据和衍生信息在监测目标、制定规划、跟踪进展等多方面发挥重要作用，并帮助各国和利益相关者做出明智的决策和计划，进行持续调整，从而为实现可持续发展目标做出贡献。各国充分利用人口、统计数据和观测资源，对情况进行分析和建模，创建地图和其他可视化效果，评价跨部门和地区的影响，以一致化和标准化的方式监测一段时间内发生的变化，并改善问责制。

① 编者译。

气候行动

与气候行动相关的地球观测并不限于天气或气候，其范围更加广阔，包括陆地和社会经济变量。地球观测组织提供地球观测服务，支持在气候变化适应、减缓和其他具体方面提供有效的政策响应，并与合作伙伴联手，提升气候行动的全球观测系统。

气候变化是世界面临的最大挑战之一，也是地球观测最大有可为的领域之一，因为地球观测有能力在一系列空间、光谱和时间分辨率上捕捉环境和社会经济数据。

所有人都面临气候变化的影响，但贫困弱势的社区和群体受到的影响最大。GEO 在应对气候变化影响的同时支持可持续发展议程，体现了该组织各优先事项之间的相互联系。

GEO 聚集其成员和关键合作伙伴，如《联合国气候变化框架公约》(*UNFCCC*)、政府间气候变化专门委员会(IPCC)、世界气象组织(WMO)、联合国环境规划署(UNEP)和地球观测卫星委员会(CEOS)等，引领各国、区域和全球的气候行动。

GEO 和《联合国气候变化框架公约》

在全球努力应对气候变化，实施《联合国气候变化框架公约》及《巴黎协定(2015)》的过程中，地球观测界能够发挥关键作用。从地球观测中获得的数据和知识有助于区域、国家和地方政府以及其他利益相关者在诸多领域做出反应，包括气候变化减适以及《巴黎协定》的其他具体规定，并通过全球盘点的方式为这一进程提供建议。

地球观测提供近实时的温室气体浓度和排量数据，为碳核算提供参考，有助于减缓气候变化。将一个地方在一定时间段的地球观测数据与其他关键的社会经济信息结合起来，除了可以更好地开展影响、薄弱性和风险评估，制定提高韧性的措施外，还可以更好地监测适应对策取得的进展。

GEO 和政府间气候变化专门委员会

地球观测在政府间气候变化专门委员会(IPCC)的工作中起着重要作用。IPCC 为《联合国气候变化框架公约》缔约方会议和公约机构，特别是附属科学技术咨询机构(SBSTA)，提供科学意见。SBSTA 更加强调"系统观测"的价值。在《联合国气候变化框架公约》背景下，"系统观测"包含地球观测。地球观测资源，特别是卫星数据，提供了对变量的基准测量，有助于提高气候模型和预测的准确性，为决策提供依据。

IPCC 最新特别报告中的发现得益于更好地使用了地球观测数据，而且仍有提升空间。例如，使用卫星观测监测几十年来海洋热浪的发生频率，以及包括冰流在内的其他变量。卫星数据也用于监测植被绿化及褐化。开放地球观测数据集可以提高监测质量，有助于解决 IPCC 指出的问题。

减少灾害风险

地球观测可协助开展防灾、减灾和救灾。GEO 加强地球观测的协调,提高应对灾害的能力,预测防范灾害,减少损失,更好地开展灾后管理和重建。

GEO 支持构建灾后恢复力,降低灾害风险。地球观测组织工作计划目前正在开展二十多项活动,利用地球观测进行备灾、防灾和减灾,提高灾后管理和重建水平。加强卫星和地面数据的合作与共享,管控火灾、洪灾、地震和其他灾害的风险,可以大幅减少人员死亡和财产损失。获取更准确的信息将有助于提高各方对灾害风险的认识。

5. 要点讲解

在本篇英译汉的翻译过程中,译员通常采用增译和省译的翻译方法,使得译文更加流畅。在增译方面,例如将 monitoring targets, **planning**, tracking progress 译为"监测目标、**制定规划**、跟踪进展"(增加词组使其成为四字格,进而达到前后语言形式的一致性);将 supporting sustainable development agendas while tackling the effects of climate change illustrates 译为"**GEO** 在应对气候变化影响的同时支持可持续发展议程,体现了……"(增加主语)等。而从省译的角度来看,例如将 the global **community** 译为"全球"(省略重复信息的翻译);将 GEO supports… by increasing… **to** forecast and prepare for disasters 译为"GEO 加强……提高……预测防范灾害"(省略表目的的介词 to 的翻译)等。

同时,译员通过调整语序的方法使得译文更加流畅,使其更加符合中文的使用习惯。例如将 monitor change over time in a consistent and standardized manner 译为"以一致化和标准化的方式监测一段时间内发生的变化"(状语提前);将 earth observations relevant to climate action 译为"与气候行动相关的地球观测"(定语提前)等。

此外,面对源语中的长难句,译员多采用拆句法,将其拆分成汉语分句,使其更加浅显易懂。例如将长句"Earth observations contribute near real-time data on greenhouse gas (GHG) concentrations and emissions for carbon accounting in relation to mitigation responses."译为"地球观测提供近实时的温室气体浓度和排量数据,为碳核算提供参考,有助于减缓气候变化。"译员梳理了源语中的信息,将源语中的主语、目的及作用译为三个短句,先翻译主语,再解释作用及意义,使译文逻辑清晰明了。

（二）杨利伟在第27届太空探索者协会年会闭幕式上的致辞

1. 词汇表

太空探索者协会	Association of Space Explorers (ASE)
中国载人航天工程办公室	China Manned Space Agency (CMSA)
执委会议	executive committee meeting
技术分会	technical sessions
职业素养	professionalism
载人航天	manned spaceflight
中国航天员中心	the Astronaut Center of China (ACC)

2. 背景知识

(1) 中国载人航天工程办公室

1992年9月21日，中国载人航天工程由中国政府批准实施。为了加强对工程的管理，设立了“中国载人航天工程办公室”(CMSA)，代表政府行使管理职能。

中国载人航天工程办公室负责中国载人航天工程的规划计划、总体技术、工程研制、基本建设和技术改造、经费等工作，按照工程总指挥、总设计师联席会议的决定，对工程的计划、技术、质量等进行全系统、全过程的管理，统筹协调工程各部门、各系统的工作。办公室设主任一名、副主任两名，下设科技计划局、工程建设局和工程总体室三个业务局室[①]。

(2) 中国航天员科研训练中心

中国航天员科研训练中心，简称中国航天员中心，成立于1968年4月1日，是中国载人航天领域内医学与工程相结合的综合型研究机构。自成立以来，中国航天员中心肩负起了航天员选拔训练和相关产品研制的艰巨任务，实现了载人航天飞行9类关键技术的重大突破，包括建造了航天飞行模拟器、模拟失重水槽、舱外航天服试验舱等十余个大型地面训练试验设施；设计并完成了近百项有人参与的空间科学试验；设计了企鹅服、太空跑台等多项失重防护设备等，为中国航空航天事业的发展

---

① 中国载人航天. 中国载人航天工程组织管理[EB/OL]. (2015-08-02)[2023-08-14]. https://www.cmse.gov.cn/art/2015/8/2/art_24_11273.html.

做出了突出贡献[①]。

3. 原 文

在第 27 届太空探索者协会年会闭幕式上的致辞[②]

中国载人航天工程办公室副主任 杨利伟

2014 年 9 月 15 日

尊敬的道瑞恩主席、王永志院士，女士们、先生们、朋友们：

从 9 月 9 日的执委会议至今天的闭幕式，我和来自世界各国的同行们共同度过了 7 天的美好时光。这 7 天，我像从太空顺利返回一样快乐，因为我们在一起，看到朋友们再次欢聚一堂，我感到无比欣慰、由衷幸福！

本届年会在全体参会代表和工作人员的共同努力下，圆满完成了预定的各项议程，即将落下帷幕。首先，我要代表中国载人航天工程办公室，代表本届年会组委会，向年会的圆满成功表示热烈祝贺。在过去的 7 天时间里，来自 18 个国家的 91 位航天员，先后举行了开幕式、主题会议、三次技术分会和两次全体航天员大会，走进了北京、天津、西安、深圳的多所学校、科研院所和企业。在平等融洽的氛围中，全体与会代表积极探讨、踊跃发言，展示了良好的职业素养和人文精神，让我们对载人航天的美好未来充满了希望，对未来的国际合作充满了期待。在年会中，全世界的航天员相互了解、相互认识，结下了深厚的友谊。可以说，本届年会无愧为一次载人航天的盛会、和平的盛会、友谊的盛会！"合作：共圆人类航天梦"已成为大家的共识，这正是本届年会所希望取得的成果，也是载人航天向更高更远发展的力量所在。

在祝贺的同时，我还要表示感谢。感谢道瑞恩先生为太空探索者协会及本届年会做出的重要贡献，感谢全体航天员不辞辛劳出席会议、参加活动、积极发言，尤其要感谢安迪先生为会议顺利举办所做的大量协调工作，也要感谢我们的航天员家属们，你们是美丽的使者，一定会把中国的文化、美食、胜景向世界传播。我要感谢来自中外新闻机构的记者朋友们，没有你们对人类探索太空事业广泛而深入的报道，就不会赢得世界各地人们对这项事关全人类福祉的伟大事业的关注和支持。同时，我还要感谢大会的每一位工作人员，感谢中国载人航天工程办公室和航天员中心的诸位同事，感谢默默奉献的会务公司员工、安保人员、医护人员、志愿者和友谊宾馆工作人员，没有你们就没有本次大会的成功，相信每一位与会代表都会记住你们的微笑、记住你们的热情、记住你们的辛勤，谢谢你们！

---

① 中国军网. 中国航天员中心成立 53 周年，生日快乐[EB/OL]. (2021-04-02)[2023-08-14]. http://www.81.cn/kt/jdt_208603/10015615.html.

② 中国载人航天. 杨利伟：愿浩瀚太空真正成为人类的共同财富[EB/OL]. (2014-09-16)[2023-07-10]. http://www.cmse.gov.cn/art/2014/9/16/art_1615_26107.html.

除了祝贺与感谢之外，我们更满怀期待。我们期待太空探索者协会在促进载人航天国际合作与交流方面发挥愈来愈重要的作用，我们愿积极支持协会工作，在载人航天领域为全世界做出积极贡献。我们期待载人航天的明天更加美好，浩瀚无垠的太空将真正成为人类的共同财富，太空探索事业会为全人类带来更多实实在在的福祉。我们期待着友谊长存，载人航天比任何其他事业都更加需要同心协力、同舟共济，希望以本届年会为契机，大家既是事业上的伙伴，也成为生命中的朋友。当然，我们也期待着第 28 届年会会更加精彩！

朋友们，年会即将闭幕，但梦想才刚刚上路，合作会越来越紧密。中国的大门永远向你们敞开，古老而美丽的中国永远欢迎你们！

谢谢大家！

4. 参考译文

Remarks at the Closing Ceremony of the 27th Planetary Congress of the Association of Space Explorers①

Mr. Yang Liwei

Deputy Director of China Manned Space Engineering Office

15 September 2014

Distinguished President Dumitru-Dorin Prunariu,

Academician Wang Yongzhi,

Ladies and Gentlemen, Friends,

I have spent seven wonderful days with colleagues from all over the world since the opening of the Executive Committee Meeting on September 9. For me, the joy of reunion with friends is as much as the joy upon the completion of the space mission. My happiness is beyond words.

With the joint efforts of all the participants and staff members, this congress is drawing to a successful close, with all the agenda items completed. First of all, on behalf of the China Manned Space Agency and the Organizing Committee, I wish to extend warm congratulations on the success of the congress. Over the past seven days, we have brought together 91 astronauts from 18 countries to join us for the opening ceremony, one keynote meeting, three technical sessions and two plenary sessions for astronauts. We also organized visits to universities, research institutes and enterprises in Beijing, Tianjin, Xi'an and Shenzhen. All the participants engaged in enthusiastic and

---

① 编者译。

friendly discussions on an equal footing. You have demonstrated professionalism and humanism, inspiring hope and confidence in a bright future of manned spaceflight and international cooperation. During the congress, astronauts from all over the world got to know each other and built friendship. It's fair to say this congress is a grand event of manned spaceflight, peace, and friendship. It has been agreed that by working together, we can fulfil the humanity's space dream. This is exactly what we aim for through this congress. It also empowers the manned spaceflight to pursue higher and further development.

After expressing my congratulations, I want to express appreciation. I wish to thank Mr. Prunariu for his great contribution to the Association of Space Explorers (ASE) and to this year's congress. I also want to extend gratitude to all the astronauts for participating in the activities and sharing your thoughts. I appreciate Andy's coordination efforts, which makes the congress possible. My thanks also go to the family members of astronauts. You are messengers that will spread the culture, food, and beauty of China to the world. I would like to thank journalists from China and around the world. Without your extensive and in-depth reporting of human space exploration, we would not have gained attention and support from across the globe for this great cause that concerns humankind's well-being. I am also indebted to all those who have made the congress possible, including the colleagues from the China Manned Space Agency and the Astronaut Center of China, and the conference service company, the security staff, the medical staff, the volunteers and the staff of Beijing Friendship Hotel. It's your joint efforts that have made this congress a success. I believe that everyone attending the congress is impressed by your smiles, your warmth, and your hard work. Thank you all.

In addition to congratulations and appreciation, my colleagues and I are also full of expectation. We hope ASE will play an increasingly important role in promoting international cooperation and exchange in manned spaceflight, and we are ready to support ASE and make contributions to the world in this field. We look forward to a better tomorrow for manned spaceflight, when the vast space will truly be shared by all and space exploration will bring more tangible benefits to humankind. We expect lasting friendship, as manned spaceflight calls for more solidarity and collaboration than any other undertakings. We hope this opportunity could cement partnership and friendship. We also look forward to a more exciting gathering at the 28th Planetary Congress of ASE.

Dear friends,

This congress is coming to an end, but our dream just begins. Our cooperation will become closer than ever. The door of China is always wide open. An ancient and beautiful China always welcomes you. Thank you.

5. 要点讲解

例 1:

在平等融洽的氛围中,全体与会代表积极探讨、踊跃发言,//展示了良好的职业素养和人文精神,让我们对载人航天的美好未来充满了希望,对未来的国际合作充满了期待。

All the participants engaged in enthusiastic and friendly discussions on an equal footing. **You** have demonstrated professionalism and humanism, **inspiring** hope and confidence in a bright future of manned spaceflight and international cooperation.

为了便于目标语听众理解,译员在翻译长句时,经常采用拆句法,把一个长句拆译成几个短句。这里,译员根据源语的意群切分,在"全体与会代表积极探讨、踊跃发言"后断开,独立成句。译员给第二个短句增加了主语 You,符合源语的意思,也使演讲人的话语更有针对性和互动性。最后,译员采用合并法,使用非谓语动词 inspiring 将后两个短句整合在一起作状语,形成一个逻辑清晰的长句。

例 2:

可以说,本届年会无愧为一次载人航天的盛会、和平的盛会、友谊的盛会!

It's fair to say this congress is a grand event of manned spaceflight, peace, and friendship.

这句话包含工整的并列结构,使汉语传递的信息更加有力,但在口译的时候,受时间限制,译员不必拘泥于中文的并列结构,只保留一个"grand event"即可,将"载人航天"、"和平"和"友谊"三个修饰词整合在一起,使英文输出简洁明了,避免重复。

例 3:

我们期待载人航天的明天更加美好,浩瀚无垠的太空将真正成为人类的共同财富,太空探索事业会为全人类带来更多实实在在的福祉。

We expect a better tomorrow for manned spaceflight, when the vast space will truly become the wealth shared by all and space exploration will bring more tangible benefits to all humanity.

在翻译这一句时,译员通过"when"将后两个分句整合在一起,形成时间状语从句,修饰前面宾语从句中的"载人航天的明天"。这样,源语三个并列的从句有机整合成了一个逻辑清晰、结构紧凑的长句,符合英文句子重"形合"的特点。

二、实战训练篇

（一）美国国家航空航天局和欧洲航天局

1. 词汇表

National Aeronautics and Space Administration (NASA)	美国国家航空航天局
civil servant	公务员
contractor	承包商
academia	学术界
fiscal year	财年
aeronautics	航空学
electric propulsion	电力推进
supersonic flight	超声速飞行
Moon to Mars exploration approach	“月球到火星”探测计划
Artemis missions	阿尔忒弥斯计划
European Space Agency (ESA)	欧洲航天局
spacecraft	航天器
ground facility	地面设施
probe	探测器

2. 原　文

About NASA①

The National Aeronautics and Space Administration is America's civil space program and the global leader in space exploration. The agency has a diverse workforce of just under 18,000 civil servants, and works with many more U. S. contractors, academia, and international and commercial partners to explore, discover, and expand knowledge for the benefit of humanity. With an annual budget of $23.2 billion in Fiscal Year 2021, which is less than 0.5% of the overall U. S. federal budget, NASA supports more than 312,000 jobs across the United States, generating more than $64.3 billion in total economic output (Fiscal Year 2019).

---

① National Aeronautics and Space Administration. About NASA[EB/OL]. (2023-01-27)[2023-07-10]. https://www.nasa.gov/about/index.html.

At its 20 centers and facilities across the country—and the only National Laboratory in space—NASA studies Earth, including its climate, our Sun, and our Solar System and beyond. We conduct research, testing, and development to advance aeronautics, including electric propulsion and supersonic flight. We develop and fund space technologies that will enable future exploration and benefit life on Earth.

NASA also leads a Moon to Mars exploration approach, which includes working with U.S. industry, international partners, and academia to develop new technology, and send science research and soon humans to explore the Moon on Artemis missions that will help prepare for human exploration of the Red Planet. In addition to those major missions, the agency shares what it learns so that its information can make life better for people worldwide. For example, companies use NASA discoveries and technologies to create new products for the public. To ensure future success for the agency and the nation, NASA also supports education efforts in STEM with an emphasis on increasing diversity in our future workforce.

European Space Agency (ESA)[①]

ESA is an intergovernmental organisation, created in 1975, with the mission to shape the development of Europe's space capability and ensure that investment in space delivers benefits to the citizens of Europe and the world.

ESA has 22 Member States: Austria, Belgium, the Czech Republic, Denmark, Estonia, Finland, France, Germany, Greece, Hungary, Ireland, Italy, Luxembourg, the Netherlands, Norway, Poland, Portugal, Romania, Spain, Sweden, Switzerland and the United Kingdom. Slovenia is an Associate Member.

ESA has established formal cooperation with six Member States of the EU. Canada takes part in some ESA programmes under a Cooperation Agreement.

By coordinating the financial and intellectual resources of its members, ESA can undertake programmes and activities far beyond the scope of any single European country.

ESA develops the launchers, spacecraft and ground facilities needed to keep Europe at the forefront of global space activities.

Today, it launches satellites for Earth observation, navigation, telecommunications and astronomy, sends probes to the far reaches of the Solar System and cooperates in the human exploration of space.

---

① IAF. European Space Agency [EB/OL]. (2023-02-15) [2023-07-10]. https://www.iafastro.org/membership/all-members/european-space-agency-esa.html.

(二) 国家主席习近平发表二〇二三年新年贺词

1. 词汇表

神舟十三号、十四号、十五号	Shenzhou-13, Shenzhou-14 and Shenzhou-15
第三艘航母"福建号"	China's third aircraft carrier Fujian
首架 C919 大飞机	C919, China's first large passenger aircraft
白鹤滩水电站	Baihetan hydropower station
党的二十大	the 20th National Congress of the Communist Party of China (CPC)
中华民族伟大复兴	the great rejuvenation of the Chinese nation
脱贫攻坚	poverty elimination
乡村振兴	advanced rural revitalization
北京冬奥会、冬残奥会	the Beijing Olympic and Paralympic Winter Games
自由贸易试验区	pilot free trade zones
海南自由贸易港	Hainan Free Trade Port
"一国两制"	One Country, Two Systems

2. 原　文

国家主席习近平发表二〇二三年新年贺词①

大家好！2023 年即将到来，我在北京向大家致以美好的新年祝福！

2022 年，我们胜利召开党的二十大，擘画了全面建设社会主义现代化国家、以中国式现代化全面推进中华民族伟大复兴的宏伟蓝图，吹响了奋进新征程的时代号角。

我国继续保持世界第二大经济体的地位，经济稳健发展，全年国内生产总值预计超过 120 万亿元。面对全球粮食危机，我国粮食生产实现"十九连丰"，中国人的饭碗端得更牢了。我们巩固脱贫攻坚成果，全面推进乡村振兴，采取减税降费等系列措施为企业纾难解困，着力解决人民群众急难愁盼问题。

疫情发生以来，我们始终坚持人民至上、生命至上，坚持科学精准防控，因时因势优化调整防控措施，最大限度保护了人民生命安全和身体健康。广大干部群众特别是医务人员、基层工作者不畏艰辛、勇毅坚守。经过艰苦卓绝的努力，我们战胜了前所未有的困难和挑战，每个人都不容易。目前，疫情防控进入新阶段，仍是吃劲的时候，大家都在坚忍不拔努力，曙光就在前头。大家再加把劲，坚持就是胜利，团结就是

---

① 中华人民共和国中央人民政府. 国家主席习近平发表二〇二三年新年贺词[EB/OL]. (2022-12-31)[2023-07-10]. https://www.gov.cn/gongbao/content/2023/content_5736705.htm.

胜利。

2022年,江泽民同志离开了我们。我们深切缅怀他的丰功伟绩和崇高风范,珍惜他留下的宝贵精神财富。我们要继承他的遗志,把新时代中国特色社会主义事业不断推向前进。

历史长河波澜壮阔,一代又一代人接续奋斗创造了今天的中国。

今天的中国,是梦想接连实现的中国。北京冬奥会、冬残奥会成功举办,冰雪健儿驰骋赛场,取得了骄人成绩。神舟十三号、十四号、十五号接力腾飞,中国空间站全面建成,我们的"太空之家"遨游苍穹。人民军队迎来95岁生日,广大官兵在强军伟业征程上昂扬奋进。第三艘航母"福建号"下水,首架C919大飞机正式交付,白鹤滩水电站全面投产……这一切,凝结着无数人的辛勤付出和汗水。点点星火,汇聚成炬,这就是中国力量!

今天的中国,是充满生机活力的中国。各自由贸易试验区、海南自由贸易港蓬勃兴起,沿海地区踊跃创新,中西部地区加快发展,东北振兴蓄势待发,边疆地区兴边富民。中国经济韧性强、潜力大、活力足,长期向好的基本面依然不变。只要笃定信心、稳中求进,就一定能实现我们的既定目标。今年我去了香港,看到香港将由治及兴十分欣慰。坚定不移落实好"一国两制",香港、澳门必将长期繁荣稳定。

今天的中国,是赓续民族精神的中国。这一年发生的地震、洪水、干旱、山火等自然灾害和一些安全事故,让人揪心,令人难过,但一幕幕舍生取义、守望相助的场景感人至深,英雄的事迹永远铭记在我们心中。每当辞旧迎新,总会念及中华民族千年传承的浩然之气,倍增前行信心。

今天的中国,是紧密联系世界的中国。这一年,我在北京迎接了不少新老朋友,也走出国门讲述中国主张。百年变局加速演进,世界并不太平。我们始终如一珍视和平和发展,始终如一珍惜朋友和伙伴,坚定站在历史正确的一边、站在人类文明进步的一边,努力为人类和平与发展事业贡献中国智慧、中国方案。

党的二十大后我和同事们一起去了延安,重温党中央在延安时期战胜世所罕见困难的光辉岁月,感悟老一辈共产党人的精神力量。我常说,艰难困苦,玉汝于成。中国共产党百年栉风沐雨、披荆斩棘,历程何其艰辛又何其伟大。我们要一往无前、顽强拼搏,让明天的中国更美好。

明天的中国,奋斗创造奇迹。苏轼有句话:"犯其至难而图其至远",意思是说"向最难之处攻坚,追求最远大的目标"。路虽远,行则将至;事虽难,做则必成。只要有愚公移山的志气、滴水穿石的毅力,脚踏实地,埋头苦干,积跬步以至千里,就一定能够把宏伟目标变为美好现实。

明天的中国,力量源于团结。中国这么大,不同人会有不同诉求,对同一件事也会有不同看法,这很正常,要通过沟通协商凝聚共识。14亿多中国人心往一处想、劲往一处使,同舟共济、众志成城,就没有干不成的事、迈不过的坎。海峡两岸一家亲。衷心希望两岸同胞相向而行、携手并进,共创中华民族绵长福祉。

明天的中国，希望寄予青年。青年兴则国家兴，中国发展要靠广大青年挺膺担当。年轻充满朝气，青春孕育希望。广大青年要厚植家国情怀、涵养进取品格，以奋斗姿态激扬青春，不负时代，不负华年。

此时此刻，许多人还在辛苦忙碌，大家辛苦了！新年的钟声即将敲响，让我们怀着对未来的美好向往，共同迎接2023年的第一缕阳光。

祝愿祖国繁荣昌盛、国泰民安！祝愿世界和平美好、幸福安宁！祝愿大家新年快乐、皆得所愿！

谢谢！

第二章 记忆训练

记忆对口译员的重要性不言而喻。在口译过程中,记忆能力在很大程度上会影响译员口译的效果。许多译员由于在口译时进行“听、记、说”等多任务处理,容易受到干扰,记忆信息较平常更加困难,经常出现听到下一条信息便遗忘上一条信息的情况,最终导致译文不连贯,语言表达不流畅。关于如何提高记忆能力,大脑潜能思维开发培训专家刘志华认为,任何人都可以完全掌握记忆方法,拥有超级记忆力的人不是天生的,而是通过后天努力,不断训练出来的①。因此,在练习口译的过程中,进行记忆训练十分必要。提升自身瞬时记忆(instantaneous memory)和短期记忆(short-term memory)的能力以及扩展记忆广度(memory span)有助于减少失误,提高译文的质量。

第一节 记忆训练技巧

在口译工作中,口译员需要记忆大量的专业术语、常用短语、地名、人名、数字等信息,并快速、准确地将之转化为目标语言。记忆力的好坏直接影响到口译的准确性和流利程度,因此口译人员有必要学习一些常用的记忆技巧,来帮助自己有效提高记忆力,从而更加准确地完成翻译任务。刘海洋在《口译技巧》中提到:“记忆技巧是为了提高口译人员的记忆能力和记忆效率,达到减少遗忘、提高记忆量、加强记忆效果以及提高记忆速度的目的。”②本节将介绍五种常用的记忆训练方法,包括复述法、综述法、成像法、信息组块法和信息联想法,同时辅以案例分析,希望对提高译员记忆力有所助益。

一、复述法

复述法在记忆训练中十分常见,它不需要译员对原文死记硬背,而是要求译员在理解原文意思和发言人意图的前提下,通过梳理各信息之间的逻辑关系等方式,对原文信息进行加工和记忆,最后重新组织语言,将原文具体内容精准复述表达出来。根据复述的语言不同,复述可分为源语复述和目标语复述。译员在复述时对所获得的信息进行整理和加工,这个过程进一步帮助译员更好地把握原文的含义,从而加深对

---

① 刘志华.超级记忆力训练法[M].北京:中国纺织出版社,2018:5.

② 刘海洋.口译技巧[M].北京:中国对外翻译出版公司,2011:40.

原文信息的记忆。以下为复述法的具体应用案例。

例 1：

原文：6 次载人飞行，次数并不多，时间也不长，必须要积累更多的数据和经验，让航天员能够适应太空失重环境，高效完成工作。在 38 项实验中，人的健康研究占主导，有 8 项之多，我感觉是非常重要和必要的。例如随着飞行时间的延长，航天员的视力会受到影响。

参考复述：为了让航天员能更好地适应太空失重环境，更有效率地完成工作，我们必须要积累更多的数据和经验，因此执行 6 次载人飞行任务是远远不够的。此外，人的健康研究十分重要且必要，并在实验中占据主导，38 项实验中就有 8 项研究人的健康，例如研究飞行时间长短对航天员视力的影响。

该选段中，句与句之间的衔接较为松散，令人难以抓住重点，为了使译员更好地理解该段落的信息，译员在复述的过程中可以梳理原文分句间的逻辑关系，调整个别句子的顺序，通过添加"为了""因此"等连词及增补主语等方式重新排列、串联原文信息，突出其逻辑关联，这一重新组织语言表达的过程有助于加深译员记忆。

例 2：

原文：Flying taxis are no longer in the realm of fantasy—they're coming very, very soon, and all of us here will have to start thinking about whether we'll jump into one to slip the surly bonds of traffic.

参考复述：Flying taxis are no longer far away from our lives. When they come, all of us here will consider whether we should take one to get away from the bonds of traffic on the ground.

该选段中，"the realm of fantasy""slip the surly bonds of traffic"含义较为抽象，不好记忆，复述过程中译员可以对这些较为抽象的词进行加工处理，化虚为实，提取原文实际想传递的意思，重新组织语言表达出来，如将"the realm of fantasy"转换为"far away from our lives"等。这种提取信息的过程也有助于加深译员对该段落的理解和记忆。

二、综述法

区别于对原文具体信息进行精准复述的复述法，综述法要求译员把握段落或篇章的大意，并用简短的语言对原文具体信息进行概括。综述法可以帮助译员更好地了解发言人的意图，理清发言人的思维框架，更好地抓住主要内容，从而加强译员的记忆。以下为综述法在段落记忆中具体运用的案例。

例 1：

然而，飞天没有坦途，从天空飞向太空，从一名飞了 1680 小时的飞行员转身成为一名航天员，可以说是经历了一场炼狱般的脱胎换骨。对于外人来说，航天员生活很神秘。可对我们而言却是非常紧张、非常单调，生活中仿佛只剩下了两件事：学习训练、训练学习。执行任务前整整两年多,我没有逛过一次街,没有看过一次电影,不仅要学完相当于大学本科整整四年的课程,同时还要进行包括体质、心理、航天专业技术等100多个科目的艰苦训练,每个科目挑战的都是生理和心理的极限。

运用综述法进行段落记忆时,译员着重关注段落中是否有中心句,据此提取原文信息中的关键词,对段落大意进行概括。不难看出,以上这段的中心句为第一句,后面几句话都在具体叙述成为一名航天员需要经历何种艰苦训练。因此,译员可根据段落中心句及"紧张""艰苦""100多个科目"等关键词概括出该段落的大意——为了成为一名航天员,发言人经历了无数的艰苦训练,并付出了不懈的努力。了解该段落大意后,译员便可从该段落的大框架中得到提示,从而更有条理地记忆段落中的细节。

例2:

The innovations are not going to slow down. That is a challenge for the FAA and other civil aviation regulators around the world. The FAA is committed to finding a way forward that not only protects the public but allows innovation and opportunity to move forward, and we'll work hand-in-hand with our international partners, including CAAC, to make sure we address together these constantly changing challenges and to learn from one another.

除了通过提取段落中心句了解段落大意以外,译员在听的过程中也要注意段落中的连词和多次重复的词语,以免遗漏重要信息。在这一段中,FAA、innovation和challenge出现了多次,通过这三个词语可以发现,该段落的中心词为FAA,它希望将来继续推动创新,应对接下来的挑战。此外在段落的后半部分还有一个连词and,衔接了句子"we'll work hand-in-hand…",表明FAA希望通过合作的方式来推动创新和应对挑战。因此,通过整合以上提到的关键词,该段落的大意可概括为"The FAA will cooperate with its partners to boost innovation and address the challenges."译员在提取关键词并对段落具体信息进行概括的过程中加深了对原文含义的理解,强化了对该段落的记忆。

三、成像法

成像法是译员通过自身想象,将听到的信息在脑海中以图画的形式呈现出来,进行图像记忆。这种方法多用于帮助译员记忆叙述性强或故事情节丰富的材料①。以下为成像法具体运用于段落记忆的例子。

例1:

许多特因环境训练更是残酷,其中有一项训练叫转椅,当飞行员时转椅只需坐两分钟,可当了航天员优秀标准却提高到了15分钟。我自认为前庭功能还是不错的,

---

① 任文.英汉口译教程[M].北京:外语教学与研究出版社,2011:73.

可当第一次转椅坐到五分钟的时候，那种突如其来的眩晕、恶心，瞬间让我脸色苍白、满头是汗，下来后整整一天吃不下饭。

这部分是发言人对其接受转椅训练的叙述，发言人在叙述过程中用了许多形容词，例如“脸色苍白”“满头是汗”等，来描述转椅训练的强度之大。译员在记忆此段落时，可以想象一位航天员从转椅上下来，由于眩晕无法站稳，满头大汗的画面，通过将信息视觉化的方式来帮助自己更快地记住段落中的更多细节。

例 2：

The Ehang operating concept is right out of a science fiction novel: A passenger boards the autonomous air vehicle. He or she selects from a list of destinations on a touchscreen display and clicks a button to depart and another button to land. The vehicle flies the route autonomously but is under continuous surveillance and monitoring from a command and control center on the ground.

该段描述了乘客乘坐全自动飞行器的过程，译员在听的过程中可想象自己坐上了该飞行器，选择目的地后按按钮起飞，再在抵达的时候按按钮着陆的画面，这样就能将操控全自动飞行器的过程完整地记忆下来。

四、信息组块法

信息组块法指的是将相似或相关的信息片段进行整合归类，减少需要记忆的信息组块的数量，用较少的组块囊括更多的信息，从而减少译员短时记忆的压力①。以下为信息组块法具体运用于段落记忆的例子。

例 1：

当我最终以优异的成绩通过全部考核，被确定为神九任务的女航天员时，那些搬着双腿上楼，连做梦都是实验和操作、技能与原理的日子，显得那样地意义非凡、那样地弥足珍贵。那一刻，一种巨大的幸福感像暖流一般在心中流淌，这是被祖国需要、被任务挑选、被人民信任的幸福，是个人梦想融入祖国荣耀的幸福！

本例最后一句话是一个并列句，发言人在描述为何在自己被确定为神九任务的女航天员时会感到如此幸福。译员在记忆时可将这些让发言人感到快乐的原因归为一个信息组块，记忆关键词“幸福”。合并信息组块进行记忆有助于为译员腾出更多时间来记忆其他重要信息，拓宽记忆内容的广度。

例 2：

As a result, today, across the United States, we are building and expanding hundreds of solar panels, wind turbine, electric vehicle, and battery manufacturing plants.

该例句选自美国副总统哈里斯 2023 年 12 月在联合国气候变化大会上的演讲，

---

① 邹德艳. 口译的记忆训练[M]. 北京：中央编译出版社，2016：10.

它提到了许多设备,如“solar panel”“wind turbine”等,通过进一步梳理信息间的联系,不难发现这些设备都与清洁能源有关,因此可依据它们的共同点,将这些设备归为一个组块进行记忆,有利于让译员更有条理地记忆该句信息。

五、信息联想法

信息联想法指的是译员将听到的信息与生活中熟悉的事物或之前记忆的旧信息建立联系,从而起到旧信息提示新信息的作用,进一步强化译员对新信息的记忆①。以下为信息联想法具体运用于段落记忆的例子。

例 1:

第一次参加舱外服水下试验,我在 120 多公斤的服装中才工作了四五个小时,手就已经不停地颤抖,而将来真正的出舱活动训练,一次就要连续工作七八个小时。没有捷径,没有巧工,只有练,练,练,加量地练! 学,学,学,玩命地学。

本段出现了几个描述工作服重量和工作时间的数字,译员记忆这些数字时可尝试与生活中熟悉的事物建立联系来强化记忆,例如译员可从“120”联想到中国大陆急救电话号码,“四五个小时”联想到工作半天的时间,从“七八个小时”联想到工作一天的时间。通过联想与这些数字建立联系,译员可以更快地将这些信息记忆下来。

例 2:

As of last month, the FAA had registered almost 1.4 million UAS, nearly 400,000 of which were for commercial operations. That's four times as many registered drones than there are manned aircraft, and we've been registering drones for less than four years.

这一段的数字非常多且十分密集,译员如果想通过死记硬背的方式将这些数字记忆下来,较为困难。译员可通过将这些数字与生活中熟悉的事物建立联系的方式进行记忆,例如从“1.4 million”联想到中国人口 1.4 billion,但要特别记忆计数单位不同。此外,也可在之前记忆的旧信息和新信息之间寻找共同点,建立联系,再进行记忆。例如译员可先记忆“1.4 million”,再将下一数字“400,000(0.4 million)”与前一数字建立联系,如联想到 1.4 million 减去 1 million 即为 0.4 million。如此,译员便可更准确地记忆同一段落中密集的数字信息。

---

① 任文.英汉口译教程[M].北京:外语教学与研究出版社,2011:74.

第二节　篇章讲解与练习

一、重点解析篇

(一) 利用蘑菇在火星上为人类建造未来家园

1. 词汇表

Space Technology Mission Directorate	空间技术任务局
NASA's Innovative Advanced Concepts (NIAC) Program	美国国家航空航天局创新先进概念计划
mycelia	菌丝
myco-architecture	菌制建筑
fungal mycelium	真菌菌丝体
reinforced concrete	钢筋混凝土
false positive reading	假阳性读数
spore	孢子
cyanobacteria	蓝细菌

2. 背景知识

(1) 美国国家航空航天局

1958 年 7 月 29 日，美国总统艾森豪威尔签署了《美国公共法案 85-568》。1958 年 10 月 1 日，美国国家航空航天局正式成立，其作为美国联邦政府的一个行政性科研机构，负责制定、实施美国的太空计划，并开展航空科学暨太空科学的研究。NASA 是世界上最权威的航空航天科研机构，与许多国内及国际上的科研机构分享其研究数据，其领导项目包括阿波罗登月计划、建立"天空实验室(Skylab)空间站"等①。

(2) 美国国家航空航天局创新先进概念(NIAC)计划

美国国家航空航天局创新先进概念(NIAC)计划创立于 1998 年，原名为"NASA 先进概念研究所"，于 2011 年更名为"NASA 创新先进概念计划"。该计划旨在支持

---

①　新华网. 背景："神机构"NASA 从何来？美国瞄准太空探索[EB/OL]. (2015-09-28)[2023-07-10]. http://www.xinhuanet.com//world/2015-09/28/c_128276493.htm.

可能从根本上改变未来的早期空间技术研究，是一个具有远见和深远意义的航空航天计划，它有可能为未来可能的太空任务创造突破性技术。目前，NIAC 计划已资助如“PI 卫星防御”“太空制药厂”等多个项目①。

3. 原　文

Using Mushrooms to Build Future Homes on Mars for Humans②

Have you ever wondered what could happen if Earth becomes uninhabitable?

After Earth, there's only one planet where humans can survive—MARS.

As phrased by Therese Griebel, deputy associate administrator for programs at NASA's Space Technology Mission Directorate, "People have always shown they will do what is necessary to survive." "If the Earth becomes unsustainable, there will be a reason to go somewhere else."

And NASA has started projecting their focus on Mars.

Currently, NASA is investigating the possibilities of finding technologies that can grow habitable structures on Mars. Surprisingly, these technologies include mycelia and fungi, the underlying thread that makes up a whole fungus.

The traditional route of building habitats on Mars is like the turtle, carrying homes in their backs wherever they go. Despite being a reliable plan, it can cause huge energy costs, says Lynn Rothschild, the principal investigator on the early-stage project of building homes out of mushrooms on Mars. However, Lynn suggests we can harness mycelia to grow the habitats ourselves once we all get to the planet.

According to astronauts, it might be possible that one day we might all be living under the revolutionary concept called "myco-architecture." This concept demonstrates how fungal mycelium is extremely stronger than reinforced concrete can grow and repair itself.

Will future homes on Mars be built with fungi?

NASA is already on it; the project is part of NASA's Innovative Advanced Concepts program that considers different aspects of life to be a technology. Astronauts can bring in a much reliable and compact habitat built from lightweight materials that are

---

① 网易. 盘点 NASA 十大“特别”的太空计划[EB/OL]. (2022-12-10)[2023-07-10]. https://www.163.com/dy/article/HO7LRNEB0553VYNH.html.

② NASA. Could Future Homes on the Moon and Mars Be Made of Fungi[EB/OL]. (2020-01-14)[2023-07-10]. https://www.nasa.gov/centers-and-facilities/ames/could-future-homes-on-the-moon-and-mars-be-made-of-fungi/.(文字有校改)

embedded with fungi. This can survive long-term spaceflight. Now when these materials touch the surface, the only thing astronauts will be needing to do is to activate the fungi, using water. This habitat will not only protect humans but the lunar of the Martian surface as well since the fungi will be contained within the structure.

Further on, the mycelia will be genetically altered to remain sustainable even when separated from the habitat. This prevents the surface of Mars from getting contaminated. Besides this, it will also help in detecting false positive reading for life on the surface of Planet Mars. These structures will then be forced to reinforce its structure and further prevent contamination.

These fungi feed on organic material to produce spores. Inside the spores are the mycelia that disguise like the roots to help build the fungi. This further spreads out into millions of mushrooms.

Planet Mars will neither be a harsh environment for humans nor for the fungi. However, the fungi will need cyanobacteria for survival. The cyanobacteria, with the help of solar energy, could convert carbon dioxide and water into food and oxygen. A domed habitat is designed with three layers—the outside layer consisting of the frozen water acts like a roadblock between radiation and the astronauts. The third layer provides water for the second layer, i.e. the cyanobacteria, that further converts into oxygen. The final layer, made up of mycelia, gathers nutrients from the cyanobacteria layer.

According to what Rothschild says, "When we design for space, we're free to experiment with new ideas and materials with much more freedom than we would on Earth." Once the prototypes complete the design for other worlds, they can directly be brought to our planet.

4. 参考译文

美国国家航空航天局利用蘑菇在火星上为人类建造未来家园[①]

你是否曾想过,如果地球不再适合人类居住,将会发生什么?

除地球外,适合人类居住的星球仅剩一个——火星。

美国国家航空航天局(NASA)空间技术任务局(STMD)副局长帮办特蕾莎·格里贝尔(Therese Griebel)表示,"人类无数次地证明:只要能生存下去,什么必要的事情都愿意做。如果地球变得不可持续,应当另寻他处。"

NASA 已经开始计划给予火星更多关注。

目前,NASA 正在研究是否能够找到某些技术,让火星上"长出"可居住建筑。

---

① 编者译。

令人惊讶的是，菌丝和真菌便符合该技术要求，这些位于地下的丝状体能够构成一个真菌建筑。

在火星上建造栖息地的传统思路类似乌龟，无论走到哪里都要把家背在背上。林恩·罗斯柴尔德(Lynn Rothschild)致力于探索在火星利用蘑菇建造房屋的可能性，是其早期项目的主要研究人员。他表示尽管传统计划很可靠，但也会产生巨大的能源消耗。不过，林恩认为，一旦我们到了火星，不妨利用菌丝体来"种出"自己的栖息地。

宇航员们认为，或许真的有一天我们都会生活在"菌制建筑"这一革命性概念之下。这个概念表明真菌菌丝体要比钢筋混凝土坚固得多，而且能够自我生长修复。

未来真的可以用真菌在火星上建造房屋吗？

NASA已经启动了相关项目。该项目隶属于NASA创新先进概念计划，该计划将生命的不同方面视为一种技术。宇航员可以使用植入真菌的轻质材料建造一个更为可靠、更为紧密的栖息地。栖息地可以经受长期的太空飞行。这些材料运到火星表面后，宇航员唯一需要做的就是用水激活真菌。因为真菌保存在该结构中，这样的栖息地不仅能够保护人类，还能够保护火星的卫星。

之后，专家还会对菌丝体进行基因编辑，使其即便与栖息地分离也能够保持可持续性，防止火星表面受到污染。除此之外，它还将有助于检测火星表面生命的假阳性读数。编辑后的菌丝体基因会加固整体结构，并防止进一步污染。

这些真菌以有机材料为食，产生孢子。孢子里面是菌丝体，它们形如植物根系，能帮助构建真菌，从而进一步扩散为数以百万计的蘑菇。

火星对人类和真菌来说，环境都不算恶劣。不过，真菌的存活离不开蓝细菌。蓝细菌在太阳能的帮助下可以将二氧化碳和水转化为食物和氧气。栖息地上罩穹顶，并采用三层式设计，最外层由冷冻水构成，为宇航员阻隔外界辐射，第三层为第二层——蓝藻层提供水分，使其进一步转化为氧气。最后一层由菌丝体组成，从蓝藻层中收集营养物质。

罗斯柴尔德表示，"当我们为太空设计时，可以自由地尝试新的想法和材料，比在地球上自由得多。"只要完成了对其他太空领域的设计并创造出雏形，我们就能直接将其应用在地球上。

5. 要点讲解

本篇材料语言简练，信息量大，且常出现了mycelia、cyanobacteria等专业术语，译员在听信息的过程中理解部分段落的难度较大。在记忆本篇材料时，译员可先做好译前准备，熟悉文章中常用的科技术语，再运用综述法、复述法和成像法等方法辅助记忆。

听本篇材料时，译员可先运用综述法，把握整篇文章的框架。该材料并没有出现可以概括整篇文章内容的总领句和中心句，因此，译员在听的过程中需要提取每段的关键词，并发挥主观能动性将关键词串联起来得出段落大意。通过提取uninhabitable、MARS、NASA、focus等关键词可以得出，材料前四段主要交代了

NASA 在火星种植真菌计划的背景。通过 technologies、habitable structures、turtle、myco-architecture、stronger 等词，译员可推断出 5 至 7 段主要是对菌制建筑的介绍。最后，通过 project、protect、genetically altered、neither be a harsh environment 和 freedom 等词和短语，译员可推断出剩下的内容主要交代了该真菌种植计划操作的原理和可行性，以及其建成的好处。经过对以上关键词的提取，译员便可对该材料的组织框架有更清晰的认知，进一步帮助其记忆。

对于该材料具体段落的细节，译员可在理解原文的基础上，运用复述法，通过对长难句断句、重新组织语序等方式对原文进行加工处理，并用自己的话将原文含义表述出来以强化记忆。同时也可运用成像法进行记忆。段落中有许多对内容描述得十分具体的地方，例如在第 6 段，材料将在火星种植真菌作为人类栖息地的做法和乌龟联系起来，译员可通过想象在头脑中生成乌龟背着龟壳随处走动的图像，以便更好地理解和记忆人类生活在真菌建筑里的情景。在记忆"The cyanobacteria, with the help of solar energy… The final layer, made up of mycelia, gathers nutrients from the cyanobacteria layer."这段时，译员可在脑海中想象段落中提到的内外三层式建筑——最外层为冷冻水，第二层铺上了蓝藻，最里面的一层种上了菌丝体。想象出这样的画面有利于译员强化自己对该建筑的印象及记忆。

（二）习近平主席在"领导人气候峰会"上的讲话

1. 词汇表

2023 年可持续发展议程	2030 Agenda for Sustainable Development
碳中和	carbon neutrality
《应对气候危机联合声明》	*Joint Statement Addressing the Climate Crisis*
《〈蒙特利尔议定书〉基加利修正案》	*Kigali Amendment to the Montreal Protocol*
气候遥感卫星	remote sensing satellites for climate monitoring
低碳示范区	low-carbon demonstration zone
节能灯	energy-efficient light
煤电项目	coal-fired power generation project

2. 背景知识

(1) 领导人气候峰会

领导人气候峰会由美国总统拜登发起，由拜登总统和副总统哈里斯主持，于 2021 年 4 月 22—23 日以视频方式举行，聚焦气候变化挑战、应对气候变化解决方案、资金援助、创新等议题。美国白宫发布声明称，此次峰会邀请40位国家和国际组织领导人参加，包括中国、俄罗斯等38个国家的领导人以及欧盟委员会主席和欧洲理事会主席。此次峰会包含5项议程，讨论的主题涉及提高应对气候变化的决心、气候适应能力与韧性、气候安全、应对气候变化的创新技术、应对气候变化带来的经济机会等。出席会议的发展中国家领导人呼吁发达国家在应对气候变化方面展现更大的决心和行动，切实履行气候变化融资承诺，为发展中国家提供更多资金、技术、能力建设等方面的支持①。

(2) 碳中和

碳中和一般是指国家、企业、产品、活动或个人在一定时间内直接或间接产生的二氧化碳或温室气体排放总量，通过植树造林、节能减排等形式，以抵消自身产生的二氧化碳或温室气体排放量，实现正负抵消，达到相对"零排放"。2023年9月，《2023全球碳中和年度进展报告》在北京正式发布。报告系统评价了全球197个国家在碳中和承诺、低碳技术、气候投融资、国际气候合作等方面的进程，为推动各国深化碳中和转型、弥合全球碳中和进展与《巴黎协定》温升目标的差距提供了重要信息和参考②。

3. 原　文

共同构建人与自然生命共同体

——习近平主席在"领导人气候峰会"上的讲话③

2021年4月22日　北京

尊敬的拜登总统，

尊敬的各位同事：

很高兴在"世界地球日"到来之际出席领导人气候峰会，感谢拜登总统的邀请。借此机会，我愿同大家就气候变化问题深入交换意见，共商应对气候变化挑战之策，共谋人与自然和谐共生之道。

人类进入工业文明时代以来，在创造巨大物质财富的同时，也加速了对自然资源的攫取，打破了地球生态系统平衡，人与自然深层次矛盾日益显现。近年来，气候变化、生物多样性丧失、荒漠化加剧、极端气候事件频发，给人类生存和发展带来严峻挑战。新冠肺炎疫情持续蔓延，使各国经济社会发展雪上加霜。面对全球环境治理前

---

① 新华网. 综合消息：发展中国家领导人呼吁发达国家为应对气候变化提供更多支持[EB/OL]. (2021-04-23)[2023-11-14]. http://www.xinhuanet.com/world/2021-04/23/c_1127363475.htm? baike.

② 中国青年报.《2023全球碳中和年度进展报告》发布[EB/OL]. (2023-09-25)[2023-11-14]. https://baijiahao.baidu.com/s? id=1777994674123873577&wfr=baike.

③ 中国政府网. 习近平在"领导人气候峰会"上的讲话(全文)[EB/OL]. (2021-04-22)[2023-11-14]. https://www.gov.cn/xinwen/2021-04/22/content_5601526.htm.

所未有的困难,国际社会要以前所未有的雄心和行动,勇于担当,勠力同心,共同构建人与自然生命共同体。

——坚持人与自然和谐共生。“万物各得其和以生,各得其养以成。”大自然是包括人在内一切生物的摇篮,是人类赖以生存发展的基本条件。大自然孕育抚养了人类,人类应该以自然为根,尊重自然、顺应自然、保护自然。不尊重自然,违背自然规律,只会遭到自然报复。自然遭到系统性破坏,人类生存发展就成了无源之水、无本之木。我们要像保护眼睛一样保护自然和生态环境,推动形成人与自然和谐共生新格局。

——坚持绿色发展。绿水青山就是金山银山。保护生态环境就是保护生产力,改善生态环境就是发展生产力,这是朴素的真理。我们要摒弃损害甚至破坏生态环境的发展模式,摒弃以牺牲环境换取一时发展的短视做法。要顺应当代科技革命和产业变革大方向,抓住绿色转型带来的巨大发展机遇,以创新为驱动,大力推进经济、能源、产业结构转型升级,让良好生态环境成为全球经济社会可持续发展的支撑。

——坚持系统治理。山水林田湖草沙是不可分割的生态系统。保护生态环境,不能头痛医头、脚痛医脚。我们要按照生态系统的内在规律,统筹考虑自然生态各要素,从而达到增强生态系统循环能力、维护生态平衡的目标。

——坚持以人为本。生态环境关系各国人民的福祉,我们必须充分考虑各国人民对美好生活的向往、对优良环境的期待、对子孙后代的责任,探索保护环境和发展经济、创造就业、消除贫困的协同增效,在绿色转型过程中努力实现社会公平正义,增加各国人民获得感、幸福感、安全感。

——坚持多边主义。我们要坚持以国际法为基础、以公平正义为要旨、以有效行动为导向,维护以联合国为核心的国际体系,遵循《联合国气候变化框架公约》及其《巴黎协定》的目标和原则,努力落实2030年可持续发展议程;强化自身行动,深化伙伴关系,提升合作水平,在实现全球碳中和新征程中互学互鉴、互利共赢。要携手合作,不要相互指责;要持之以恒,不要朝令夕改;要重信守诺,不要言而无信。

中方欢迎美方重返多边气候治理进程。中美刚刚共同发布了《应对气候危机联合声明》,中方期待同包括美方在内的国际社会一道,共同为推进全球环境治理而努力。

——坚持共同但有区别的责任原则。共同但有区别的责任原则是全球气候治理的基石。发展中国家面临抗击疫情、发展经济、应对气候变化等多重挑战。我们要充分肯定发展中国家应对气候变化所作贡献,照顾其特殊困难和关切。发达国家应该展现更大雄心和行动,同时切实帮助发展中国家提高应对气候变化的能力和韧性,为发展中国家提供资金、技术、能力建设等方面支持,避免设置绿色贸易壁垒,帮助他们加速绿色低碳转型。

各位同事!

中华文明历来崇尚天人合一、道法自然,追求人与自然和谐共生。中国将生态文明理念和生态文明建设写入《中华人民共和国宪法》,纳入中国特色社会主义总体布局。中国以生态文明思想为指导,贯彻新发展理念,以经济社会发展全面绿色转型为引领,以能源绿色低碳发展为关键,坚持走生态优先、绿色低碳的发展道路。

去年，我正式宣布中国将力争2030年前实现碳达峰、2060年前实现碳中和。这是中国基于推动构建人类命运共同体的责任担当和实现可持续发展的内在要求作出的重大战略决策。中国承诺实现从碳达峰到碳中和的时间，远远短于发达国家所用时间，需要中方付出艰苦努力。中国将碳达峰、碳中和纳入生态文明建设整体布局，正在制定碳达峰行动计划，广泛深入开展碳达峰行动，支持有条件的地方和重点行业、重点企业率先达峰。中国将严控煤电项目，"十四五"时期严控煤炭消费增长、"十五五"时期逐步减少。此外，中国已决定接受《〈蒙特利尔议定书〉基加利修正案》，加强非二氧化碳温室气体管控，还将启动全国碳市场上线交易。

作为全球生态文明建设的参与者、贡献者、引领者，中国坚定践行多边主义，努力推动构建公平合理、合作共赢的全球环境治理体系。中方将在今年10月承办《生物多样性公约》第十五次缔约方大会，同各方一道推动全球生物多样性治理迈上新台阶，支持《联合国气候变化框架公约》第二十六次缔约方会议取得积极成果。中方秉持"授人以渔"理念，通过多种形式的南南务实合作，尽己所能帮助发展中国家提高应对气候变化能力。从非洲的气候遥感卫星，到东南亚的低碳示范区，再到小岛国的节能灯，中国应对气候变化南南合作成果看得见、摸得着、有实效。中方还将生态文明领域合作作为共建"一带一路"重点内容，发起了系列绿色行动倡议，采取绿色基建、绿色能源、绿色交通、绿色金融等一系列举措，持续造福参与共建"一带一路"的各国人民。

各位同事！

"众力并，则万钧不足举也。"气候变化带给人类的挑战是现实的、严峻的、长远的。但是，我坚信，只要心往一处想、劲往一处使，同舟共济、守望相助，人类必将能够应对好全球气候环境挑战，把一个清洁美丽的世界留给子孙后代。

谢谢大家。

4. 参考译文

For Man and Nature: Building a Community of Life Together
—Remarks at the Leaders Summit on Climate[①]

H. E. Xi Jinping

President of the People's Republic of China

Beijing, 22 April 2021

Honorable President Joe Biden,

Honorable Colleagues,

It is a great pleasure to join you at the Leaders Summit on Climate on Earth Day.

---

① Ministry of Foreign Affairs of the People's Republic of China. For Man and Nature: Building a Community of Life Together[EB/OL]. (2021-04-22)[2023-11-14]. https://www.fmprc.gov.cn/mfa_eng/wjdt_665385/zyjh_665391/202104/t20210422_9170542.html.

I wish to thank President Biden for the kind invitation. It is good to have this opportunity to have an in-depth exchange of views with you on climate change, and to discuss ways to tackle this challenge and find a path forward for man and Nature to live in harmony.

Since time of the industrial civilization, mankind has created massive material wealth. Yet, it has come at a cost of intensified exploitation of natural resources, which disrupted the balance in the Earth's ecosystem, and laid bare the growing tensions in the human-Nature relationship. In recent years, climate change, biodiversity loss, worsening desertification and frequent extreme weather events have all posed severe challenges to human survival and development. The ongoing COVID-19 pandemic has added difficulty to economic and social development across countries. Faced with unprecedented challenges in global environmental governance, the international community needs to come up with unprecedented ambition and action. We need to act with a sense of responsibility and unity, and work together to foster a community of life for man and Nature.

—We must be committed to harmony between man and Nature. "All things that grow live in harmony and benefit from the nourishment of Nature." Mother Nature is the cradle of all living beings, including humans. It provides everything essential for humanity to survive and thrive. Mother Nature has nourished us, and we must treat Nature as our root, respect it, protect it, and follow its laws. Failure to respect Nature or follow its laws will only invite its revenge. Systemic spoil of Nature will take away the foundation of human survival and development, and will leave us human beings like a river without a source and a tree without its roots. We should protect Nature and preserve the environment like we protect our eyes, and endeavor to foster a new relationship where man and Nature can both prosper and live in harmony.

—We must be committed to green development. Green mountains are gold mountains. To protect the environment is to protect productivity, and to improve the environment is to boost productivity—the truth is as simple as that. We must abandon development models that harm or undermine the environment, and must say no to shortsighted approaches of going after near-term development gains at the expense of the environment. Much to the contrary, we need to ride the trend of technological revolution and industrial transformation, seize the enormous opportunity in green transition, and let the power of innovation drive us to upgrade our economic, energy and industrial structures, and make sure that a sound environment is there to buttress sustainable economic and social development worldwide.

—We must be committed to systemic governance. Mountains, rivers, forests as well as farmlands, lakes, grasslands and deserts all make indivisible parts of the ecosystem. Protecting the ecosystem requires more than a simplistic, palliative approach. We need to follow the innate laws of the ecosystem and properly balance all elements and aspects of Nature. This is a way that may take us where we want to be, an ecosystem in sound circulation and overall balance.

—We must be committed to a people-centered approach. The environment concerns the well-being of people in all countries. We need to take into full account people's longing for a better life and a good environment as well as our responsibility for future generations. We need to look for ways to protect the environment, grow the economy, create jobs and remove poverty all at the same time, so as to deliver social equity and justice in the course of green transition and increase people's sense of benefit, happiness and security.

—We must be committed to multilateralism. We need to work on the basis of international law, follow the principle of equity and justice, and focus on effective actions. We need to uphold the UN-centered international system, comply with the objectives and principles laid out in the *UN Framework Convention on Climate Change* (*UNFCCC*) and its *Paris Agreement*, and strive to deliver the 2030 Agenda for Sustainable Development. We need to each take stronger actions, strengthen partnerships and cooperation, learn from each other and make common progress in the new journey toward global carbon neutrality. In this process, we must join hands, not point fingers at each other; we must maintain continuity, not reverse course easily; and we must honor commitments, not go back on promises.

China welcomes the United States' return to the multilateral climate governance process. Not long ago, the Chinese and US sides released a Joint Statement Addressing the Climate Crisis. China looks forward to working with the international community including the United States to jointly advance global environmental governance.

—We must be committed to the principle of common but differentiated responsibilities. The principle of common but differentiated responsibilities is the cornerstone of global climate governance. Developing countries now face multiple challenges to combat COVID-19, grow the economy, and address climate change. We need to give full recognition to developing countries' contribution to climate action and accommodate their particular difficulties and concerns. Developed countries need to increase climate ambition and action. At the same time, they need to make concrete efforts to help developing countries strengthen the capacity and resilience against climate change, support them in financing, technology, and capacity building, and refrain from creating green trade barriers, so as to help developing countries accelerate the transition to green and low-carbon development.

Colleagues,

The Chinese civilization has always valued harmony between man and Nature as well as observance of the laws of Nature. It has been our constant pursuit that man and Nature could live in harmony with each other. Ecological advancement and conservation have been written into China's Constitution and incorporated into China's overall plan for building socialism with Chinese characteristics. China will follow the Thought on Ecological Civilization and implement the new development philosophy. We will aim to achieve greener economic and social development in all aspects, with a special focus on developing green and low-carbon energy. We will continue to prioritize ecological conservation and pursue a green and low-carbon path to development.

Last year, I made the official announcement that China will strive to peak carbon dioxide emissions before 2030 and achieve carbon neutrality before 2060. This major strategic decision is made based on our sense of responsibility to build a community with a shared future for mankind and our own need to secure sustainable development. China has committed to move from carbon peak to carbon neutrality in a much shorter time span than what might take many developed countries, and that requires extraordinarily hard efforts from China. The targets of carbon peak and carbon neutrality have been added to China's overall plan for ecological conservation. We are now making an action plan and are already taking strong nationwide actions toward carbon peak. Support is being given to peaking pioneers from localities, sectors and companies. China will strictly control coal-fired power generation projects, and strictly limit the increase in coal consumption over the 14th Five-Year Plan period and phase it down in the 15th Five-Year Plan period. Moreover, China has decided to accept the Kigali Amendment to the Montreal Protocol and tighten regulations over non-carbon dioxide emissions. China's national carbon market will also start trading.

As a participant, contributor and trailblazer in global ecological conservation, China is firmly committed to putting multilateralism into action and promoting a fair and equitable system of global environmental governance for win-win cooperation. China will host COP15 to the Convention on Biological Diversity this October and looks forward to working with all parties to enhance global governance on biodiversity. We support COP26 to the *UNFCCC* in achieving positive outcomes. As we in China often say, "It is more important to show people how to fish than just giving them fish." China has done its best to help developing countries build capacity against climate change through various forms of results-oriented South-South cooperation. From remote sensing satellites for climate monitoring in Africa to low-carbon demonstration zones in Southeast Asia and to energy-efficient lights in small island countries, such cooperation has yielded real, tangible and solid results. China has also made ecological cooperation a key part of Belt and Road cooperation. A number of green action initiatives have been launched, covering wide-ranging efforts in green infrastructure, green energy, green transport and green finance, to bring enduring benefits to the people of all Belt and Road partner countries.

Colleagues,

As we say in China, "When people pull together, nothing is too heavy to be lifted." Climate change poses pressing, formidable and long-term challenges to us all. Yet I am confident that as long as we unite in our purposes and efforts and work together with solidarity and mutual assistance, we will rise above the global climate and environment challenges and leave a clean and beautiful world to future generations.

Thank you.

5. 要点讲解

本篇演讲稿语言平实,表达逻辑清晰,科技类术语数量较少,译员理解该发言的难度不大。在收听该发言稿时,译员可运用综述法、成像法、信息组块法、复述法及信息联想法等方法来辅助记忆。

译员可先运用综述法,把握整篇演讲稿的叙述框架,了解发言人的意图。不难发现,该发言稿的中心句在文章第一段的最后一句"……我愿同大家就气候变化问题……人与自然和谐共生之道",因为接下来的内容都围绕着这句话展开。习近平主席先是提出了目前气候问题带来的一些挑战,接着提出了几点应对对策。发言人在阐述每个对策时,都有一句主题句,如"坚持绿色发展"等,通过这些主题句,译员也可快速了解和记忆每个对策的大致内容。紧接着,习近平主席还提到了中国在过去应对气候变化问题所采取的行动和未来需要实现的目标。最后,他呼吁各国携手合作,共同应对气候变化问题。经过以上对发言内容的概括和梳理,译员即可更好地掌握该讲话的总体框架。

除了运用综述法梳理演讲的具体框架以外,译员还可以运用成像法、信息组块法、信息联想法和复述法等方法记忆段落细节。习近平主席在发言中用了很多比喻，十分生动形象，译员可尝试运用成像法对这些比喻进行记忆。例如，在记忆“绿水青山就是金山银山”这句话时，译员可想象金银珠宝堆积如山的场景，加强对绿水青山重要性的认识，对该句内容形成图像记忆。在记忆“……采取绿色基建、绿色能源、绿色交通，绿色金融……”时，译员可运用信息组块法将这句话中提到的各种绿色举措合并为一个组块，减少记忆的压力。原文中还提到了一些数字，如“今年 10 月”“2030 年前”“‘十四五’”等。遇到这种情况，译员可善用信息联想法，将这些数字与生活中熟悉的概念建立联系，如将 10 月份和庆祝国庆节联系起来等。习近平主席在发言时还引用了不少诗句，在记忆“众力并，则万钧不足举也”等诗文时，译员可运用复述法，在充分理解该诗文含义的前提下，对信息进行加工，并转化为自己的语言表达出来，如表达为“只要各国携手合作，就没有闯不过的难关”等。

二、实战训练篇

（一）美国联邦航空局(FAA)前代理副局长卡尔·E·伯利森(Carl E. Burleson)在中国民航发展论坛上的讲话

1. 词汇表

CAAC Flight 981	中国民航 981 次航班
Boeing 747SP	波音 747SP 机型
unmanned aircraft systems (UAS)	无人机系统
manned aircraft	载人飞行器
China Civil Aviation Development Forum	中国民航发展论坛
Acting Deputy Administrator	代理副局长
Civil Aviation Administration of China (CAAC)	中国民用航空局
Federal Aviation Administration (FAA)	美国联邦航空管理局
International Air Transport Association (IATA)	国际航空运输协会
DJI	大疆公司
Ehang	亿航公司
UAS Integration Pilot Program (IPP)	无人机系统融合试点计划
higher-density urban area	高密度城区

2. 原　文

China Civil Aviation Development Forum[①]

Carl E. Burleson
Former Acting Deputy Administrator
Federal Aviation Administration
1 May 2019

Thank you, for the warm welcome from the Civil Aviation Administration of China and for the opportunity to join you at this impressive event. It is an honor to be here in Beijing representing the Federal Aviation Administration (FAA).

As we gather here this week, we find both of our countries—and all countries for that matter—on the edge of a bold new era of innovation, one that will require even closer collaboration among us as new entrants shatter old norms with their game-changing technologies. I know that we have a partner in that journey with CAAC. You share a similar safety philosophy and our bilateral cooperation helps amplify our safety messages.

Let's reflect for a moment on the FAA/CAAC partnership over the past 40 years and on how far we've come since the United States and China resumed diplomatic relations in 1979. Our work supported the restart of commercial air service in January 1981. The first was CAAC Flight 981, a Boeing 747SP that flew from Beijing to San Francisco.

Think about this—starting with Flight 981 in 1981, there were on average about two daily flights between Chinese mainland and the United States. Based on the latest data from the International Air Transport Association, there are now 100 daily flights.

That stellar growth in flight activity is also matched by what our two countries have done in practically all other aspects of aviation, including operations, aircraft certification, air traffic management, and always most importantly, aviation safety. We thank you for being our partner in this journey and look forward to the next 40 years. One thing is for sure, aviation will look much different by that time, but our bedrock principles on safety will not.

So how do we and the world's aviation regulators introduce these new and radically different users into the safest form of transportation on the planet? How do we do it

---

① Federal Aviation Administration. China Civil Aviation Development Forum[EB/OL]. (2019-05-01) [2023-07-10]. https://www.faa.gov/speeches/china-civil-aviation-development-forum.(文字有校改)

wisely, safely, efficiently and in a reasonable timeframe?

In thinking about the challenges, I'm reminded of a conversation I had with my sons on a stormy evening long ago.

It was a sticky August night in Washington D. C. and a raging thunderstorm had knocked the electrical power out. I took the boys out to our screened-in porch to watch the driving rain, the wind and the lightning show. My adventurous 12-year old pleaded, "Hey Dad, can I go out and play in this?" My younger son was sitting very quietly in my lap for the longest time. I finally asked him what he thought about the storm. He says, "Dad, is this any way for a seven-year-old to die?" That's reality—two boys, from the same genetic stock, having two polar opposite views of the same phenomena they were observing.

Their divergent views of the world is a perfect analogy for the challenge we in the aerospace industry face with the rapid development of new technologies that enable increasingly autonomous operations. Some see an incredible opportunity for development and new markets and services. Others see this as potentially undermining safety and adding new levels of risk into the system. As a regulator, our job is to find the pathway forward.

One area where we're having to find that way—and quickly—is unmanned aircraft systems, or UAS, and urban air mobility vehicles, also known as flying taxis.

As of last month, the FAA had registered almost 1.4 million UAS, nearly 400,000 of which were for commercial operations. That's four times as many registered drones than there are manned aircraft, and we've been registering drones for less than four years. We already issued about 130,000 remote pilot licenses. A good many of these are small, semi-autonomous systems that you fly with your finger and a smart phone, the type made so popular by China-based DJI, now the world's largest manufacturer of civilian drones.

Flying taxis are no longer in the realm of fantasy—they're coming very, very soon, and all of us here will have to start thinking about whether we'll jump into one to slip the surly bonds of traffic.

Ehang, based here in China, is a frontrunner in the emerging urban air mobility market. It was the first to carry passengers on test flights of its fully automated, two-seat, electrically powered flying taxi. Although not yet certified by the CAAC, Ehang is delivering vehicles for a variety of demonstrations here in China, in the United States, Canada and elsewhere. Other nascent air mobility companies like Uber Elevate and Kitty Hawk are also racing quickly to deploy these vehicles to the market.

The Ehang operating concept is right out of a science fiction novel: A passenger boards the autonomous air vehicle. He or she selects from a list of destinations on a touchscreen display and clicks a button to depart and another button to land. The vehicle flies the route autonomously but is under continuous surveillance and monitoring from a command and control center on the ground. And this is already happening in tests.

Not hard to see what might make regulators nervous. Certifying the safety of a vehicle that carries passengers yet has no pilot on board; figuring out how to integrate its operations with traditional air traffic; and regulating numerous fleets of these vehicles.

Let's face it—these are big challenges compared to the manned aircraft we've managed very successfully for decades—not just in terms of use but in terms of product cycles can generally be measured in months, not years.

So let me touch on how we're charting a course toward integration of these new entrants—gathering the data, creating the framework, and building the infrastructure, and doing it all safely.

Our risk management approach, how we set regulations, the policies we promote—and most importantly—how we've gotten to the level of safety in traditional aviation is all about having access to data to properly understand and manage risk. That is why the UAS Integration Pilot Program or IPP that Derek Kan talked about this morning is so important.

The IPP is allowing us to work with a variety of users and government organizations to safely test and validate advanced operations of drones. This will help us gain the data to develop the UAS regulations, policy and guidance through practical applications. And as he noted, it has allowed us to grant the first air carrier certification to Alphabet Inc's Wing Aviation for package deliveries in rural Blacksburg, Virginia. We anticipate several other companies will gain this approval over the next few months.

The FAA is focused on enabling an ever-expanding universe of UAS operations. In order to allow for such operations to be conducted safely and securely, we've moved forward with a number of rulemakings for small UAS, including external markings, flights over people and at night, and safe and secure operational concepts and will soon be issuing a rule on remote ID. Together, these rules will form our framework for advancing this new industry.

We won't just need regulatory framework, we will need an air traffic management system capable of managing manned and unmanned vehicles, where many of the latter will be highly autonomous. As Derek Kan noted this morning, we are working closely with NASA and industry to develop the concepts and tools we'll need to accomplish this task. What we learn from these increasingly complex demonstrations in higher-density urban areas will help inform the requirements for providers of these services going forward.

How fast can we bring new innovations and technologies into service? In commercial aviation, no new technology historically has premiered before its time where we have confidence it can be safely integrated in our airspace. And while I've been speaking of the future, we can't forget the need to address the present challenges.

We've had two tragedies in the last year with the Boeing 737 MAX, and I want to extend the FAA's heartfelt sympathy to the friends and families of those two accidents. Aviation is something that has broken down borders and opened opportunity across the world. But it also means that we all share the sadness of each other's losses.

We continue to review any and all evidence from the ongoing accident investigations. We became an international leader in aviation by taking action based on data and addressing risk. And we'd be the first to admit that we have not always gotten it perfect. That is why we are relentless advocates for transparency and continuous improvement. In fact, it's one of the reasons we have invited Director's General from around the world to Dallas this month to provide a comprehensive overview of the steps that have been taken since the accidents. And let me be clear—the only timeline FAA has in returning the 737 MAX to service is simple—when we are confident that the issues arising from these accidents have been addressed and it is safe to operate.

When we look to the future, and the new users and new technologies coming into aviation, it's a certainty that there will be differing opinions on whether any particular technology or innovation will help or hurt-just like my sons' impression of the lightning storm that night.

The innovations are not going to slow down. That is a challenge for the FAA and other civil aviation regulators around the world. The FAA is committed to finding a way forward that protects the public but allows innovation and opportunity to move forward, and we'll work hand-in-hand with our international partners, including CAAC, to make sure we address together these constantly changing challenges and to learn from one another.

Thank you so much for your kind attention.

（二）国际科技合作倡议

1. 词汇表

科技创新	science, technology and innovation (STI)
大变局	profound changes
开放共享	openness and sharing
时代挑战	challenges of the times
科学无国界	develop science for the benefit of all regardless of borders
知识产权保护	protection of intellectual property rights
绿色低碳转型	green and low-carbon transition
开放科学精神	open science spirit
人才交流合作	talent exchange and cooperation
平等参与	equal participation
科技合作政治化	politicization of science and technology cooperation
团结协作	solidarity and coordination
全球科技创新合作新模式	new model of global cooperation on STI

2. 原　文

国际科技合作倡议[①]

科技创新是人类社会发展的重要引擎，是应对全球性挑战的重要手段。当前，世界百年未有之大变局加速演进，人类发展面临越来越多的重大挑战。人类社会比以往任何时候都更需要国际合作和开放共享，通过科技创新合作探索解决全球性问题，共同应对时代挑战，共同促进和平发展。为倡导并践行开放、公平、公正、非歧视的国际科技合作理念，坚持“科学无国界、惠及全人类”，携手构建全球科技共同体，中国提出国际科技合作倡议：

——坚持崇尚科学。以科学的态度对待科学，以真理的精神追求真理。坚持科研诚信，尊重科研伦理，塑造科技向善理念，完善全球科技治理。加强知识产权保护，加强对新兴技术发展的包容与审慎管理。

---

① 科技部．国际科技合作倡议[EB/OL]．(2023-11-07)[2023-11-15]．https://www.most.gov.cn/kjbgz/202311/t20231107_188728.html.

——坚持创新发展。加强全球科技创新协作,共建全球创新网络,促进新兴技术推广应用,加强企业间创新和技术合作,为世界经济复苏和发展注入新动能。各国应该携手推动数字时代互联互通,加快全球绿色低碳转型,实现全人类可持续发展。

——坚持开放合作。秉承无国界、无障碍的开放科学精神,坚持科技创新人员和资源等在全球范围内自由流动,加强人才交流合作,构建开放自由的国际科技合作生态。坚决反对限制或阻碍科技合作、损害国际社会共同利益。

——坚持平等包容。秉承互相尊重、公正平等、非歧视的合作理念,倡导各个国家和科学研究实体平等参与国际科技合作。坚决反对将科技合作政治化、工具化、武器化,反对以国家安全为借口实施科技霸权霸凌。

——坚持团结协作。面对气候变化、卫生健康、环境保护、能源安全、粮食安全等人类社会迫切需要解决的全球性问题挑战,各国要同舟共济,加强科技创新主体深度协作、互学互鉴,推进实施国际大科学计划和大科学工程,共同突破关乎人类未来命运的重大科技难题。

——坚持普惠共赢。要坚持真正的多边主义,探索互利共赢的全球科技创新合作新模式,促进科技创新成果互惠互享。中国将面向全球设立科学研究基金,并加大对发展中国家科技援助,让科技进步惠及全人类。

(三) 航天员大队航天员陈冬在载人航天工程应用成果情况介绍会上的讲话

1. 词汇表

载人航天工程	manned spaceflight project
天宫二号	Tiangong-2
实验舱	experimental cabin
机械臂人机协同在轨操作维修技术	human-machine coordination in-orbit operation and maintenance technology of the robotic arm
植物栽培验证试验	plant cultivation verification experiment
神舟十一号	Shenzhou-11
在轨飞行时间	on-orbit flight time
中国人民解放军航天员大队	Chinese People's Liberation Army Astronaut Corps

2. 原 文

在载人航天工程应用成果情况介绍会上的讲话①

航天员大队航天员　陈冬

2018 年 9 月 26 日

天宫二号已经发射两周年了。33 天太空驻留是神舟系列飞船历次任务中时间最长的一次，祖国投入了大量的人力、物力、财力送我们上太空，为了利用好这宝贵的机会，作为航天员，我们都是争分夺秒、全力以赴、高标准完成多达 38 项各类实验。这也充分体现了人在航天活动中的地位、作用和价值。

这些实验，每个都有很强的指导性和实用性。前期的构想、方案设计修改，设备选材加工，大家花费了很多的精力，我们在太空中利用失重环境完成这些实验，飞行后，我们的专家花大量的时间去分析、处理这些实验数据。这是一个漫长的过程，最终才获得了宝贵的成果。在这里，我也对每个实验的团队、专家的辛苦付出，表示感谢！

从三个方面谈谈自己的感受。

一是保障人。6 次载人飞行，次数并不多，时间也不长，必须要积累更多的数据和经验，让航天员能够适应太空失重环境，高效完成工作。在 38 项实验中，人的健康研究占主导，有 8 项之多，我感觉是非常重要和必要的。例如，随着飞行时间的延长，航天员的视力会受到影响。在实验舱内，工作生活始终是在噪音环境下，听力发生变化，有些变化是不可逆的，而且对睡眠、心情会有影响，进而对工作产生影响。这些实验都是为了研究人在失重环境下身体各功能发生的某些变化，从而提出有效的预防和保护措施。人是载人航天的主体，将来走出地球寻找新的家园也是人类，所以要持续关注人的健康保障问题，把相关实验继续做下去。

二是利用人。地面科学家设计了非常好的各种实验，而我们在太空就相当于他们的眼睛和手臂，所以一定要利用好。例如，这次我们向习主席展示了机械臂人机协同在轨操作维修技术，体现了人与机器的完美结合；我们还进行了植物栽培验证试验，如播种、浇水、施肥、间苗，等等，记录每一天植物的生长情况，直到最后带回九棵绿油油的生菜。我们可以骄傲地说，通过我们的双手实现了地面科技人员的太空梦想。另一方面，我们的载人航天工程进行科学试验、技术验证的同时，对科普工作也非常重视。这次任务中也完成了很多的科普教育活动。例如，香港小朋友设计的太

---

① 陈冬. 航天技术为创造美好生活发挥更大作用[EB/OL]. (2018-09-26)[2023-07-10]. http://www.cmse.gov.cn/art/2018/9/26/art_18_32662.html.（文字有校改）

空养蚕实验，就引起了中小学生的浓厚兴趣。通过航天员的展示，可以让更多的人了解航天、认识航天，最终参与航天。

三是相信人。有人说机器人也可以当我们的眼睛和手，未来是否可以用机器人代替？但我想说，谁也不能低估人的主观能动性。后续空间站任务，系统会更加庞大，出现故障进行维修是不可避免，就像神舟十一号任务，我们专门开展了在轨维修实验。在随机应变、化险为夷方面，我感觉人比机器人更有优势。还有，也请科技工作者相信航天员，我们都是竭尽全力做好实验，拿出最好的状态，取得完美的数据。

有些实验需要很强的专业知识，对我们来说有些难度，但可以根据我们独有的失重经验提出建议，例如程序的优化、工具的改进，实际上也是参与设计。而且我们也在加强学习，不断提高自身能力素质，努力成为科研型航天员。后续我们还会有第三批航天员的加入，他们都有很强的专业背景。空间站任务在轨飞行时间长，实验项目会更多，我们也希望自己可以独立设计实验，将来在太空再去亲手完成它，那种感觉一定非常奇妙。

我相信，在天地的共同努力配合下，我们一定能完成好实验，并利用好实验成果，服务于大众，让航天技术为创造美好生活发挥更大作用。

第三章　笔记训练

上一章介绍了记忆在口译中的作用，译员需要在听辨的基础上，抓住主要内容，理清思路，并借助复述法、综述法、成像法、信息组块法和信息联想法等辅助手段，将源语信息印在脑海里。但交替传译发言人一次讲话时间可能过长，如果仅靠脑记，译员输出时难免会丢失数字、人名、地名和机构名称等重要细节。另外，听到语法结构复杂的复句时，译员的记忆系统无法对整个语句进行综合、分析加工，这就需要辅以适当的笔记，将冗长的语句以合适的形式记录下来[①]。

第一节　笔记训练技巧

在口译中，主要靠脑记，笔记只起辅助作用。口译笔记不是速记，亦不是课堂笔记和会议记录。因为口译笔记的目的不是记"原话"，而是记录信息要点和关键词语[②]。常听到口译初学者关于口译笔记的疑惑"自己的笔记记得很详尽，为什么翻的时候大脑一片空白?"问题可能出在听的时候记得过多，没有分配足够的精力分析整理信息，听辨环节没有做好，在看笔记输出的环节，自然不能脱口而出，给出逻辑清晰、表达通顺的产出。所以，初学口译笔记的同学，不要心急，要边听边分析，捋清逻辑关系，笔记越简约，输出才能越流畅。

一、记什么

那么译员需要记什么？译员需要记下有助于重现源语意思的信息。

1. 句子主干

句子主干需要记，包括主语、谓语（含动词时态）和宾语等。当然要视具体情况而定，如果译员很熟悉讲话内容，主语和宾语等主干信息可以省略。方向性和定性的内容一定要记准确，是"上升"还是"下降"，是"赞同"还是"反对"，一定不能记反了。

例 1：

A number of stakeholders and partners, including representatives of States members of the Committee, were interviewed for the evaluation and this indeed brought

---

① 鲍刚．口译理论概述[M]．北京：中国对外翻译出版有限公司，2011：149.

② 鲍刚．口译理论概述[M]．北京：中国对外翻译出版有限公司，2011：152.

valuable insight to the work and performance of the Office.

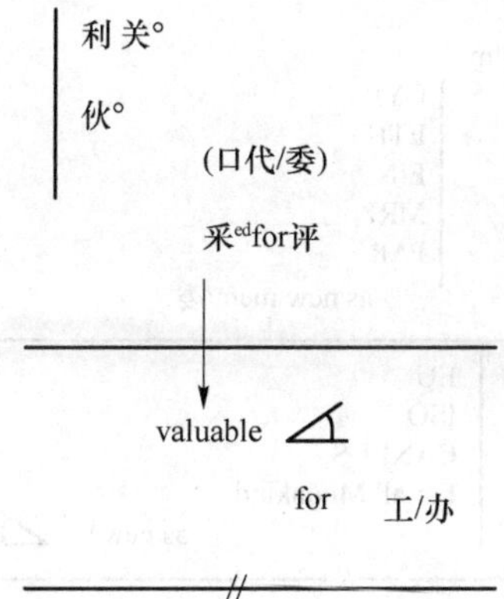

这个并列句中第一个简单句的主语较长，两个简单句的动词词组也较复杂，需要记下来。

例 2：

On behalf of the Office for Outer Space Affairs, I thank you for the opportunity to address this session on the work of the Office. I would like to assure you Mr. Chair of our commitment to assist you in running the meetings.

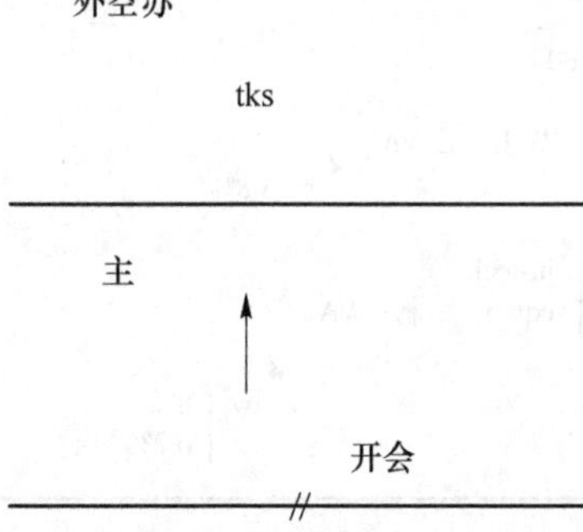

这两句是开场白的套话，结构清晰，内容简单，译员记忆没有压力，做笔记的时候可以简化，只记机构名称和承诺的对象及内容。

2. 容易遗忘的细节

一些靠脑记难以复盘的细节需要靠笔记，包括时间、地名、组织机构、人名、数字和单位等。

例 1：

At the outset, allow me to welcome Cyprus, Ethiopia, Finland, Mauritius and Paraguay as new States members of the Committee, and the European Union, International Organization of Standardization, CANEUS-International and For All

Moonkind, as the newest observer organizations to the Committee.

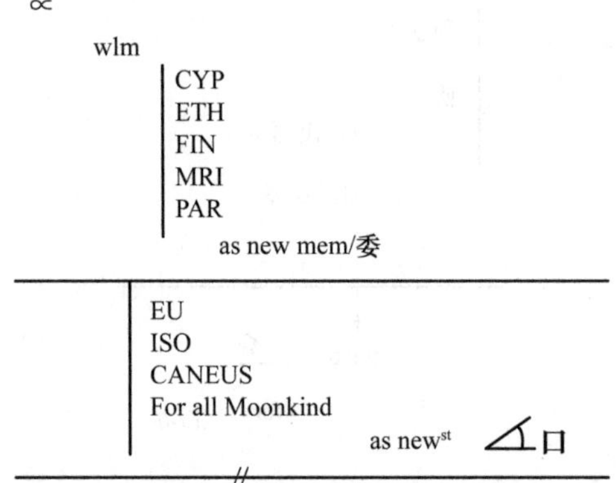

此例中，多个国家、国际组织名称也会对译员造成较大的短时记忆压力，需要记清楚。

例 2：

2022 年 1—11 月，我国规模以上工业增加值同比增长 3.8%，高技术制造业和装备制造业增加值同比分别增长 8%和 6.2%。制造业占 GDP 比重基本稳定，发挥了工业稳定宏观经济大盘的"压舱石"作用。与此同时，发展质量稳步提升，新能源汽车、大型客机、光伏等领域取得一批标志性成果。

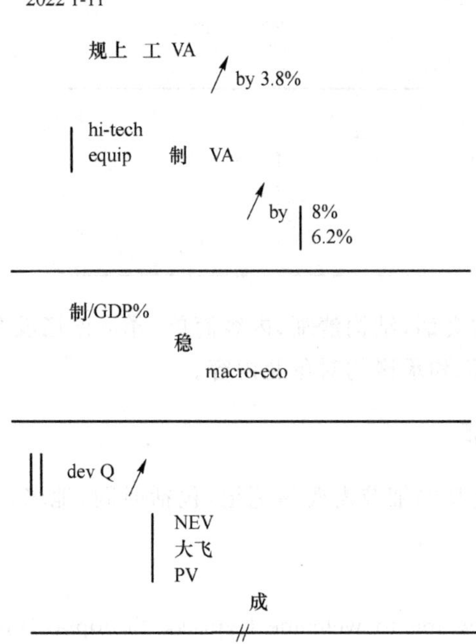

此例中数字较多，要准确记录数字及其对应的内容。如果只记数字，不记下与数

字相关或对应的概念或指标，就无法准确地进行传译。

3. 逻辑关系词

在记录冗长的复句时，不能忽视逻辑关系词，要醒目地标记好各语法结构之间的关系。

例 1：

I would like to thank the ones among you who took the time to answer the questions of OIOS and presented such a positive opinion on the performance of the Office. While some recommendations can, and will, be implemented and that the Office has already taken actions in the proper direction, others cannot be implemented due to a lack of human resources in the Office which has been reported in the formal response to OIOS, available as annex Ⅰ and Ⅱ of the evaluation report.

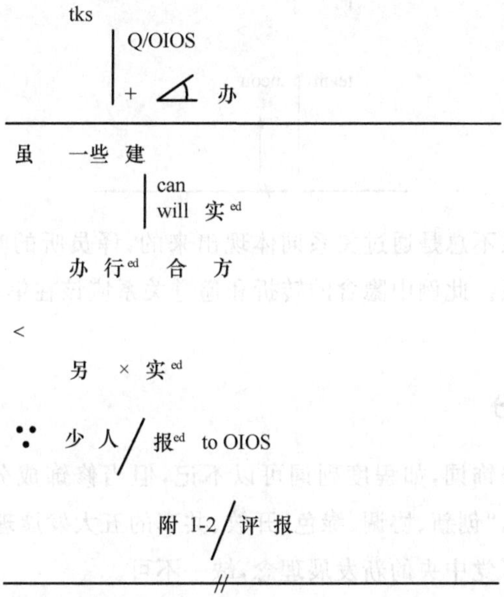

tks
| Q/OIOS
| + ⊿ 办

虽　一些 建
| can
| will 实ed

办 行ed 合 方

<

另 × 实ed

∵ 少 人 / 报ed to OIOS

附 1-2 / 评 报

此例逻辑关系复杂，表示转折和因果关系的关系词要写在笔记的最左侧，便于译员抓住逻辑脉络。

例 2：

人说："智者乐水，仁者乐山"，我们航天员都比较"乐天"，这么长的时间，我们难免也会有感到孤独寂寞、疲惫和想家的时候，这次任务我们除了有专业的地面心理支持手段外，我还可以和家人、战友通话，视频连线，我们乘组之间也互相鼓励，互相支持。

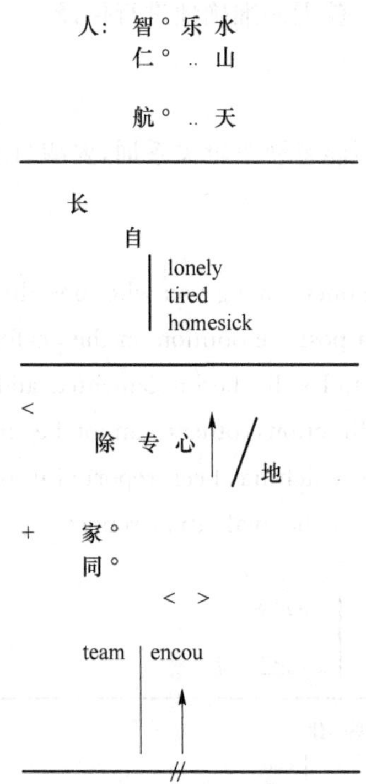

中文的逻辑关系不总是通过关系词体现出来的，译员听的时候，要集中精力，分辨出潜在的逻辑关系。此例中隐含的转折和递进关系应该在笔记本左侧清楚地标记出来。

4. 重要修饰成分

通常不重要的修饰词，如程度副词可以不记，但当修饰成分关乎对内容的理解时，需要记全。例如，"创新、协调、绿色、开放、共享的五大发展理念"中五个并列的修饰词每一个都体现了党中央的新发展理念，缺一不可。

例 1：

载人航天也是系统最复杂、科技最密集、创新最活跃的科技活动。这项工程不仅综合性强、协作面广、技术难度高、风险大，而且研制周期短，任务十分艰巨。

此例中，"系统最复杂、科技最密集、创新最活跃"以及"综合性强、协作面广、技术难度高、风险大"等并列修饰成分文字虽然不多，但涉及方面多，信息量大，是对载人航天工程特点的重要概括，需要在笔记中记全，便于译员忠实、全面地译出源语的意思。

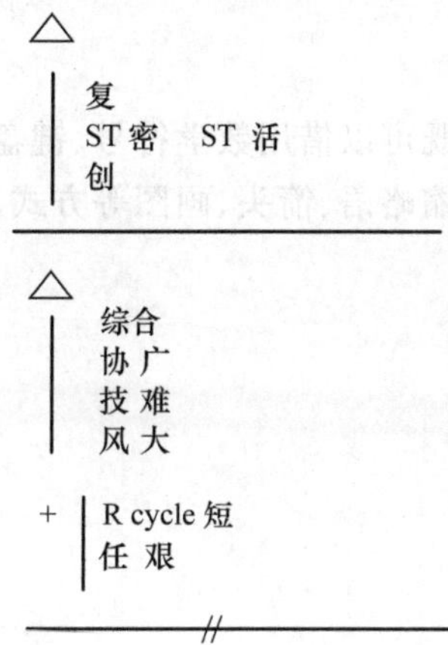

二、怎么记

（一）竖记法

为了便于译员迅速抓住核心意思，口译员惯用的方法是竖着记，并且缩进。

（二）关系词的记法

关系词记在本子的左侧，一目了然，便于译员迅速梳理源语逻辑关系。

（三）意群的记法

每个意群要醒目分隔，可以用"//"或"#"，因人而异。讲话人讲完一段停顿后，译员可以在笔记下面画一条长线，下一段讲话的内容就记在长线下面。每次翻译的时候，都从前一个长线开始，不会因找不到起始点而陷入慌乱的状态。如果讲话人停顿后，译员还埋头看笔记，迟迟不开口，就会影响听众对译员的信任。所以准确定位从哪里开始译十分关键。

（四）笔记的语言

关于口译笔记用什么语言记没有严格的规定，既可以用源语缩略语也可以用目标语缩略语，只要便于记录和输出即可。例如，"一带一路"可以用目标语缩略语记为"BRI"。在听辨和记录过程中，如果有能力，可多用目标语来记，这样会降低输出时语言转换的压力。

（五）时间的记法

时间可以用字母加"."的方法来作笔记。明天用"d."，后天用"d.."，去年用".y"，前年用"..y"。

(六) 笔记符号

笔记符号可以多种多样,既可以借用数学符号、键盘符号、标点符号、货币符号和元素符号,也可以采用中英文缩略语、箭头、画图等方式,记录听到的意思。

1. 数学符号

= 等于、等同
+ 额外、递进、另外
> 大于、高于、胜于
< 小于、低于、弱于
% 表示百分比关系
/ 表示除法关系,也可以接后置定语
∵ 表示原因
∴ 表示结果

例如,“一国两制”可以记为“1+2”。

2. 键盘符号

& 表示并列,例如,“改革开放”倡议可写成“R&O”
@ 表示 at 或者 according to

3. 标点符号

? 可以表示“问题”
! 可以表示“强调、重视”
( ) 可以表示区间,例如,(1978—2023)
: 可以表示认为、建议、推测……

4. 货币符号

¥ 人民币
$ 美元
£ 英镑
€ 欧元

5. 元素符号

CO_2 二氧化碳
CO 一氧化碳
NO_2 二氧化氮

Na 钠

6. 中文缩写

天 航天
火 火箭
卫 卫星
宇 宇航员
斗 北斗
合 合作

7. 英文缩写

min minimum
max maximum
gov government
org organization
info information
rfm reform

8. 箭 头

口译笔记中箭头使用频率高,可以表示逻辑关系和发展趋势。

→ lead to, contribute to
← come, originate from
↑ go up, rise, climb, increase
↓ go down, fall, decrease, reduce

9. 国字框

口 nation
口al national (*adj.*)
口ze to nationalize
口tn nationalization
口o national (*n.*), citizen

10. 画 图

☺ happy, pleased, excited
☹ unhappy, sad, gloomy

✈ plane, aviation

同学们可以参考一些教程,学习笔记书写的一些基本原则和方法,但不建议照搬别人的符号。笔记训练的关键是要通过大量练习,形成自己能够熟练掌握的符号体系。

第二节 篇章讲解与练习

一、重点解析篇

(一) IAF主席介绍联合会发展历程

1. 词汇表

International Astronautical Federation (IAF)	国际宇航联合会
supersonic flight	超声速飞行
space flight	太空飞行
Gesellschaftf ür Weltraumforschung (GfW)	德国太空研究协会
International Astronautical Congress (IAC)	国际宇航大会
global space governance	全球空间治理
"3G" (Geography, Generation, Gender) Diversity	"3G"多样性
butterfly effect	蝴蝶效应

2. 背景知识

(1) 国际宇航联合会

国际宇航联合会成立于1951年,是一个非盈利性的非政府组织,在78个国家拥有513名成员,包括各种政府组织、研究所、博物馆、工业专业协会及学术团体。国际宇航联合会鼓励促进对空间的了解以及开发和应用空间资产为人类造福,在国际宇航界具有广泛的影响力,多年来与中国航天机构建立了良好的合作关系①。

① International Astronautical Federation. HISTORY AND MISSIONS[EB/OL]. (2019-07-25)[2023-09-10]. https://www.iafastro.org/about/history-and-missions.html.

(2) 国际宇航大会

国际宇航大会(IAC)是由国际宇航联合会成员举办的年度会议,吸引全球航天界广泛参与,享有"宇航奥林匹克"之誉。IAC层次高、规模大、影响广、具有商业价值,每年吸引大约3000名来自世界各地的政府官员、企业家、科技人员、学生和媒体人士参加会议。会议同期举办航天展览会,世界知名航天企业和卫星公司、主要航天国家的航天机构都积极参展①。

3. 原 文

The Story of the International Astronautical Federation②

Pascale Ehrenfreund

IAF President

The origins

During the mid-20th Century, the rapid development of military rockets during and after World War Ⅱ and their potential led to a growing public interest in rocketry and in the possible use of rocket power for the attainment of space flight. Flight outside the Earth's atmosphere has been a dream of human beings for generations. The advancing rocket technology, together with other rapid innovation in electronics, supersonic flight, and atomic power, led professional scientists and engineers to a startling conclusion: space flight could be achieved within their generation.

When dealing with rocket technology, those pioneers were able to see beyond the use for merely military purposes but aimed at something much greater: the realization of space flight and space research as a peaceful enterprise. Although national pride and competition were the main drivers of space research, many experts in astronautics soon realized that the implementation of space flights was a scientific, technical, and economic challenge of such magnitude that it could not be solved by one single nation.

Some space enthusiasts started to look for an international platform where they could meet, share, connect, and exchange their studies, their thoughts, their visions beyond the limits of national ambitions and politics.

The initiative leading to the foundation of the International Astronautical

---

① 新华丝路.国际宇航大会是指什么?[EB/OL].(2020-01-06)[2023-09-10].https://www.imsilkroad.com/news/p/397554.html.

② International Astronautical Federation. 70 YEARS - Connecting @ll Space People[EB/OL].(2022-03-13)[2023-07-10].https://www.iafastro.org/assets/files/publications/iaf-70-year-anniversary-booklet-2022-03-13-full-document-final-online-v2.pdf.

Federation (IAF) dates back to the summer of 1949, when the Board of Directors of the German Gesellschaftfür Weltraumforschung (GfW) adopted a resolution stating that "the push into interplanetary space and the future research by space flight is an international task" and recommended an international meeting of all societies researching in this field to foster friendly relations and an exchange of experiences and ideas. This resolution was circulated to the astronautical societies around the world, to immediately find positive feedback, which led to the organization of their first public gathering in Paris in 1950, known as the first International Astronautical Congress. At this first Congress the delegates unanimously voted a resolution in which they expressed wishes to create an international federation to keep contact and secure future international cooperation.

The positive outcome of this experience led to a second Congress in 1951 in London, where the original participants, along with new delegations from other countries, officially founded the International Astronautical Federation (IAF).

The founding members from Argentina, Austria, Denmark, France, Germany, Italy, Spain, Sweden, Switzerland, United Kingdom, and United States, committed to organize the International Astronautical Congress on a yearly basis and meet regularly. During the Cold War period, the IAF Congress soon became the one and only forum where space enthusiasts from all over the world could meet.

Today

Since its establishment, the IAF has been promoting dialogue and cooperation, establishing itself as a key actor in the space community. Today, with its 433 members from 72 countries, the IAF is truly the world's leading space advocacy body. Following its theme—"A space-faring world cooperating for the benefit of humanity" and its motto "Connecting @ll Space People"—the Federation keeps advancing knowledge about space and fostering the development and application of space assets by promoting global cooperation. As organizer of the world's premier space event, the International Astronautical Congress (IAC), and other gatherings on specific topics, the IAF is at the heart of the international space dialogue.

Capitalizing on its many years of expertise and relying on its worldwide network of volunteers, the IAF over the past years has been expanding its scope and reach, focusing also on inclusiveness in the space sector, through its "3G" (Geography, Generation, Gender) Diversity initiative, creating a butterfly effect resonating within the global space community.

The future

Bringing with us our 70 years of history, we look now to the future: a future that will surely be exciting but also challenging, with many revolutionary developments in technology, science, society, and geopolitics.

I am convinced that the IAF will continue to play a major role in the global space arena by supporting global space governance, propelling the global space economy, and fostering global space advocacy.

No matter what the future will hold, the IAF is committed to connecting all space people for a space-faring world cooperating for the benefit of humankind.

4. 参考译文

国际宇航联合会的故事①

国际宇航联合会主席 帕斯卡·埃伦弗朗德

IAF 的诞生

20 世纪中期,军用火箭在第二次世界大战期间和战后快速发展,潜力巨大,公众对火箭技术以及可能使用火箭动力实现太空飞行产生了愈加浓厚的兴趣。在地球大气层外飞行一直是几代人的梦想。先进的火箭技术,加上电子技术、超声速飞行和原子能技术的飞速创新,使科学家和工程师们得出了一个惊人的结论:太空飞行可以在他们这一代实现。

在发展火箭技术的过程中,这些先驱们不仅能看到其在军事上的应用,还着眼于更伟大的目标,即实现太空飞行,并将空间研究作为一项和平事业。尽管民族自豪感和竞争是空间研究的主要动力,但许多航天专家很快意识到,实施太空飞行是一项巨大的科学、技术和经济挑战,不可能由一个国家来解决。

一些太空爱好者开始寻求一个国际平台,在这里他们能够互相结识并分享、联络、交流他们的研究、想法和愿景,且不受国家野心和政治的限制。

国际宇航联合会成立的倡议可以追溯到 1949 年夏天,彼时德国太空研究协会董事会通过了一项决议,指出"进军星际空间和通过太空飞行开展未来研究是一项国际任务"。董事会建议召开一个由该领域所有研究团体参加的国际会议,发展友好关系、交流经验想法。这项决议传送到了世界各地的宇航学会,收到了积极反馈,也促成了 1950 年在巴黎召开第一次公开会议,史称"第一届国际宇航大会"。在大会上,代表们一致投票通过一项决议,即希望能够建立一个国际联盟,保持联系,确保未来开展国际合作。

有了第一届大会取得的丰硕成果,第二届大会随即于 1951 年在伦敦举行,上一

① 编者译。

届的与会者和其他国家的新代表团正式成立了国际宇航联合会。

来自阿根廷、奥地利、丹麦、法国、德国、意大利、西班牙、瑞典、瑞士、英国和美国的创始成员决定,每年举行一次国际宇航大会,并定期举行会议。在冷战时期,IAF大会很快成了唯一能让世界各地的太空爱好者会面的论坛。

IAF 的现况

自成立以来,IAF 始终推进对话合作,成为了空间领域的关键力量。如今,IAF 拥有来自 72 个国家的 433 名成员,是真正世界一流的太空倡导机构。IAF 以"世界航天合作造福人类"为主题,以"连接所有航天人"为座右铭,持续推动全球合作,不断提高对太空的认识,促进空间资产的开发和应用。作为全球最为重要的宇航界盛会活动——IAC 以及其他特定主题活动的组织者,IAF 已成为国际空间对话的核心。

利用其多年的专业知识并依靠其全球志愿者网络,国际宇航联合会在过去几年里一直在扩大其范围和影响力,注重太空行业的包容性,并通过其地理、时代、性别,即"3G"多样性倡议,在全球航天界引发了"蝴蝶效应"。

IAF 的未来

我们有着 70 年的历史,我们的未来注定令人兴奋,也注定充满挑战,因为在技术、科学、社会和地缘政治等方面,我们都取得了许多革命性的进展。

我相信,通过支持全球空间治理、推动全球空间经济发展、促进全球对空间发展的重视,IAF 将继续在全球空间领域发挥重要作用。

纵使未来风云万变,IAF 都将致力于连接所有航天人,让世界航天合作造福人类。

5. 要点讲解

(1) 缩进式笔记有助于译员清晰地浏览所记内容、轻松地梳理各成分之间的语义逻辑关系,以提取关键信息,提高翻译的准确性。

例 1:

原文:When dealing with rocket technology, those pioneers were able to see beyond the use for merely military purposes **but** aimed at something much greater: the realization of space flight **and** space research as a peaceful enterprise. **Although** national pride and competition were the main drivers of space research, many experts in astronautics soon realized that the implementation of space flights was a scientific, technical, and economic challenge of such magnitude that it could not be solved by one single nation.

通过观察,我们可以发现,该段文字中存在许多逻辑连接词、同位语成分和定语成分等,若想快速理清句子与句子之间、各句子成分之间的关系,最好采用缩进式笔记法。

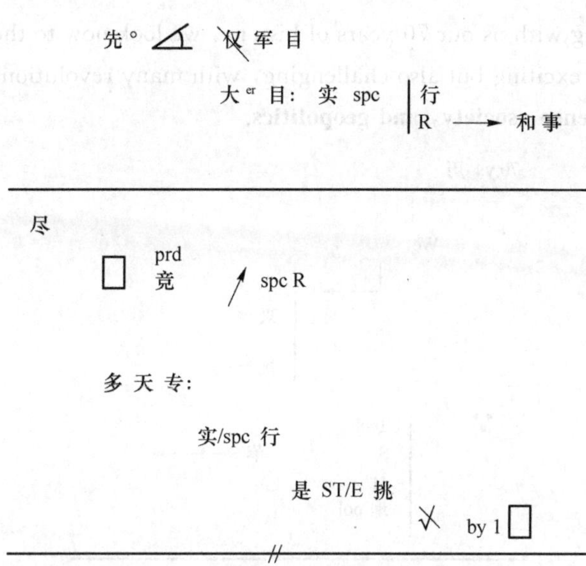

(2) 当发言人列举某些事物时，译员通常采用叠加式笔记法，这样在翻译时便可以一眼看出这些所列举的事物之间是并列关系。通常，译员还会在这一系列并列事物左侧画一条竖线来明确此关系。

例 2：

原文： The founding members from **Argentina**, **Austria**, **Denmark**, **France**, **Germany**, **Italy**, **Spain**, **Sweden**, **Switzerland**, **United Kingdom**, **and United States**, committed to organize the International Astronautical Congress on a yearly basis and meet regularly.

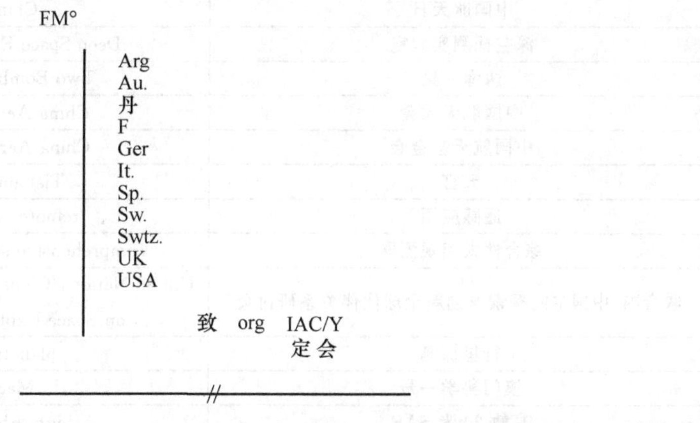

通过分析可知，本句话的主语中心词是“founding members”，表示其来源地的国家名词(from Argentina, Austria, Denmark…)地位相同，属于并列关系，宜采用叠加式笔记加以记录。

例 3:

原文: Bringing with us our 70 years of history, we look now to the future: a future that will surely be exciting but also challenging, with many revolutionary developments in **technology, science, society, and geopolitics.**

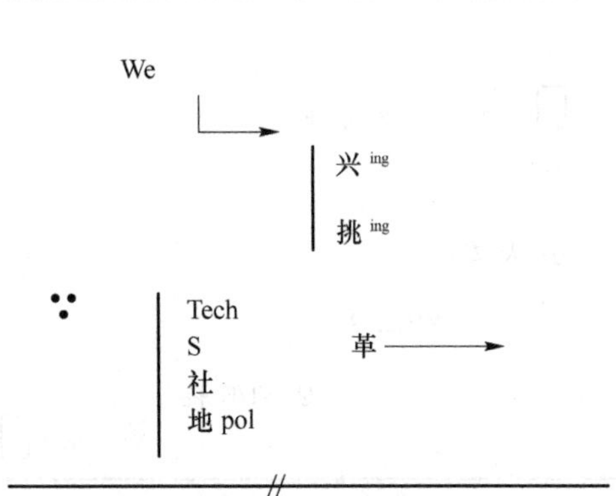

发言者表示,未来将取得多方面的进展,而这些方面"technology, science, society, and geopolitics"彼此之间属于并列关系,使用叠加式笔记法便能毫不费力地辨识清楚这种关系,避免与句子的其他成分相混淆。

(二) 2023 中国航天日新闻发布会(节选一)

1. 词汇表

中国国家航天局	China National Space Administration
中国航天日	China Space Day
深空探测实验室	Deep Space Exploration Laboratory
两弹一星	Two Bombs and One Satellite
中国航天大会	China Aerospace Conference
中国航天基金会	China Aerospace Foundation
天宫	Tiangong space station
遥感应用	remote sensing application
综合性太阳观测网	comprehensive solar observation network
联合国/中国空间探索与创新全球伙伴关系研讨会	United Nations/China Global Partnership Workshop on Space Exploration and Innovation
行星探测	planetary exploration
澳门科学一号	Macao Science 1
高轨 20 米 SAR	high-orbit 20-meter SAR
新一代海洋水色	new-generation marine color
工业和信息化部	Ministry of Industry and Information Technology
中国科学技术大学	University of Science and Technology of China

2. 背景知识

(1) 中国航天日

为了纪念中国航天事业取得的成就,发扬中国航天精神,2016 年 3 月 8 日,国务院批复同意将每年 4 月 24 日设立为"中国航天日",激发全民尤其是青少年崇尚科学、探索未知、敢于创新的热情,为实现中华民族伟大复兴的中国梦凝聚强大力量①。2023 年 4 月 24 日,第八个"中国航天日"主题是"格物致知,叩问苍穹"。

(2) 建设航天强国

2016 年 4 月 24 日,习近平总书记在首个"中国航天日"作出重要指示,指出"探索浩瀚宇宙,发展航天事业,建设航天强国,是我们不懈追求的航天梦"。

航天梦是强国梦的重要组成部分。近年来,中国航天坚持自主创新,不断突破关键核心技术,奋力实现高水平科技自立自强,以载人航天、月球探测、火星探测等为代表的重大工程连战连捷。我国从航天大国加快向航天强国迈进②。

3. 原 文

2023 年"中国航天日"新闻发布会③

(节选一)

2023 年 4 月 18 日

张涛:

各位来宾、记者朋友们:大家下午好!

国家航天局 2023 年"中国航天日"新闻发布会现在开始。"中国航天日"自 2016 年经党中央批准、国务院批复设立以来,得到社会各界广泛关注和踊跃参与,先后在北京、西安、哈尔滨、长沙、福州、南京、海口七地线上线下成功举办主场活动,已成为"弘扬航天精神,传播航天文化,普及科学知识,凝聚航天力量"的重要平台和窗口。特别感谢媒体界和公众,一直以来关注航天,参与航天,感受航天,与中国航天发展,与航天日活动同频共振共鸣共情。

今年是全面贯彻落实党的二十大精神开局之年,也是国家航天局成立 30 周年,

---

① 人民网. 国务院批准:四月二十四日为中国航天日[EB/OL]. (2016-03-22)[2023-08-14]. http://politics. people. com. cn/n1/2016/0322/c1001-28215823. html.

② 央广网,海外网. 新思想引领新征程丨我国从航天大国加快向航天强国迈进[EB/OL]. (2023-04-25)[2023-08-14]. https://baijiahao. baidu. com/s? id=1764124411758148008&wfr=spider&for=pc.

③ 国家航天局. 2023 年"中国航天日"新闻发布会[EB/OL]. (2023-04-18)[2023-07-10]. https://www. cnsa. gov. cn/n6758967/n6758969/n10008374/index. html. (文字有校改)

第八个“中国航天日”主场活动将于 4 月 24 日在安徽省合肥市举办，由工业和信息化部、国家航天局、安徽省人民政府共同主办，合肥市人民政府、中国科学技术大学、深空探测实验室承办。

今天我们非常荣幸邀请到国家航天局系统工程司吕波副司长，安徽省政府办公厅张亚伟副主任，合肥市委袁飞常委，中国科学技术大学傅尧党委常委、副校长，深空探测实验室关锋执行主任出席发布会，分别介绍一年来中国航天成就，安徽省产业特点和航天日主场特色活动等相关情况。我是国家航天局新闻宣传办公室主任张涛。下面请吕波副司长发布 2023 年“中国航天日”活动主题、活动安排、宣传海报等内容。

吕波：

各位朋友：下午好！我是国家航天局系统工程司副司长吕波。感谢大家一直以来对中国航天事业、对“中国航天日”的关注和支持。今年是全面贯彻落实党的二十大精神开局之年，也是国家航天局成立 30 周年，工业和信息化部、国家航天局、安徽省人民政府共同主办的 2023 年“中国航天日”主场活动将于 4 月 24 日在合肥市举办。现将有关情况作简要介绍。

一、关于今年“中国航天日”主题

今年航天日主题是“格物致知、叩问苍穹”，以此为题，意在勉励广大航天人在党的二十大开局之年，继续秉承“两弹一星”精神、载人航天精神、探月精神和新时代北斗精神，怀着探索未知的决心，加快航天强国建设，积极构建外空领域人类命运共同体，倡导广大公众特别是青少年，行而致知、知而促行，不断探索宇宙奥秘。每年航天日我们都面向全国公众，组织开展主题海报征集活动，今年共收到海报作品 1058 幅。经过评审，由来自福建商学院范治鸣老师设计的作品被选定为 2023 年“中国航天日”宣传海报。该作品创作灵感源于今年航天日主题，包含了“格物致知”“叩问苍穹”“绵延赓续”“礼敬安徽”四个层面，将中华传统文化、航天典型元素以及徽派特色有机融合，具有较强视觉冲击力。海报可从“中国的航天”微信公众号、国家航天局网站下载。

二、关于今年“中国航天日”主场活动情况

今年航天日主场活动主要包括：启动仪式、中国航天大会、第一届深空探测天都国际会议、空天信息产业与商业航天发展研讨会、航天科普系列展览、中国航天文化艺术论坛等。4 月 24 日上午，将于合肥市滨湖国际会展中心举办 2023 年“中国航天日”启动仪式，首播今年航天日主题宣传片；公布 2023 年“中国航天公益形象大使”；颁发 2022 年度中国航天基金会钱学森最高成就奖、钱学森杰出贡献奖、航天创新团队奖、航天国际合作奖；发布首次火星探测相关成果。同期，还将举办以“走向世界的中国航天”为主题的航天系列展览。届时，将有来自中央和国家机关有关部委、中科院、高等院校、航天企业的领导和院士专家，以及来自 30 多个国家的航天机构、驻华使馆、高校科研机构和 7 个国际组织的外宾，参加航天日主场活动。

三、关于今年"中国航天日"全国系列活动情况

今年航天日我们鼓励各单位结合自身情况开展"中国航天日"系列活动。初步统计，截至4月14日下午，国家有关部委、各地方国防科技工业管理部门、航天企事业单位以及有关高校、社团，将在4月24日前后一段时间举办航天开放日、科普讲堂、知识竞赛、有关交流研讨等400多项活动。相关航天展馆、航天设施将集中向社会公众和大中小学生开放；一批院士专家走进校园，为青少年进行科普宣讲；一些航天科普电视节目或线上科普讲座将在未来一段时间陆续播出。我们将在国家航天局网站及"中国的航天"微信公众号及时公布有关情况，欢迎大家积极参与。

回望过去的一年，中国航天以高质量创新引领高质量发展，全年实施宇航发射任务64次，再创历史新高；中国空间站全面建成，梦圆"天宫"；我国首次在月球上发现新矿物"嫦娥石"；中国再添"太空印记"；高分专项完成天基部分建设，遥感应用服务华夏，惠及全球；"羲和""夸父"双星逐日，我国初步建立综合性太阳观测网；"句芒"就位，遥瞰祖国绿水青山，助力低碳减排；成功召开联合国/中国空间探索与创新全球伙伴关系研讨会，推动构建新型空间伙伴关系，开创空间探索和创新合作新局面。

2023年，中国航天踏上新征程。探月工程四期、行星探测重大任务推进工程研制；中国空间站转入应用与发展阶段；澳门科学一号、高轨20米SAR、新一代海洋水色等卫星今年将发射入轨。中国航天将以习近平新时代中国特色社会主义思想为引领，深入贯彻党的二十大精神，踔厉奋发、勇毅前行，加快建设航天强国，推进空间科学、空间技术、空间应用全面发展，继续与国际社会一道，为探索宇宙奥秘、保护地球家园、增进民生福祉、服务人类文明进步作出新的更大贡献。

谢谢大家！

4. 参考译文

Press Conference of Space Day of China 2023①

(Excerpt 1)

18 April 2023

Zhang Tao:

Good afternoon, guests and colleagues from the media.

We are here for the press conference of Space Day of China 2023 organized by China National Space Administration. Since its launch in 2016 with the approval of the Central Committee of the Communist Party of China and the State Council, Space Day of China has received extensive attention and active participation from the public. It was hosted in seven cities including Beijing, Xi'an, Harbin, Changsha, Fuzhou,

① 编者译。

Nanjing, and Haikou both online and offline. It has become an important platform to carry forward the spirit of China's earlier development of atomic bomb and space technology, and that of manned spaceflight, lunar exploration, and BeiDou Navigation Satellite System in the new era. Here, I wish to thank the media and the public for closely following China's space exploration and engaging in China's space endeavors and activities on the Space Day.

This year marks the beginning of the full delivery of the 20th National Congress of the Communist Party of China and also the 30th anniversary of the establishment of the China National Space Administration. The eighth Space Day of China will be launched in Hefei, Anhui Province on April 24th. It is jointly sponsored by the Ministry of Industry and Information Technology, the China National Space Administration, and the People's Government of Anhui Province. The Hefei Municipal People's Government, University of Science and Technology of China, and Deep Space Exploration Laboratory are co-organizers of the events.

Today we are very honored to have with us Lü Bo, Deputy Director-General of the System Engineering Department of the China National Space Administration, Zhang Yawei, Deputy Director-General of the Anhui Provincial Government Office, Yuan Fei, Standing Member of the Hefei Municipal Committee, Fu Yao, Vice President of the University of Science and Technology of China, and Guan Feng, Executive Director of the Deep Space Exploration Laboratory. They will share with us China's aerospace achievements over the past year, Anhui's industrial strengths and the activities on the Space Day. I am Zhang Tao, Director-General of the News and Communications Office of the China National Space Administration.

Next, let's invite Deputy Director-General Lü Bo to release the theme, schedule, and promotional posters of the Space Day of China 2023.

Lü Bo:

Hello everyone! Good afternoon! I am Lü Bo, Deputy Director-General of the System Engineering Department of the China National Space Administration. Thank you for following and supporting China's aerospace industry and Space Day of China. This year marks the 30th anniversary of the China National Space Administration. Jointly sponsored by the Ministry of Industry and Information Technology, the China National Space Administration, and the People's Government of Anhui Province, the Space Day of China 2023 will be held in Hefei City on April 24th. I will give you a brief picture about the event.

Ⅰ. The theme of this year's Space Day is "learning through investigation and exploration"

With this theme, it aims to carry forward the spirit of China's earlier development of "Two Bombs, One Satellite", manned spaceflight, lunar exploration, and BeiDou system. With the determination to explore the unknown, efforts will be made to expand the aerospace industry and build a community with a shared future for mankind in outer space. The public, especially young people, will be encouraged to explore the physical world, and discover the secrets of the universe. Every year, we called for poster designs for the Space Day from the public. This year we received a total of 1, 058 poster designs. After review, the design by Professor Fan Zhiming from Fujian Business College was selected as the promotional poster for 2023 China Space Day. With inspiration coming from this year's Space Day theme, this poster covers four aspects: learning through investigation, making space exploration, passing on the legacy, and paying tribute to Anhui. It brings together Chinese traditional culture, typical aerospace elements, and Huizhou characteristics with strong visual impact. The poster can be downloaded from the WeChat public account "China's Aerospace" and the website of the National Space Administration.

Ⅱ. Main events of this year's Space Day

The main venue activities of this year's Space Day include: the opening ceremony, China Aerospace Conference, the first Deep Space Exploration International Conference, the Sky Information Industry and Commercial Space Development Seminar, Aerospace Science Popularization Series Exhibitions, and the China Aerospace Culture and Art Forum. On the morning of April 24th, the Space Day of China 2023 will be opened at the Binhu International Convention and Exhibition Center in Hefei. The theme promotional film of this year's Space Day will be premiered; the "China Space Public Welfare Image Ambassador" for 2023 will be announced; the Qian Xuesen Highest Achievement Award of China Aerospace Foundation, Qian Xuesen Outstanding Contribution Award, Aerospace Innovation Team Award, and Aerospace International Cooperation Award for 2022 will be presented; and the first Mars exploration-related achievements will be released. In the meanwhile, a space exhibition with the theme of "China's space going global" will also be held. At that time, leaders and academicians from central and state organs and departments, Chinese Academy of Sciences, universities, and aerospace enterprises, as well as foreign guests from aerospace agencies from more than 30 countries, embassies in China, university research institutions and seven international organizations will participate in the main venue activities of Space Day.

Ⅲ. About this year's Space Day events across the country

For this year's Space Day, we encourage participation in a series of activities on Space Day of China. According to preliminary statistics, as of the afternoon of April 14th, more than 400 activities such as space open days, science lectures, knowledge competitions and seminars will be held by relevant national departments and commissions, local science and technology authorities, aerospace enterprises and institutions, relevant universities and associations around April 24th. Relevant aerospace exhibition halls and facilities will be opened to the public and students of all ages. A group of academicians and experts will go to schools to give lectures to teenagers; some aerospace TV programs or online science lectures will be broadcast in the near future.

We will brief you on the latest information of the event on the website of the National Space Administration and the WeChat public account "China's Aerospace", and look forward to your participation.

Looking back on the past year, we are proud to see the high-quality development of China's aerospace industry driven by high-quality innovation. 64 space launch missions were completed throughout the year, setting a new record. China's Tiangong space station has been fully built. China discovered a new mineral Chang'e Stone on the moon for the first time, making another space mark. China High-resolution Earth Observation System has completed part of the space-based system, providing remote sensing application services to China and benefiting the world. With the launching of the Xihe satellite and the solar probe Kuafu-1, China has initially set up a comprehensive solar observation network. Goumang is in place, overlooking the green mountains and lucid waters of our motherland and enabling low-carbon emission reduction. The United Nations/China Global Partnership Workshop on Space Exploration and Innovation was held to promote a new type of space partnership and facilitate space exploration and collaborative innovation.

In 2023, China's aerospace embarks on a new journey. The fourth phase of the lunar exploration project and the major task of planetary exploration are well underway. China's space station has entered the application and development stage. Macao Science 1, high-orbit 20-meter SAR, a new-generation marine color satellite will be launched into orbit this year. China's aerospace industry will take Xi Jinping Thought on Socialism with Chinese Characteristics for a New Era as its guide, follow up on the 20th National Congress of the Communist Party of China, work hard and forge ahead with courage, build a country strong on aerospace, improve space science, space technology and space applications, continue to work with the international community to uncover mysteries of the universe, protect our home planet, promote well-being and contribute to social progress.

Thank you!

5. 要点讲解

例 1：

原文：今年航天日主题是"格物致知、叩问苍穹"，以此为题，意在勉励广大航天人在党的二十大开局之年，继续秉承"'两弹一星'精神、载人航天精神、探月精神和新时代北斗精神"，怀着探索未知的决心，加快航天强国建设，积极构建外空领域人类命运共同体，倡导广大公众特别是青少年，行而致知、知而促行，不断探索宇宙奥秘。

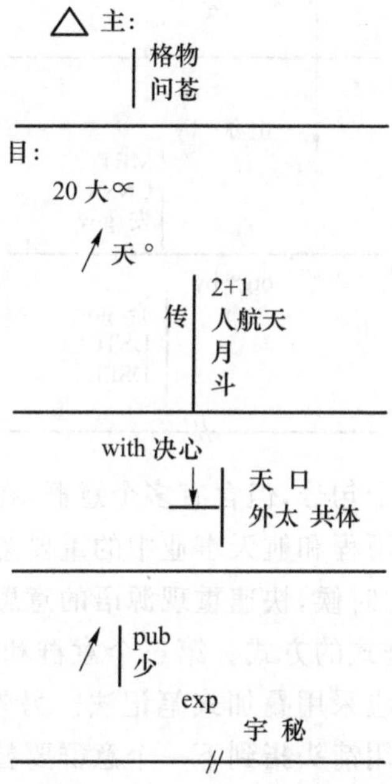

这一段发言可以分为四个意群，每个意群用直线分隔。首先介绍航天日的主题，可以用"△"指代"今年航天日"，节约时间。"格物致知、叩问苍穹"可以采用叠加式的方法记。第二个意群中，"在党的二十大开局之年"可以记作"20 大∝"，其中"∝"在希腊字母中排序第一个，可表示开始。"勉励"可借用向上的箭头。"广大航天人"采用中文缩略语"天"+"°"。传承的四个精神采用叠加式，垂直列出，一目了然。接下来是一系列无主句，如果不加处理，全按照并列句记录，翻译的时候译员理不清逻辑，听众也听得一头雾水。如果能够分析短句之间的逻辑关系，则会减少译员在输出时的压力。第三个意群中，可将"怀着探索未知的决心"记成"with 决心"，作为状语。紧接着"建设"和"构建"均可用组合箭头"⊥↓"指代。从"倡导广大公众特别是青少年"开始，因为对象发生变化，所以作为第四个意群。

例 2:

原文:今年是全面贯彻落实党的二十大精神开局之年,也是国家航天局成立 30 周年,第八个"中国航天日"主场活动将于 4 月 24 日在安徽省合肥市举办,由工业和信息化部、国家航天局、安徽省人民政府共同主办,合肥市人民政府、中国科学技术大学、深空探测实验室承办。

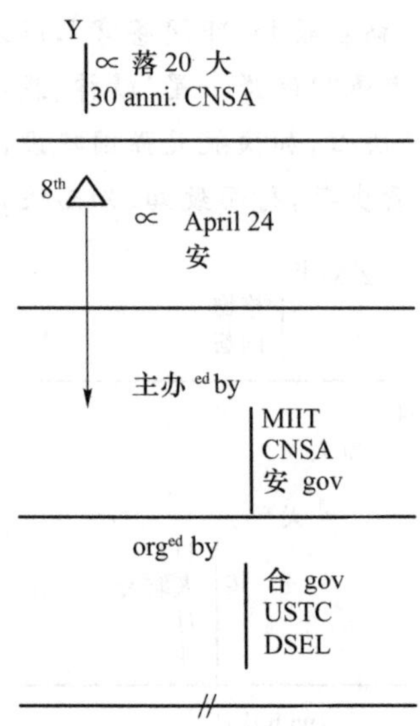

这一段中文虽然只有一个句号,但含有多个意群,在记录时要按意群拆分。第一个意群介绍今年在党的发展历程和航天事业中的重要意义,可采用叠加式方法来记,很清晰,便于译员在读笔记的时候,快速重现源语的意思。第二个意群介绍活动主办的时间和地点,可以采用缩进式的方式。第三个意群和第四个意群介绍活动的主办方和承办方,在做笔记时,均应采用叠加式笔记法。另外,第二个意群和第三个意群主语一致,不必再写一遍,可用箭头指到下一个意群要替代的位置。这种用箭头将上文中出现过的信息引过来的方法在口译笔记中十分常用,节约时间。

二、实战训练篇

(一) 国际民航组织秘书长在国际民航组织创新博览会上的闭幕词

1. 词汇表

airframe	机身
aeronautics	航空学
Urban Air Mobility	城市空中交通
Civil Aviation Authority	民航局

续表

cleaner energy propulsion	清洁能源推进
blockchain	区块链
machine learning	机器学习
emerging industry	新兴产业
national incubator programme	国家孵化器计划
national identification management system	国家身份管理系统
secure backbone	安全骨干网
United States Transport Security Administration	美国运输安全管理局
interoperable secure digital credentials for health and identity	健康和身份的可互操作安全数字凭证
intelligent data sharing	智能数据共享
reorientation of operational processes	操作流程重新定位
United Nations Summit of the Future	联合国未来峰会
Pact for the Future	未来公约

2. 原　文

Closing Remarks at the 2022 ICAO Innovation Fair①

Mr. Juan Carlos Salazar
Secretary General of the
International Civil Aviation Organization (ICAO)
Montréal, 25 September 2022

Innovation has been instrumental in furthering ICAO's mandate for more than 75 years. The monitoring and understanding of the impact of technological and process evolutions are an essential pre-requisite for the establishment of international standards.

As our sector adopts and integrates the latest technologies and solutions, it also enhances the capability of our global network to perform more safely, securely, efficiently, and sustainably, and to drive enhanced socio-economic benefits for societies and businesses in every world region.

These benefits in turn augment the ability of States to invest in and achieve the

① Juan Carlos Salazar. Closing remarks by the Secretary General of the International Civil Aviation Organization (ICAO) Mr. Juan Carlos Salazar to the 2022 ICAO Innovation Fair[EB/OL]. (2022-12-25)[2023-07-10]. https://www. icao. int/secretariat/SecretaryGeneral/Documents/Addresses% 20and% 20Messages/20220925_Innovation_Fair%20concluding. pdf.

Sustainable Development Goals now being globally pursued under the United Nation's 2030 Sustainable Development Agenda.

We're living today in the midst of an incredible era of innovation in every branch of research and development.

It's important to recall in this context that our sector can benefit as much from new discoveries in cleaner energy propulsion, battery science, or digital revolutions, such as blockchain and machine learning, as it can from direct aviation innovations in terms of new airframes and aeronautics, or passenger processing technologies.

This 21st century discovery ecosystem is proving to be especially fertile and diverse, and this presents some incredible opportunities for non-traditional aircraft or engine manufacturing States.

But to fully exploit the potential of emerging industries and new entrants, governments and industry need to work closely together to create dependable and equally fertile investment environments, closely integrated with effective national incubator programmes.

This process of innovation democratization is a real paradigm shift in aviation, and indeed in all industries and sectors, and it highlights why global analysis and certification processes need to be both diligent and dynamic.

This year's Innovation Fair has highlighted an array of challenges in bringing aviation innovations from concept to reality, both for innovators and for regulators.

Many of these factors are common across jurisdictions, and the panel discussions brought this to the fore.

This highlights the need for regulators to operate in new and creative ways to productively engage with innovators, and new ways to communicate with a broader range of stakeholders.

In the area of Urban Air Mobility, for example, we heard numerous insights on how a more inclusive governance system could help pave the way to societal acceptance and deployment.

We also heard about the significant transformations now underway, with many Civil Aviation Authorities already reorganizing the way they work and how they interact with innovators.

Some governments are establishing public-private Committees to facilitate mutual understanding, which includes enhanced familiarization of the regulatory parameters within which start-up companies will have to operate.

And many incubator and accelerator initiatives are also increasingly identifying awareness—raising on regulation as an essential aspect of how they can contribute to a vibrant innovation ecosystem.

In my role as Secretary General, I am also working to transform ICAO to make it more responsive and agile, and able to process and enable innovations to benefit our sector and the countries it serves as rapidly and prudently as possible.

There are no shortcuts when assuring the safety, security, and sustainability of international aviation, but that doesn't mean we cannot be continuously more efficient at how we responsibly assess and standardize new technologies and procedures.

It is also important to recognize that these innovations are only as effective as the regulators who apply them, and ICAO will accordingly become more proactive in supporting States in this area.

Our outreach to innovators needs to be intensified, and we will also seek new ways to maximize added value for States on the basis of the clearly defined goals in our Strategy on Innovation.

We'll also be paying very close attention to assuring that the latest solutions are made accessible to ALL States, equally and rapidly, both for purposes of global harmonization, and to assure no country is left behind, as our network becomes greener, more agile, and more resilient in the years ahead.

I think that this Innovation Fair has also helped us to appreciate ICAO's role as the nexus of public and private cooperation in air transport. This makes it a natural platform for global conversation, partnership, and action on innovation.

Our status as a UN system agency additionally compels us to define a vision of aviation innovation, fully inclusive of related societal priorities, whether in terms of gender equality, socio economic development opportunities, or sustainable prosperity and resilience.

In recent years, our UN role has also seen us coordinating very closely with UN bodies concerned with combatting terrorist and criminal mobility, and contributing to their work as an important partner, given the foundational role of air transport in the international movement of people and goods.

These priorities affect planning and decision making regarding the collection and transmission of passenger data, national identification management systems inclusive of printed or digital passports, visas and health proofs, and the secure backbones needed to house and protect related data.

Some of these points were explored on the second day of the Innovation Fair, when we discussed latest developments in aviation security and facilitation.

Mr. Pekoske, Administrator of the United States Transport Security Administration, opened these engagements by highlighting the potential of a more open architecture in aviation screening, potentially helping States achieve a more secure, resilient and interoperable "system of systems" in aviation security.

These open architecture approaches also have their own sets of challenges, however, both from the development and implementation standpoints.

Further engagements in aviation security focused on recent developments in machine learning capabilities, and the interconnections in this domain between human and artificial intelligence.

There was broad agreement here that the future of aviation security will need to achieve an effective and complementary balance between the two.

Lastly, our discussions on the implementation challenges of seamless and contactless travel, focused on the need for interoperable secure digital credentials for health and identity, intelligent data sharing, consent-based use of biometric data, and a reorientation of operational processes.

Developers and policy makers from government and industry will need to work better together toward these priorities, in order to ultimately realize more efficient traveler processing at airports.

In closing now, I wish to call on all innovators and regulators to stay cognizant of the need to enhance the contributions and visibility of women in the innovation area.

An inclusive and diversity approach recognizes that the populations we serve are diverse, and will help to identify and mitigate the risk of misguided bias detrimentally affecting the Artificial Intelligence and data analytics applications which many future solutions will depend on.

In 2024, the United Nations Summit of the Future will be held in New York, and the entire UN system, States and civil society will be invited at that time to formulate and agree on a new "Pact for the Future".

The goal of this new pact will be to accelerate progress towards the achievement of the Sustainable Development Goals, and to ensure that the world is more aligned as it considers and takes the next steps to protect succeeding generations from current existential threats.

ICAO will assure that international aviation has a strong voice in these discussions, and that the expanding roles for civil society which innovation is now enabling will be appreciated by all concerned.

It has been a sincere pleasure for us to host so many brilliant speakers and engaged participants on the eve of our 41st Assembly, and I'm sure that the vibrant discussions we've heard here over the past three days will leave a lasting impression on its outcomes. I also express my appreciation for the many exhibitors of new innovations, which contributed to the success of this Innovation Fair, and the future-focused discussions among the panelists and participants. We are also grateful for the sponsors of the social events and refreshments which allowed us to remain on premises and take advantage of the rich networking opportunities.

Thank you.

（二）习近平主席在中国-海湾阿拉伯国家合作委员会峰会上的主旨讲话

1. 词汇表

遥感和通信卫星	remote sensing and communications satellite
中海联合月球和深空探测中心	China-GCC joint center for lunar and deep space exploration
中国-海湾阿拉伯国家合作委员会峰会	China-GCC Summit
古丝绸之路	ancient Silk Road
全球发展倡议	Global Development Initiative (GDI)
联合国 2030 年可持续发展议程	2030 Agenda for Sustainable Development
全球安全倡议	Global Security Initiative (GSI)
上海石油天然气交易中心平台	Shanghai Petroleum and Natural Gas Exchange platform
风电光伏	wind and photovoltaic power
中海和平利用核技术论坛	China-GCC forum on peaceful use of nuclear technology
中海核安保示范中心	China-GCC nuclear security demonstration center
中海产业和投资合作论坛	China-GCC forum on industrial and investment cooperation
多边央行数字货币桥项目	m-CBDC Bridge project
中海语言文化论坛	China-GCC language and culture forum
中海人文交流和互鉴双语文库	bilingual library for people-to-people and cultural exchanges and mutual learning

2. 原 文

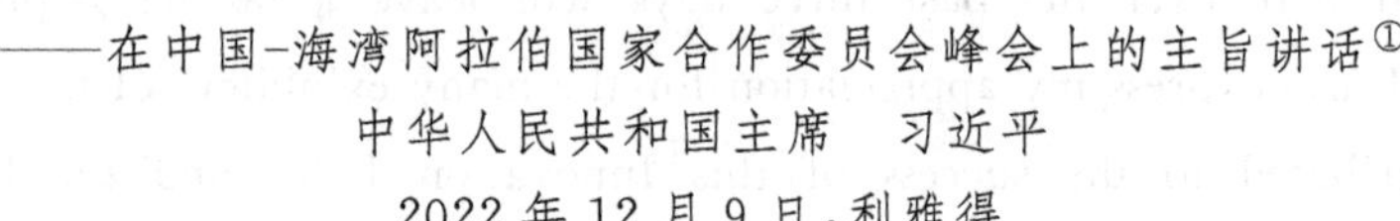

继往开来，携手奋进
共同开创中海关系美好未来

——在中国-海湾阿拉伯国家合作委员会峰会上的主旨讲话①

中华人民共和国主席 习近平

2022 年 12 月 9 日，利雅得

尊敬的各位同事，

纳伊夫秘书长：

大家好！

首先，我谨对沙特阿拉伯王国为举办首届中国-海湾阿拉伯国家合作委员会峰会作出的努力表示衷心感谢。很高兴同大家齐聚一堂，共商中海关系发展大计。

中国同海合会国家有近两千年友好交往历史。双方人民秉持重和平、尚和谐、求真知的“东方智慧”，沿着古丝绸之路往来不绝。1981 年海合会一成立，中国即同海合会建立联系。40 余年来，双方谱写了团结互助、合作共赢的灿烂篇章。

中海关系之所以实现跨越式发展，归根于深厚互信，中国和海合会国家始终相互支持彼此主权独立，尊重各自发展道路，坚持大小国家一律平等，坚定维护多边主义。归根于高度互补，中国拥有广阔消费市场，工业体系完备；海方能源资源丰富，经济多元化发展方兴未艾，双方是天然合作伙伴。归根于民心相通，中海同属东方文明，文化价值相近，人民相知相亲。归根于患难与共，面对国际和地区风云变幻以及金融危机、新冠肺炎疫情、重大自然灾害等挑战，双方同舟共济，守望相助。

各位同事！

面对百年变局，海合会国家团结自强，克服疫情影响实现经济增长，积极推动地区热点难点问题政治解决，推动海合会成为中东海湾最具活力的地区组织，中方对此高度赞赏。站在历史的十字路口，我们要赓续中海友好传统，以建立中海战略伙伴关系为契机，充实中海关系战略内涵。

——做共促团结的伙伴。我们要不断夯实政治互信，坚定支持彼此核心利益。共同维护不干涉内政原则，携手践行真正的多边主义，维护广大发展中国家共同利益。

——做共谋发展的伙伴。我们要加强发展战略对接，发挥互补优势，培育发展动能。中方期待同各方一道推进落实全球发展倡议，落实联合国 2030 年可持续发展议

---

① 习近平. 继往开来，携手奋进 共同开创中海关系美好未来——在中国-海湾阿拉伯国家合作委员会峰会上的主旨讲话[EB/OL]. (2022-12-09)[2023-07-10]. http://www.news.cn/politics/leaders/2022-12/10/c_1129197120.htm.

程，促进地区发展繁荣。

——做共筑安全的伙伴。中国将继续坚定支持海合会国家维护自身安全，支持地区国家通过对话协商化解分歧，构建海湾集体安全架构。欢迎海合会国家参与全球安全倡议，共同维护地区和平稳定。

——做共兴文明的伙伴。我们要增进民心相通，丰富人文交流，借鉴彼此优秀文化成果，弘扬东方文明深厚精髓，为人类文明发展进步作出积极贡献。

各位同事！

未来3到5年，中国愿同海合会国家在以下重点合作领域作出努力：

第一，构建能源立体合作新格局。中国将继续从海合会国家持续大量进口原油，扩大进口液化天然气，加强油气上游开发、工程服务、储运炼化合作。充分利用上海石油天然气交易中心平台，开展油气贸易人民币结算。加强氢能、储能、风电光伏、智能电网等清洁低碳能源技术合作和新能源设备本地化生产合作。设立中海和平利用核技术论坛，共建中海核安保示范中心，为海合会国家培养300名和平利用核能与核技术人才。

第二，推动金融投资合作新进展。中国愿同海合会国家开展金融监管合作，便利海合会国家企业进入中国资本市场。同海方成立共同投资联合会，支持双方主权财富基金以多种方式开展合作。研究举办中海产业和投资合作论坛。加强数字经济和绿色发展等领域投资合作，建立双边投资和经济合作工作机制。开展本币互换合作，深化数字货币合作，推进多边央行数字货币桥项目。

第三，拓展创新科技合作新领域。中国愿同海合会国家共建大数据和云计算中心，加强5G和6G技术合作，共建一批创新创业孵化器，围绕跨境电商合作和通信网络建设等领域实施10个数字经济项目。建立中海气象科技合作机制，举办中海应对气候变化研讨会。

第四，实现航天太空合作新突破。中国愿同海合会国家在遥感和通信卫星、空间应用、航天基础设施等领域开展一系列合作项目。开展航天员选拔训练合作，欢迎海合会国家航天员进入中国空间站，同中国航天员联合飞行并进行空间科学实验。欢迎海方参与中国嫦娥和天问等航天任务的搭载合作，研究成立中海联合月球和深空探测中心。

第五，打造语言文化合作新亮点。中国将同300所海合会国家大中小学合作开展中文教育，同海合会国家合作设立300个中文智慧教室，提供3000个“汉语桥”夏（冬）令营名额，建立中文学习测试中心和网络中文课堂。举办中海语言文化论坛，共建中海人文交流和互鉴双语文库。

各位同事！

中国和海合会国家各自肩负民族发展振兴的光荣使命，中海关系既古老又年轻。让我们继往开来，携手奋进，共同开创中海关系美好未来！

谢谢大家！

（三）王毅国务委员兼外长在第十四届亚欧外长会议第一次全会上的引导性发言

1. 词汇表

第十四届亚欧外长会议第一次全会	the first Plenary Session of the 14th ASEM Foreign Ministers' Meeting
强权政治	power politics
《巴黎协定》	*Paris Agreement*
上诉机构	appellate body
技术封锁	technological monopoly
数字霸权	digital hegemony
《生物多样性公约》第15次缔约方大会	the 15th meeting of the Conference of the Parties to *the Convention on Biological Diversity*
联合国宪章	UN Charter
亚欧会议	Asia-Europe Meeting(ASEM)
欧亚经济联盟	Eurasian Economic Union
欧亚互联互通战略以及东盟互联互通总体规划	EU Strategy on Connecting Europe and Asia, and the Master Plan on ASEAN Connectivity
联合国第二届全球可持续交通大会	the second UN Global Sustainable Transport Conference

2. 原 文

践行多边主义，共创美好未来

——在第十四届亚欧外长会议第一次全会上的引导性发言①

国务委员兼外长 王毅

2019年12月16日，马德里

主席先生，各位同事：

很高兴出席第十四届亚欧外长会议，感谢东道主西班牙为这次亚欧外长会议所做的周到安排。

当今世界正经历百年未有之大变局。单边主义愈演愈烈，冲击国际规则法治。

① 王毅. 践行多边主义，共创美好未来——王毅国务委员兼外长在第十四届亚欧外长会议第一次全会上的引导性发言[EB/OL].(2019-12-17)[2023-07-10]. http://www.fmprc.gov.cn/web/gjhdq_676201/gj_676203/oz_678770/1206_679810/1209_679820/201912/t20191217_9870545.shtml.

保护主义持续蔓延，拖累世界经济增长。强权政治四处横行，威胁世界和平稳定。亚欧国家代表着近六成的世界人口和世界经济总量。一直以来，亚欧国家都是多边主义的受益者和维护者。坚持多边主义，早已成为亚欧各国领导人的普遍共识和集体行动。本次会议以“有效多边主义”为主题，反映出亚欧国家加强团结协作的共同心声。相信在主席先生领导下，53个亚欧会议成员携手共进，将弘扬多边主义的时代强音，汇聚支持多边主义的强大能量。

各位同事，

中国始终是多边主义的坚定捍卫者和积极践行者。作为首个在联合国宪章上签字的国家，中国加入了几乎所有政府间国际组织和500多项国际公约，基本实现了与国际规则的全面接轨。我们忠实信守签署过的每一个条约，全力落实作出过的每一项承诺。我们已经提前3年兑现气候变化《巴黎协定》相关承诺，提前10年完成联合国2030年可持续发展议程制定的消除贫困目标。中国将于明年彻底消除绝对贫困，全面建成小康社会。当前，单边主义和保护主义乌云正笼罩在世界上空，但只要各国团结一致，坚定有效地践行多边主义，乌云就不可能遮住太阳。我们建议：

践行多边主义，要坚持以维护和平、促进发展为目标。多边主义是应对全球性挑战的“金钥匙”。不管哪个版本的多边主义，目标都是促进世界和平与发展。亚欧各国应在捍卫多边主义上作表率，坚定维护以联合国为核心的国际体系，维护以世界贸易组织为基石的多边贸易体制。我们支持对世贸组织进行必要改革，支持各方为恢复世贸组织上诉机构运作作出的努力。新一轮科技革命应该成为共同发展的新动力和各国共享的新机遇，我们要以创新合作挖掘增长动力，反对搞技术封锁和数字霸权，反对制造科技鸿沟和发展脱钩。人工智能、5G、生物技术等前沿科技关乎全球经济发展，各国有义务为他国企业在本国投资兴业、开展合作提供公平、公正、非歧视的营商环境。马德里会议让世界再次聚焦气候变化问题，也凸显了开展全球合作的紧迫性。2020年，中国将承办《生物多样性公约》第15次缔约方大会。我们期待会议取得具有雄心的成果，为世界应对气候变化作出重要贡献。

践行多边主义，要坚持以国际法和公认的国际关系准则为基础。多边主义是现有国际秩序的重要基石。国际秩序不能肆意践踏，国际协议不能动辄退出。我们要共同维护联合国宪章宗旨和原则，恪守维护国际法治的承诺。反对将一己私利凌驾于他国之上，反对肆意歪曲国际法，反对“合则用、不合则弃”的霸权逻辑。反对长臂管辖和单边制裁的霸凌行径。中美达成阶段性贸易协议，旨在解决双方的关切，是本着相互尊重精神平等协商的结果，符合两国人民的利益，符合国际社会的愿望，有利于为世界经济注入信心，有利于为国际贸易带来稳定性。

践行多边主义，要坚持以公平正义、合作共赢为宗旨。相互尊重、平等相待是亚欧会议的重要精神。我们要摒弃意识形态偏见和冷战旧思维,尊重别国走符合自身国情的发展道路,反对将自身价值观强加于人。互利合作、共同发展是亚欧国家的共同愿望。我们应当积极推动共建"一带一路"、欧亚经济联盟、欧亚互联互通战略以及东盟互联互通总体规划等倡议之间的对接。明年,中国将承办联合国第二届全球可持续交通大会,我们希望以此为契机,为促进亚欧地区乃至全球的互联互通与共同发展注入新动力,作出新贡献。

总之,坚持多边主义是世界发展的大势所趋,符合包括亚欧国家在内世界各国的共同利益和福祉。中国愿继续同亚欧各方一道,继续高举多边主义旗帜,不断推动构建人类命运共同体,携手为各国人民创造更加美好的未来。

谢谢!

第四章　数字口译训练

数字翻译是口译工作中不可或缺的一部分，在科技、金融、医疗、教育等诸多领域应用广泛。因为英语和汉语表达和记录数字的方法并不一致，数字口译也是口译工作中的一大难点。数字转换和笔记技巧一样，只有经过大量的练习才能融会贯通。当源语中出现数字时，译员需要全神贯注、快速记录、迅速转换。译员不经过专业训练，很难做到口译数字时准确无误。能够快速、准确地传译数字正是优秀译员的标志[①]。本章将结合具体案例，重点介绍不同类别数字的表达和记录，以及相应的翻译技巧和方法。

第一节　数字口译技巧

一、基数词翻译

英汉两种语言间基数词转换的最大障碍来源于数位划分的不同。中文以四位数为计数单位，由此划分数字，如：万、亿、兆。英语则以三位数为一个计数单位，从小到大依次为：千(thousand)、百万(million)、十亿(billion)、万亿(trillion)。汉英计数单位并不能一一对应，翻译时需要作相应的数位转换，如表 4-1 所列。

表 4-1　三位以上数字的转换规则

位　数	中　文	英　文
4 位数	1 千	1 thousand
5 位数	1 万	10 thousand
6 位数	10 万	100 thousand
7 位数	100 万	1 million
8 位数	1000 万	10 million
9 位数	1 亿	100 million
10 位数	10 亿	1 billion
11 位数	100 亿	10 billion
12 位数	1000 亿	100 billion
13 位数	1 万亿	1 trillion
14 位数	10 万亿	10 trillion

① 任文.英汉口译教程[M].北京：外语教学与研究出版社，2011：163.

大数字的翻译可以分为两步，第一步记录数字，第二步翻译数字。

记录数字要讲方法。对于较为复杂的英文基数词，译员可以使用缩略字母记录，如：听到英文数字 eighty-eight million three hundred and fifty-four thousand three hundred and thirteen，可以简记为“88m354t313”，也可以采用特殊符号标记数位，将数字从左向右，每隔三位用逗号标记为“88，354，313”，从左至右的逗号依次读作billion、million、thousand。同样，译员可以采用字母或符号标记法记录汉语数字，听到中文数字“两亿八千八百三十五万四千三百一十三”，记录为“2 亿 8835 万 4313”，或“2 / 8835 / 4313”，从左至右读标记符号为“亿”“万”。

翻译数字讲技巧。翻译中文大数字时，可以先用中文数字符号标记法进行记录，再用英文标记法标记，译的过程只需照笔记读出被符号分隔的每三位数字加相应的分隔符即可。如：听到汉语数字“两亿八千八百三十五万四千三百一十三”，先用中文符号记录为“2 / 8835 / 4313”，再用英文符号从右向左标记为“2 / 88,35 / 4,313”，最后按标记的英文符号从左至右读出数字即可，即“288(million)354(thousand)313”。

翻译英文大数字时，同样可以采用这个方法。先用英文数字符号标记法进行记录，再用中文标记法标记，翻译时照笔记读出被符号分隔的每四位数字加相应的分隔符。如：听到英文数字 thirty-five million two hundred and eighty-seven thousand nine hundred and twenty-six，用英文计数法将数字从右向左，每隔三位用逗号标记为“35,287,926”，再用中文标记法标注为“35,28/ 7,926”，最后按中文符号标记读出数字。翻译中文基数词需要注意的一点是，在记录中文数字时，译员需要补零空缺的位数。

例 1：

原文： five hundred forty-nine million nine hundred forty-six thousand seven hundred and sixty-eight

记录： 549,946,768

中文标注： 5 / 49,94 / 6,768

译文： 五亿四千九百九十四万六千七百六十八

例 2：

原文： 120 万亿元

英文标注：120t

译文：one hundred and twenty trillion

二、序数词翻译

源语中出现有关名次、排名和“最”的表述时，就涉及序数词的翻译方法。“排名第 X”一般用不带定冠词的序数词来表示：be / rank (be ranked)/ come / finish＋序数词。另外，译员也可以结合具体语句，使用“(序数词＋)形容词最高级＋中心词”的结构表示“第 X”。

例 1：

原文：At the height of her career she **ranked second** in the world.

参考译文：在她事业的巅峰时期，她排名世界**第二位**。

例 2：

原文：The university is **ranked number one** in the country for engineering.

参考译文：这所大学的工程学领域，**在全国排名第一**。

例 3：

原文：在全球 129 个主要经济体中，中国对外开放指数排名从**第 62 位**上升到**第 39 位**。

参考译文：China's **ranking** in the World Openness Index has moved up from **the 62nd** to **the 39th** out of 129 major economies of the world.

例 4：

原文：中国货物出口世界**第一**，产品种类丰富、性价比高。

参考译文：China is the **largest exporter** of goods, offering a rich variety of products at competitive prices.

例 5：

原文：正是在中国共产党的坚强领导下，我们踏上了独立自主建设现代化伟大征程，从一穷二白，成长为世界**第二大**经济体、**第一**货物贸易大国、**第一**外汇储备大国、**第一**制造大国。

参考译文：It is under the CPC's strong leadership that we have embarked on the great journey of independently building a modern country. We have turned China from an impoverished and backward land into the world's **second largest** economy, **top trader** in goods, **biggest holder** of foreign exchange reserves, and **biggest manufacturer.**

例 6：

原文：我们稳步推进能源结构调整，风电、光伏装机量、发电量均居世界**第一**，新能源汽车产销量居全球**之首**。

参考译文：We have worked steadily to adjust the energy mix. Today, China has the world's **largest** installed capacity and power generation from wind and solar power, and **leads** the world in the output and sales of new energy vehicles.

三、小数翻译

小数的翻译无需赘述。小数点读作 point，小数点后的数字需逐一读出。如：0.08读作 zero point o eight。小数翻译常涉及单位换算，译员需及时反应，快速、准确地对数字和单位进行转换。

例 1：

原文：中国企业对非投资 **21.7 亿**美元。

参考译文：Chinese companies have invested **2.17 billion** US dollars in Africa.

例 2：

原文：中方已向非洲 27 国提供 **1.89 亿**剂新冠疫苗。

参考译文：China has provided **189 million** doses of COVID-19 vaccines to 27 African countries.

四、分数翻译

通常情况下，分数的分子用基数词表示，分母用序数词表示，比如，1/3 是 one-third，1/5 是 one-fifth。如果分子大于 1，分母的序数词则需要使用复数，比如，2/3 是 two-thirds，3/5 是 three-fifths。需要注意，1/2 直接用 one half 表示；1/4 既可以说成 one-fourth，也可以使用 one quarter，同理，3/4 既可以用 three fourths，也可以用 three quarters。遇到带分数时，先说整数部分，再用 and 连接分数部分，比如 $5\frac{2}{3}$ 是 five and two-thirds。

对于分子和分母都比较大的分数，可以使用介词 over 来表示分数线，比如 21/76 是 twenty-one over seventy-six，68/123 可以说成 sixty-eight over one hundred and twenty-three，或者简单读作 sixty-eight over one two three。

表 4 - 2 所列为常见分数类型的翻译示例。

表 4 - 2　常见分数类型的翻译示例

分　数	表达方式	说　明
1/2	one half	特殊情况，分子为 1，分母为 2，直接用 one half 表示
1/3	one-third	分子为 1，分母为≤10，分子用基数词，分母用序数词
1/4	one fourth/one quarter	特殊情况，分子为 1，分母为 4，有两种说法
2/3	two-thirds	分子>1，分母≤10，分母要用序数词的复数形式
$5\frac{2}{3}$	five and two-thirds	带分数，先说整数部分，再用 and 连接分数部分
21/76	twenty-one over seventy-six	分子和分母都比较大，使用介词 over 表示分数线
68/123	sixty-eight over one hundred and twenty-three	同上，也可以说成 six-eight over one two three

例 1：

原文：中国的全国增绿面积占全球**四分之一**。

参考译文：China's newly-added green coverage accounts for **one-fourth** of the global total.

例 2：

原文：60 年来，不结盟运动成员由 25 个增加到现在的 120 个，占联合国会员国的三分之二。

参考译文：Sixty years on, Non-Aligned Movement (NAM) membership has grown from 25 to 120, accounting for **two-thirds** of the UN membership.

例 3：

原文：亚洲地大物博、山河秀美，在世界三分之一的陆地上居住着全球三分之二的人口。

参考译文：Asia, this vast and beautiful continent, covers **a third** of the earth's land mass and has **two-thirds** of the world's population.

五、百分数翻译

百分数的英译需在前面的整数或者小数后加上 percent，比如 86.03%读作 eighty-six point o three percent。文字形式的"百分点"对应 percentage point，比如"5 个百分点"是 five percentage points。

例 1：

原文：我们严格控制二氧化碳排放，2020 年单位国内生产总值二氧化碳排放比 2005 年降低 **48.4%**，到 2030 年将比 2005 年下降 **65%**以上。

参考译文：We have put CO_2 emissions under strict control. As a result, CO_2 emissions per unit of GDP in 2020 was **48.4%** lower than that of 2005, and by 2030, it will be over **65%** lower than that of 2005.

例 2：

原文：中德位居世界前三大贸易国之列，两国外贸总额占世界贸易的比重近 **20%**，但双边贸易额占两国外贸总额的比重不到 **3%**，有很大发展空间。

参考译文：China and Germany are among the top three trading nations in the world, accounting for nearly **20%** of global trade. However, trade between us only accounts for less than **3%** of our total foreign trade, leaving huge space for further expansion.

例 3：

原文：刚刚修订发布的 2018 版外商投资准入负面清单，限制性条目压减 **24%**。

参考译文：In the recently released 2018 version of the negative list on access for foreign investment, the number of restricted items was reduced by **24%**.

例 4：

原文：2001—2017 年，中国货物进口额累计 20 万亿美元，年均增长 **13.5%**，高出全球平均水平**6.9个百分点**；服务进口额累计3.7万亿美元，年均增长**16.7%**，高出全球平均水平**8.8个百分点**；今年以来，中国货物和服务进口均增长**20%**左右。

参考译文：Between 2001 and 2017, China imported a total of $20 trillion worth of goods, which comes down to an annual increase of **13.5%**, **6.9 percentage points** higher than the global average; during the same period, our total import in services reached $3.7 trillion, posting an annual growth of **16.7%**, **8.8** percentage points higher than the global average. This year so far China's imports in goods and services have both increased by around **20%**.

六、倍数翻译

倍数的英语表达比较多样，"A是B的几倍"通常可以使用以下几种句型：

① A＋be动词＋倍数＋as＋形容词或副词原级＋as＋B。比如This tree is three times as tall as that tree. ②A＋be动词＋倍数＋形容词或副词比较级＋than＋B。比如This tree is three times taller than that tree. ③A＋be动词＋倍数＋the＋名词＋of＋B或者The＋名词＋of＋A＋be动词＋倍数＋that of＋B。比如The newly broadened square is four times the size of the previous one.（或者The size of the newly broadened square is four times that of the previous one.）

表4－3所列为常见倍数的翻译示例及其说明。

表4－3 常见倍数的翻译示例及其说明

倍　数	表达方式	说　明
两倍	twice/double	最常用，形容词或副词原级后面加twice或double即可
三倍	thrice/three times/triple/treble	四种表达都可以，triple和treble更口语化
四倍	quadruple/fourfold/four times	四倍及以上一般采用"数字＋fold/times"

需注意，"增长至N倍"或者"是原来的N倍"意味着"增长了N－1倍"。另外还需注意，中文中的"翻一番"是变成原来的两倍，翻两番是变成原来的四倍，以此类推。

例1：

原文：30年来贸易规模**扩大85倍**，相互累计投资超过3100亿美元。

参考译文：Over the past 30 years, our trade volume has **grown by 85 times** and two-way investment has exceeded $310 billion in cumulative terms.

例2：

原文：中国的潜在市场规模是美国的**3到4倍**。

参考译文：It is to be well expected that the Chinese market potentially is **three to four times that of** the US.

例 3：

原文：过去 40 年，中美双边货物贸易额**增长了 252 倍**。

参考译文：Over four decades, China-US trade in goods **surged by 252 times.**

例 4：

原文：这个区域目前有 28000 多处宗教场所，近 3 万名宗教教职人员，这两个数字比几十年前**增长了 10 倍**。

参考译文：There are over 28,000 religious sites in this area and close to 30,000 clerical personnel. Both figures have **increased ten fold** compared with several decades ago.

七、单位转换

单位转换是口译中经常遇到的问题。有些单位仅限于汉语，如亩、尺、斤，译员需要将他们换算成国际通用的公制单位；有些单位仅限于英语，如英尺、英亩、盎司，译员可以直译英制单位，也可以适当添加标注（“相当于……”），使目标语听众有更明确的概念。

出现在俗语中的单位有时不必追求等值翻译，如“冰冻三尺非一日之寒”可译为 It takes more than one cold day to freeze three feet of ice。有时还可以采用意译的方法，将含义表达出来即可，如“树高千尺必有根，水流万里总有源”可译为 A towering tree grows from its root, and a long river flows from its source。

例 1：

原文：牢牢守住**十八亿亩**耕地红线。

参考译文：China's total area of farmland does not fall below the redline of **120 million hectares.**

（1 亩＝1/15 公顷）

例 2：

原文：粮食产量 **1.37 万亿斤**，增产 **74 亿斤**。

参考译文：Grain output reached **685 million metric tons**, an increase of **3.7 million metric tons.**

（1 斤＝0.5 千克＝0.0005 吨）

例 3：

原文：“魔高一尺，道高一丈”，各国应与时俱进，大力提升反恐科技水平与能力。

参考译文：For justice to prevail over evil, countries need to keep pace with the times, upgrade their capabilities of counterterrorism technology.

第二节　篇章讲解与练习

一、重点解析篇

（一）机场：美国航空业的心脏

1. 词汇表

Stephen M. Dickson	斯蒂芬·M·迪克森
Airports Council International—North America (ACI-NA)	国际机场理事会北美分会
American Association of Airport Executives (AAAE)	美国机场管理人员协会
Federal Aviation Administration (FAA)	美国联邦航空管理局
U. S. Air Force	美国空军
Delta	达美航空
Hartsfield-Jackson	哈兹菲尔德-杰克逊国际机场
Great Falls	大瀑布国际机场
Eagle	伊格尔机场
JFK	肯尼迪国际机场
Monroe	门罗机场
O'Hare	奥黑尔国际机场
Air Traffic Organization	空中交通组织
Reagan National Airport	里根国家机场
Part 139	"139部分"(指2004年FAA发布的联邦机场认证规定)
Office of Airport	机场办公室
BWI Thurgood Marshall	马歇尔机场
Dulles	杜勒斯国际机场
College Park	科利奇帕克机场
Wright Brothers	莱特兄弟
U. S. Postal Air Mail service	美国邮政航空邮件服务
Leesburg Executive	利斯堡行政机场
approach	进近
departure	离场
budget request	预算申请

续表

Airport Technology Research program	机场技术研究计划
Research, Engineering & Development	研究、工程和发展
Department of Defense	美国国防部
Department of Transportation (DOT)	美国交通部
Bureau of Transportation Statistics	美国运输统计局
Airport Improvement Funding (AIP)	机场改善资金
taxiway	滑行道
737 MAX	波音 737 MAX 机型
Lion Air	狮子航空
Ethiopian Airlines	埃塞俄比亚航空
Ethiopian Airlines Flight ET302	埃航 ET302 航班
grounding	停飞
Just Culture	安全文化
Safety Management System (SMS)	安全管理系统
margin of safety	安全裕度
funneling airport	(因设置新冠病毒检查卡口,导致进机场旅客多,出机场旅客少的)漏斗机场
Center for Disease Control and Prevention (CDC)	美国疾病控制与预防中心
Customs and Border Protection (CBP)	美国海关边境保护局
Department of Health and Human Services	美国卫生与公众服务部
Department of Homeland Security	美国国土安全部
Transportation Security Administration	美国运输安全管理局
National Airspace System	国家空域系统
Noise Complaint Initiative	噪声投诉倡议
Noise Portal	噪声门户网站
Continuous Lower Energy, Emissions and Noise (CLEEN) Program	"持续降低能源、排放和噪声"计划
LAX	洛杉矶国际机场
New York Idlewild	纽约艾德威尔机场
Washington National	华盛顿国家机场
preferential runway	优先跑道
landing facility	着陆设施
hydroplaning	滑水现象
high-definition screen	高清屏幕

续 表

remote tower technology	远程塔台技术
urban air mobility	城市空中交通
pavement material	路面材料
aircraft airworthiness	飞机适航性
unmanned aircraft system (UAS)	无人驾驶飞机系统
fluorine-free alternative	无氟替代品
baseline testing	基线测试
fluorinated foam	含氟泡沫
PFAS firefighting foam	含多氟烷基物质消防泡沫
runway incursion	跑道侵入
wrong-surface takeoff and landing	错误道面起飞降落
Runway Incursion Mitigation (RIM) program	跑道侵入缓解计划
spaceport	太空港
low Earth orbit	近地轨道

2. 背景知识

① 美国联邦航空管理局(Federal Aviation Administration,简称 FAA):FAA 是美国运输部下属、负责民用航空管理的机构,和欧洲航空安全局同为世界上主要的航空器适航证颁发者。总部设在华盛顿,主要任务是保障民用航空的飞行安全,促进民航事业的发展,但不直接经营民航企业[①]。

② 139 部分("Part 139"):在美国,所有商业机场跑道均由美国联邦航空局根据联邦法规标题 14 第 139 部分"商业服务机场认证"进行认证。自从 FAA 出台 139 部分以后,该法规对规范民用机场的安全运行、改善机场安全水平起到了巨大的促进作用[②]。

③ 波音 737 MAX 机型:波音 737 MAX 是波音 737 装配新发动机的衍生机型。2018 年 10 月 29 日,一架印尼狮航的波音 737 MAX 航班起飞 13 分钟后坠毁,导致 189 人遇难。截至 2019 年 3 月 13 日,埃航空难发生后,包括中国、印尼、欧盟在内,已经有几十个国家和地区宣布停飞这款飞机。2020 年 12 月 2 日,FAA 宣布解除对

① 搜狐网.美国联邦航空管理局的无人机监管"壮举"[EB/OL].(2016-05-16)[2023-08-14]. https://www.sohu.com/a/75577543_115926.

② HISOUR.飞机场[EB/OL].(2023-07-14)[2023-08-14]. https://www.hisour.com/zh/airport-34619/.

波音 737 MAX 机型的“禁飞令”,首次载客复飞[①]。

3. 原　文

Airports: The Heart of American Aviation[②]

Former FAA Administrator Stephen M. Dickson

Federal Aviation Administration

5 March 2020

Good afternoon everyone, and thank you to the Airports Council International—North America and the American Association of Airport Executives for the invitation to be here today.

I'm a relative newcomer to the FAA, having started in the role as Administrator back in August. But I'm no stranger to airports, having spent the last 40 years as a pilot, first as a military pilot in the U.S. Air Force at home and abroad, then at Delta for 27 years.

Over the course of my career I've operated into airports as diverse as Hartsfield-Jackson in Atlanta, to Great Falls, Montana; from Eagle, Colorado, to JFK; and from Monroe, Louisiana to O'Hare. One thing I have realized is how incredibly efficient our airport system is in the US, especially compared to other airports around the world.

Through the collaboration of our airport operators, the FAA airports team, the Air Traffic Organization, the airlines and other stakeholders, we get tremendous utilization out of the investments in our airports. And we are currently seeing a tremendous amount of capital investment in our airports around the country. This is definitely a good thing for our communities and our economy.

At the FAA, I've already had the chance to take part in a Part 139 airport certification inspection at the Reagan National Airport, including an airboat water rescue demonstration. The thoroughness of these reviews and the dedication and professionalism of airport employees, working behind the scenes, always ready for contingencies, make me certain that the public is in good hands both in the air and on the ground when they travel.

As Administrator of FAA, it's important to me that we celebrate and recommit to our longstanding relationships and partnerships with the airport community. We have

① 中国新闻网.美国联邦航空局解禁波音 737MAX[EB/OL].(2020-11-19)[2023-08-14]. https://www.chinanews.com/gj/2020/11-19/9341933.shtml.

② Federal Aviation Administration. Airports: The Heart of American Aviation [EB/OL]. (2023-07-13)[2023-08-14]. https://www.faa.gov/speeches/airports-heart-american-aviation.

worked with ACI and AAAE for more than 70 years to ensure the safety, capacity, and efficiency of our nation's system of airports.

Our collaboration is vitally important, because airports are at the heart of the U. S. aviation transportation system, an economic powerhouse that is without rival anywhere in the world. Without the heart, nothing is moving. Without a healthy heart, the viability and safety of the entire system is also at risk.

I'm proud to say that the heart of our aviation system is beating strong and steady. Through congressional support and the ongoing collaboration between our Office of Airports and industry, we will continue to ensure the long-term health of our entire airport system.

The depth and breadth of the airport business—and the 19,000-plus landing facilities in our system—never ceases to amaze me. Consider that within a 25-mile radius of where we're sitting right now, there are three major international airports—BWI Thurgood Marshall, Reagan National, and Dulles—and dozens of public, private, military airports and heliports.

One of those airports, College Park, is the oldest continuously operating airport in the world and is where the Wright Brothers first demonstrated the usefulness of aviation to the military starting in 1909. College Park was also the site of the first U. S. Postal Air Mail service and the first controlled flight of a helicopter.

Also nearby, National airport, later renamed for a famous president, became the first airport to get groovy in the late 1960s—they cut grooves into the runway to reduce hydroplaning. Think about how many accidents and incidents that technology has prevented.

We are constantly planning for the future of our airports, and testing new technology. At another local airport, Leesburg Executive, controllers work with high tech computer tools and video feeds in front of high-definition screens in a dark room rather than a tower cab. The remote tower technologies and standard operating procedures that they are using are still in the testing phase, but we are making progress.

At Dulles airport, we have cameras installed at various points on the approach and departure to gather data that will potentially influence future airport design standards.

There's much more to come. In our fiscal year 2021 budget request, we are also requesting over $200 million in airport research and technology to improve airports not just today but well into the future.

This budget includes $40.6 million for the Airport Technology Research program, directed at the safe and efficient integration of new and innovative technologies into the airport environment. This includes an additional $ 1.4 million to conduct research and to develop standards related to urban air mobility—also known as flying taxis. It also includes funding for new and innovative pavement materials testing.

The budget includes $ 170 million in our Research, Engineering & Development account to continue other research at the Tech Center in areas that will ultimately benefit airports. Included are fire safety, human factors, advanced materials, aircraft airworthiness, and unmanned aircraft systems research.

You'll be interested to know that in January, we opened a new $ 5 million research facility at the Tech Center to concentrate on one on our highest safety priorities—finding fluorine-free alternatives to PFAS firefighting foams. We're making progress, and in fact have begun baseline testing fluorinated foams, the first step in developing alternatives.

Our ultimate goal is to continue protecting the safety of the traveling public while also addressing this important environmental issue in collaboration with our government partners, including the Department of Defense.

From the Administration, to Secretary Chao, to Congress, we are getting the support we need to continue to provide the safest, most efficient airports possible. Our priorities dovetail with the DOT's: Reducing transportation-related fatalities and serious injuries; investing in infrastructure; innovating, and being accountable.

This is important, because the number of people using the transportation system is growing, and the only way to continue that successful growth is to maintain or increase the safety, efficiency, and capacity of all of our nation's airports.

According to 2019 data from the Bureau of Transportation Statistics, U.S. airlines carried approximately 926 million passengers. That's up more than 4% compared to 2018 and more than 12% compared to 2016.

To keep up with growth and maintain safety and efficiency, we are working to expedite the granting of $ 3.17 billion in congressionally approved Airport Improvement Funding, or AIP, and $ 400 million in supplemental funding this year. That makes for a total of $ 3.57 billion going to airports this year, and a total of $ 1.9 billion in supplemental funding over the past three years.

This investment reflects DOT's and FAA's commitment to our nation's airport infrastructure. It supports our continued focus on capacity, efficiency, and environmental sustainability of our airports, and—most importantly—our safety related development projects, including those that reduce runway incursions and reduce the risk of wrong-surface takeoffs and landings.

Not surprisingly, the bottom line for all of our activities, investments, and research has to be this: Safety must be maintained or improved, preferably with—but not dependent on—a boost in efficiency and capacity.

That core value is nowhere more visible than our work with reducing the potential for runway incursions. Through our Runway Incursion Mitigation, or RIM, program, we've been focusing our analysis and risk assessments on runway incursions and wrong surface events.

The RIM program remains the gold standard for reducing runway incursions. The FAA has shown a reduction of more than 67% in runway incursion rates at airports where we've mitigated those problematic locations. We've completed RIM modifications to runways and taxiways at close to 50 locations, and construction is underway at another 14 locations. We have mitigations for about 100 locations in the planning or design phase.

However, as with all things related to safety, the work is never done. In particular, the Office of Airports continues to encourage industry and sponsors to address airport geometry as a primary consideration when analyzing RIM locations.

We can't discuss safety without touching on the 737 MAX situation. First off, on behalf of everyone at the FAA, I would like to, once again, extend our deepest sympathy and condolences to the families of the victims of both the Lion Air and Ethiopian Airlines Boeing 737 MAX accidents.

Our international air transportation network is a tightly woven fabric that is vital to the world's economy. When that fabric unravels, we feel the effects globally. We have to look no further than these crashes to understand this. Onboard Ethiopian Airlines Flight ET302—which crashed one year ago on March 10—were the citizens of 35 countries.

We will honor the memory of those who lost their lives by working tirelessly every day to ensure the highest possible margin of safety in the global aviation system.

For the MAX, I have been steadfast in saying that our return-to-service decision will be based solely on our assessment of the sufficiency of Boeing's proposed software updates and pilot training that address the known issues for grounding the aircraft. I realize this grounding has had an impact on certain airports due to airline schedule changes, but our course is set. We have no choice. If the public is not confident in their aviation system, they simply will not fly.

We at the FAA have welcomed the scrutiny and feedback from near and far on how we can improve our processes. There have been multiple independent reviews launched to look at the 737 MAX and the FAA's certification and delegation processes. Going forward beyond the MAX, we are ready to stand up and speak out on key themes that are emerging regarding aircraft certification, operations, processes, and pilot training not only in the U. S., but around the world.

One of those key themes and one of my main goals is to promote the adoption of a Just Culture and Safety Management Systems, or SMS, throughout the aerospace system, including at certain Part 139 airports.

I know SMS for airports has been a long time coming, but I want to assure you that we have not forgotten this important sector. I've directed our folks to take a strategic look at rolling out SMS at airports. You've provided many great comments over the past 10 years, and many of you have voluntarily implemented SMS in your organizations—thank you for that. Rest assured, there is more to come on this subject.

It's important to note that when we look broadly at what we must do to meet the public's expectations of the highest possible levels of safety globally, we have to consider everything that impacts safety, even unusual or unplanned events like the spread of infectious diseases or drones affecting airport operations.

First, I'll discuss the new Covid-19 virus.

The very connectedness that makes our industry so vital to the global economy also puts us on the front lines for protecting our citizens from outbreaks like Covid-19 within our borders. We must be proactive and strategic in our response—but tactical as necessary—as we combat the threat.

Speaking of being proactive, I would like to take this opportunity to thank all the airports that took the initiative to work across federal agencies to help with the U. S. response to the Covid-19 outbreak. In particular, I'd like to thank those 11 funneling airports who have worked closely with the Centers for Disease Control and Prevention as well as Customs and Border Protection. Your help has been invaluable and effective, and it has been noticed.

The FAA is engaged at both the national and international levels on communicable disease preparedness. Within the United States, the FAA is collaborating and coordinating daily with the Departments of Health and Human Services and Homeland Security, the Centers for Disease Control and Prevention, the Transportation Security Administration, and State and Local public health partners.

The FAA's role is essentially one of support and facilitation on this issue, but, as the focal point for aviation in the U. S. Government, we are very well positioned to bring together our civil aviation stakeholders and our international and interagency partners to work towards preventing the spread of communicable disease.

We are supporting our interagency and industry partners by facilitating operational discussions with our public health and homeland security partners. We have worked closely with CDC and CBP to develop crew health guidance and screening protocols to maximize protection of the traveling public while minimizing operational impacts to the aviation system, including airports.

We must be sure that we maintain the highest levels of safety for airports, whether we are responding to the novel coronavirus or working to integrate emergent technology and innovative new ideas that are reshaping our industry. Consider the meteoric rise in unmanned aircraft operations. In the U.S., we've registered about 1.5 million of these aircraft, that's already about five-times as many drones as manned aircraft in our registry.

The Office of Airports is actively working with the various FAA lines of business to integrate UAS into the airport environment, protecting aviation safety, while enabling airport operators to use drones for key functions. As you know, we're conducting research on UAS Integration at airports to evaluate how they can be used to perform airport-centric operations, such as wildlife monitoring, aircraft rescue and firefighting operations, surveying, and pavement and infrastructure inspection.

We are also finalizing a research plan for evaluating UAS detection and mitigation technologies and establishing performance standards through the Tech Center, as well as reviewing proposals from airports looking to install UAS detection systems.

Since we're talking new entrants, I'll also mention the rise in commercial space and spaceports. The FAA is making rapid progress in our regulatory role in commercial space transportation by paving the way for easier access to low Earth orbit through the National Airspace System.

We're doing this by streamlining the rules for commercial launch and re-entry while at the same time protecting national security and public safety. There's really not much choice—given that Commercial space launch activity in general has ramped up tenfold in just a few years, we either innovate and move forward, or risk being left behind.

We understand some airports embrace this new technology, but others are concerned about how it will impact their operations. All FAA lines-of-business are working together to develop operating procedures to minimize conflicts in our National Airspace System and better ways of coordinating with all of our stakeholders.

And speaking of stakeholders—which includes communities—let's talk about noise.

Over the past two years, the FAA has implemented a standard, repeatable process to ensure productive and effective community involvement for new or modified air traffic procedures. We have also put in place the Noise Complaint Initiative, with a system called the Noise Portal, to more effectively and efficiently track and respond to noise complaints. We have been using the system internally since 2018 and anticipate opening this portal to the public by the end of March.

Of course the FAA will continue to pursue technological improvements to reduce noise, fuel burn and emissions under our Continuous Lower Energy, Emissions and Noise (CLEEN) Program, which will continue to be funded in FY21. In addition to technological advancements, the FAA is assessing take-off and landing operational procedures in order to reduce aircraft noise near airports.

Historically, the FAA's noise strategy has been to hold local community roundtables with residents, airport management, government officials, and industry, to try to develop solutions where there are concerns.

In the future, we'd like to develop tool kits tailored to address specific concerns of individual communities, prepare historical traffic analyses, and evaluate the feasibility of changes proposed through these roundtables to performance based navigation procedures. Our FY21 budget request includes $4.3 million for this work.

I'll close out by going back into the history books on this topic of noise. One month from now—April 4—will mark the 60th anniversary of the very first regulations the FAA issued to minimize aircraft noise at major airports, starting with LAX, New York Idlewild—later to become JFK—and Washington National.

The rules were clear—safety was the highest priority—but where possible, pilots and controllers could use procedural methods—minimum altitudes, preferential runways, and approach and departure routes over the least populated areas—to offer relief to communities.

Obviously, aircraft these days are much quieter and environmentally friendlier, but the sheer number of machines in the air 24/7/365 makes the issue of noise—and other elements of our air transportation system—a continuing concern not only for communities, but for airports and other stakeholders.

I'm here to tell you that we were listening then, in 1960, and we're still listening now.

And thank you for listening! I appreciate having the chance to speak with you today.

4. 参考译文

机场:美国航空业的心脏[①]

美国联邦航空管理局前局长 斯蒂芬·M·迪克森

2020 年 3 月 5 日

大家下午好,感谢国际机场理事会北美分会,即 ACI-NA 以及美国机场管理人员协会,即 AAAE,邀请我参加今天的会议。

我去年 8 月刚开始担任美国联邦航空管理局,即 FAA 局长,所以在局里相对还是个新人。但是我对机场并不陌生。此前我当了 40 年飞行员,先是在国内外的美国空军担任飞行员,然后又在达美航空工作了 27 年。

在我的飞行员生涯中,我曾在亚特兰大哈兹菲尔德-杰克逊国际机场(Hartsfield-Jackson)和蒙大拿州的大瀑布国际机场、科罗拉多州的伊格尔机场和肯尼迪国际机场、路易斯安那州的门罗机场和奥黑尔国际机场等多个机场工作过。我意识到,我们美国的机场系统是多么的高效,远比世界上其他机场要出色。

通过机场运营商、联邦航空局机场团队、空中交通组织、航空公司和其他利益攸关方的合作,我们最大限度利用了机场投资。同时,我们也看到全国各地的机场获得了大量资本投资,这对我们的社区和经济来说绝对是一件好事。

在 FAA,我曾参加了在里根国家机场进行的 139 部分机场认证检查,并参与了汽艇水上救援演示。这些审查彻底、细致,机场员工甘于奉献,专业水平高,这些幕后工作者们,时刻准备着应对突发情况,这使我有理由相信,公众在空中和地面的旅途都有良好的保障。

作为联邦航空局局长,对我来说,重要的是我们要赞颂并继续坚守我们与机场界的长期伙伴关系。我们已经与 ACI 和 AAAE 合作了 70 多年,确保了我们国家机场系统的安全、容量和效率。

我们的合作极其重要,因为机场是美国航空运输系统的心脏,是世界上任何地方都无法匹敌的经济动力源。没有心脏,一切都将停滞。没有一个健康的心脏,整个系统的生存能力和安全也会受到威胁。

我很自豪地说,我们航空系统的心脏正在强劲而稳定地跳动着。通过国会的支持以及我们的机场办公室和行业之间的持续合作,我们会继续保持整个机场系统的长期健康发展。

我们的机场系统涵盖了 19000 多个着陆设施,机场业务的深度和广度一直在刷新我的认知。我们现在所在会场的 25 英里半径内,就有三个主要的国际机场,即马

① 编者译。

歇尔机场、里根国家机场和杜勒斯国际机场,此外还有几十个公共、私人、军用机场和直升机机场。

而其中的科利奇帕克机场,是世界上最古老的仍在持续运营的机场,也是莱特兄弟于1909年首次向军方展示航空效用的地方。该机场还是首个美国邮政航空邮件服务点和进行首次直升机受控飞行的地点。

同时还有旁边的国家机场,之后因里根总统更名为里根国家机场。该机场在20世纪60年代末成为了第一个跑道上设有凹槽的机场,以减少滑水现象。大家可以想象这项技术避免了多少事故和意外。

我们一直在为机场的未来做规划,并尝试新技术。在当地的另外一个机场,也就是利斯堡行政机场,控制员并不在塔台工作,而是在一个黑暗的房间里使用高科技的计算机工具,在高清屏幕前查看视频资料。他们正在使用的远程塔台技术和标准操作程序仍处于测试阶段,不过我们正在取得一定进展。

在杜勒斯国际机场,我们在飞机进近和离场的各个节点安装了摄像头,收集关乎未来机场设计标准的数据。

等待我们的还有更多。在我们的2021财年的预算申请中,我们还申请了2亿多美元的资金用于机场研究技术发展,这不仅是为了机场的现在,更是为了改善机场的未来。

该预算中的4060万美元将用于机场技术研究计划,旨在将新研发的创新技术安全有效地应用于机场环境。预算还包括额外的140万美元,用于开展城市空中交通,即空中的士的研究,并制定相关标准。预算还将提供资金,测试新研发的创新路面材料。

预算中的1.7亿美元将存入我们的研究、工程和发展账户,以便继续开展技术中心的其他研究,这些领域的研究最终将使机场受益。其中包括消防安全、人为因素、先进材料、飞机适航性和无人驾驶飞机系统研究。

有意思的是,今年1月,我们在技术中心启动了一台价值500万美元的新研究设备,集中研究我们的最高安全事项之一——寻找PFAS消防泡沫的无氟替代品。我们正在努力取得进展,事实上我们已经开始对含氟泡沫进行基线测试,这是开发替代品的第一步。

我们的最终目标是继续保护旅客安全,同时与国防部等政府机构合作,解决这一关键的环境问题。

从联邦航空局到赵部长,再到国会,我们正在争取我们所需的支持,以继续提供尽可能安全有效的机场服务。我们的优先事项与美国交通部的优先事项一致,即减少与交通有关的死亡和严重伤害、投资基础设施、开展创新,以及积极承担责任等。

这点很重要,因为使用交通系统的人数正在增长,如果想要让这种成功的增长持续,唯一途径就是保持或提高我们国家所有机场的安全、效率和能力。

根据运输统计局2019年的数据,美国航空公司运送了大约9.26亿乘客。与2018年相比增长超4%,与2016年相比增长超12%。

为了保持增长势头并保证安全和效率,我们正在努力加快发放国会批准的31.7亿美元的机场改善资金，即 AIP，以及今年 4 亿美元的补充资金。这使得今年用于机场的资金总额达到 35.7 亿美元，使得过去三年的补充资金总额达到 19 亿美元。

这项投资表明了交通部和 FAA 对改善我们国家机场基础设施的承诺。这能够支持我们继续关注机场容量、效率和环境可持续性等方面的问题，最重要的是能让我们拓展与安全相关的发展项目，包括减少跑道侵入、降低错误道面起飞降落风险等。

毫无疑问，我们所有的活动、投资和研究都要守住一个底线，即维持或提升安全水平，最好还能够提升效率和负载能力，但这不是决定性因素。

这一核心价值最明显的体现就是我们在减少跑道侵入风险方面所做的工作。我们开展了“跑道侵入减缓”计划，即“RIM”计划，一直把分析和风险评估的重点放在跑道侵入和错误道面事故上。

RIM 计划仍然是减少跑道侵入的黄金标准。FAA 显示，在对有问题地点实行风险减缓的机场，跑道侵入率减少了 67% 以上。我们已经完成了近 50 个地点的跑道和滑行道的 RIM 改造，另外 14 个地点的改造施工正在进行。还有大约 100 个地点的减缓计划处于规划或设计阶段。

然而，与所有与安全有关的事务一样，这是一个永无止境的工作。特别是，机场办公室继续鼓励行业和赞助商在分析 RIM 位置时，将机场的几何形状作为首要考虑因素。

讨论安全问题就不能不提到 737 MAX 所发生的事件。首先，我想代表联邦航空局的每个人，再次向狮子航空和埃塞俄比亚航空波音 737 MAX 事故的遇难者家属致以最深切的哀悼和慰问。

我们的国际航空交通网络是一个紧密编织的结构，对世界经济至关重要。一旦这一结构分崩离析，全球各地都会受到影响。单单是这两起坠机事件就能说明这一点。埃航 ET302 航班于去年 3 月 10 日坠毁，机上搭载了来自 35 个国家的公民。

我们日夜不停地工作，最大程度上提高全球航空系统的安全裕度，以此告慰那些罹难者。

对于波音 MAX 飞机，我的答案是确定的，飞机复飞决定的做出完全取决于以下流程：我们会对波音公司的软件更新提议以及飞行员培训的充分性进行评估，衡量其是否解决了导致飞机停飞的已知问题。我也看到，此次停飞对某些机场产生了影响，导致了航空公司时间表的变动。但我们的流程是雷打不动的，没有别的选择。如果公众对他们的航空系统失去了信心，他们就不会再乘坐飞机了。

联邦航空局欢迎来自各方的监督和反馈意见，指导我们改善工作流程。我们已经启动了多项独立审查，对 737 MAX 和联邦航空局的认证和授权流程进行了调查。展望未来，除波音 MAX 之外，我们已经做好了准备，针对新兴关键课题积极发表看法，包括飞机认证、运营、流程和飞行员培训等，范围不仅在美国，更要包括全世界。

其中一项关键课题，也是我的主要目标之一，便是促进安全文化和安全管理系统，即 SMS，在整个航空航天系统中的应用，包括部分经过 139 部分认证的机场。

将 SMS 应用于机场的这一天早该到来，但我向你们保证，我们从未有忘记这个重要的系统。我已经告诉我的各位同仁,要用战略性的眼光看待 SMS 在机场的推广。在过去的 10 年中,你们提供了许多优秀的意见,而且大家有很多人已经在各自组织中自愿实施了 SMS。感谢你们。请放心,我们在这个问题上还会有更多的进展。

值得注意的是,我们常从大体上思考自己势必要做的事情,以达到全球最高安全水平,从而满足公众的期望。可我们也应考虑到每一件会影响航空安全的事情,即便是那些罕见或着计划之外的事件,比如传染病、无人机等,都会影响到机场的运营。

首先,我想要谈谈新冠病毒。

航空业的高度关联性一方面让我们成为了全球经济的生命线,另一方面也使我们处在了抵抗新冠疫情、保卫公众健康的最前线。我们在疫情防控中必须把握主动,采取战略性措施,但必要时也要采取战术性措施。

谈到把握主动,我想借此机会感谢所有主动跨联邦机构合作,帮助美国应对新冠疫情的机场。我还要特别感谢 11 家漏斗机场同美国疾病控制与预防中心,即 CDC 和美国海关边境保护局,即 CBP 的密切合作,你们的帮助非常重要且有效,大家都看在眼里。

FAA 在国家和国际层面都参与了传染病的预防工作。在美国,FAA 每天都在与美国卫生与公众服务部和国土安全部、疾控中心、运输安全管理局以及州和地方的公共卫生合作伙伴进行协调合作。

在这个问题上,FAA 的作用本质上是支持和促进,但是,作为美国政府的航空协调中心,我们也有能力将我们民用航空的利益攸关方和我们的国际及机构间合作伙伴聚集在一起,努力防止疫情的传播。

我们正在通过促进与公共卫生和国土安全伙伴的业务讨论,以支持我们的机构间和行业伙伴。我们与 CDC 和 CBP 密切合作,制定机组人员健康指南和检查协议,以最大限度地保护旅客,同时最大限度地减少对航空系统,包括机场的运营影响。

我们必须确保维持机场的最高安全水平,在抗击新冠疫情上是如此,在努力整合、重塑航空业新兴技术和创新型新想法方面也是如此。无人机业务正在飞速发展。在美国,我们已经注册了约 150 万架无人机,无人机注册数量是有人驾驶飞机的五倍。

机场办公室正在积极与 FAA 各业务部门合作,将 UAS,即无人机系统整合入机场环境,以确保航空安全,同时使机场运营商能够使用无人机执行关键任务。大家都知道,我们正在开展研究,将 UAS 同机场相整合,以评估如何使用 UAS 执行机场中心型任务,包括野生动物监测、飞机救援和消防任务、测量,以及路面和基础设施检查等。

我们还在最后确定一项研究计划,评估 UAS 的探测和减缓技术,并通过技术中心建立性能标准,同时审查机场申请安装 UAS 的提案。

既然我们在谈论行业的新成员,我便想提一提商业航天和太空港的崛起。FAA 在商业航天运输监管方面正在取得快速进展,借助国家空域系统为更容易进入近地轨道打好基础。

为此,我们正在简化商业航天器发射和重返大气层的规则,同时保护国家安全和公共安全。我们必须这样做,因为商业航天发射活动在短短几年内大体上增加了10倍,我们要么创新和前进,要么就要承担被抛弃的风险。

我们了解到一些机场欢迎这项新技术,但也有一些机场担心这项技术会影响他们的运营。FAA的所有业务部门都在合作制定操作程序,以尽量减少国家空域系统的冲突,并更好地协调所有利益攸关方。

我们说的利益攸关方包括居民社区,所以我们来讲讲噪声。

在过去的两年里,FAA已经实施了一个标准的、可重复的程序,以确保社区可以参与到新的或修改的空中交通程序中,并产出富有成效的结果。我们还实施了噪声投诉倡议,有一个名为"噪声门户网站"的系统,以更有效且高效地跟踪和回应噪声投诉。自2018年以来,我们一直在内部使用该系统,并预计在3月底前向公众开放这一门户。

当然,FAA将继续追求技术改进,以减少噪声、燃料消耗和温室气体排放,以执行"持续降低能源、排放和噪声"计划,该计划将在2021财年继续得到资助。除了技术进步外,FAA正在评估起飞和降落的操作程序,以减少机场附近的飞机噪声。

历史上,FAA的噪声策略是与居民、机场管理部门、政府官员和工业界举行当地社区圆桌会议,试图在有问题的地方制定解决方案。

未来,我们还希望开发工具包,以量身定制解决个别社区的具体问题,准备历史交通分析,并评估通过这些圆桌会议提出的基于性能的导航程序变化的可行性。2021财年的预算申请中有430万美元就是用于这项工作。

最后,我想回顾一下关于噪声这个话题的历史。一个月之后,即4月4日,是FAA第一批法规发布的60周年,这些法规旨在最大限度地减少主要机场的飞机噪声,首先是洛杉矶国际机场、纽约艾德威尔机场——也就是后来的肯尼迪机场和华盛顿国家机场。

规则很明确:安全是最高优先事项。但如果情况允许,飞行员和管制员可以使用程序性方法,即选用最低高度、优先跑道,并选择人口最稀少地区上空的进近和离场路线,为社区居民提供便利。

显然,现在的飞机更加安静,也更加环保,但空中的机器太多,而且全年无休、日夜不停,使得噪声问题以及航空运输系统的其他问题被社区以及机场及其他利益攸关方持续关注。

今天我站在这里,想要告诉各位,1960年我们在听,现在我们仍然在听。

谢谢大家,很高兴能有机会同各位交谈!

5. 要点讲解

这篇演讲中涉及的数字翻译形式多样,包括基数词、序数词、小数、百分数、倍数和单位转换,对译员的数字翻译能力提出了较高要求。

本演讲中出现频率最高的是基数词和小数。绝大部分基数词与金额有关,比如 \$200 million、\$5 million、\$170 million 等;绝大部分小数也是金额,比如 \$40.6 million、

$1.4 million、$3.17 billion 等。面对较高频次和较高密集度的金额翻译，译员需要在头脑中及时挪动小数点，并对单位进行换算。此外，本演讲还涉及设施数量（如19,000-plus landing facilities）和人数（如926 million passengers）的翻译，译员必须快速地对这些数字进行换算，同时保证准确率。

本演讲中出现的序数词翻译较为简单，如 the 60th anniversary of the very **first** regulations the FAA issued 可译为"FAA 第一批法规发布的60周年"，the **first** airport to get groovy in the late 1960s 可译为"第一个跑道上有凹槽的机场"。

本演讲中百分数的翻译都涉及与"升降"的有关表达，如 That's **up more than 4%** compared to 2018 and more than **12%** compared to 2016 和 The FAA has shown **a reduction of more than 67%** in runway incursion rates。译员除了要听清具体的数字和上升/下降之外，也要注意比较的基准（比如具体是和哪一年相比）、是否有限制范围（比如 more than/less than、nearly、about/around）等细节，尽可能保证信息传达的准确度和完整性。

本演讲出现了两处倍数翻译，如 that's already about **five-times** as many drones as manned aircraft in our registry，该句使用的是 as… as… 结构，可译为"无人机注册数量是有人驾驶飞机的**五倍**"；commercial space launch activity in general has ramped up **tenfold** in just a few years，该句使用的是"数字＋fold"的表达，可译为"商业航天发射活动在短短几年内大体上增加了 **10 倍**"。

本演讲还出现了一处英制单位，译员可以将 within a **25-mile** radius of where we're sitting right now 直译为"我们现在所在会场的 **25 英里**半径内"。如果时间允许，可以将"25 英里"换算为"4 万多米"（1 英里＝1609.344 米），这样一来，目标语听众就可以对这一距离有更直观的认识。

（二）时任国务院总理李克强在第九届中德经济技术合作论坛上的致辞

1. 词汇表

电动汽车	electric vehicle
智能制造	smart manufacturing
新能源汽车	new energy vehicle
自动驾驶	autonomous driving
自动网联驾驶领域	intelligent and connected vehicle (ICV)
第九届中德经济技术合作论坛	the Ninth China-Germany Economic and Technological Cooperation Forum
首届中国国际进口博览会	the First China International Import Expo
产业合作园区	joint industrial park

续 表

安全审查	security review
金融监管	financial regulation
2018 版外商投资准入负面清单	Special Administrative Measures for Foreign Investment Access (Negative List) (2018 version)
零和博弈	zero-sum game
强刺激	massive stimulus
供给侧结构性改革	supply-side structural reform
宏观调控	macro regulation
改革开放	reform and opening-up
社会主义市场经济体制	socialist market economy
知识产权	intellectual property
侵权假冒行为	infringement and counterfeiting

2. 背景知识

(1) 中德经济技术合作论坛(China-Germany Economic and Technological Cooperation Forum)

中德经济技术合作论坛,由中国国家发展和改革委员会与德意志联邦经济和能源部共同举办。该论坛自 1995 年由两国领导人发起成立以来,目前已成为促进中德经济技术合作的重要平台。2018 年 7 月 9 日,第九届中德经济技术合作论坛在德国柏林召开,时任国务院总理李克强和时任德国总理默克尔出席论坛闭幕式并发表演讲①。

(2) 中德政府磋商(China-Germany Inter-governmental Consultation)

2010 年 7 月,时任德国总理默克尔访华期间与时任总理温家宝商定建立中德政府间的磋商机制。自启动以来,中德政府磋商机制对不断充实中德战略伙伴关系的内涵、拓展中德务实合作、从顶层规划两国关系等方面发挥了重要作用。当地时间 2018 年 7 月 9 日中午,时任国务院总理李克强在柏林总理府与默克尔总理共同主持第五轮中德政府磋商②。

---

① 新华网.李克强与默克尔共同出席第九届中德经济技术合作论坛并发表讲话[EB/OL].(2018-07-10)[2023-08-14].http://www.xinhuanet.com/politics/2018-07/10/c_1123106231.htm.

② 新华网.李克强与德国总理默克尔共同主持第五轮中德政府磋商[EB/OL].(2023-06-15)[2023-08-14].http://www.xinhuanet.com/world/2018-07/10/c_1123101043.htm.

3. 原 文

在第九届中德经济技术合作论坛上的致辞[①]

时任国务院总理 李克强

2018 年 7 月 9 日，柏林

尊敬的默克尔总理，

女士们，先生们：

很高兴同默克尔总理一道，出席第九届中德经济技术合作论坛闭幕式。中德两国新一届政府组建后不到半年，我与总理女士就实现互访，充分反映出两国的良好关系和加强合作的愿望。本次论坛以"保持和增强创新力与竞争力"为主题，把握了时代脉搏，契合双方需求，很有意义。我谨代表中国政府，对论坛的成功举办表示热烈祝贺，对支持和关心中德友好合作的各界人士致以崇高敬意！

今年是中国改革开放 40 周年，也是中德两国政府签署科技合作协定 40 周年。40 年来，中德合作保持全面稳定健康发展态势，规模不断扩大，领域不断拓展，层次不断提升。2017 年双边贸易额达 1681 亿美元，是 1978 年的 100 倍，占中欧贸易总额的近三成，中国已连续两年成为德国最大贸易伙伴。双向投资累计超过 400 亿美元，7000 多家德国企业、2000 多家中国企业在对方国家投资兴业。现在双方合作已远远超越了贸易范畴，涵盖了金融、科技、环保、能源、人文等领域，还结成强有力的创新伙伴关系。可以说，中德合作成长性好、稳定性高、实效性强，长期引领中欧合作，堪称大国合作的典范，为两国乃至世界经济发展注入了动力。中德合作能取得今天这样的成就，关键在于双方坚持相互尊重、平等相待、互利共赢、开拓创新，也得益于两国工商界朋友的长期辛勤耕耘。

去年，习近平主席对德国进行成功访问，同德方领导人共同规划两国关系发展。今天上午，我与默克尔总理共同主持第五轮中德政府磋商，达成了一系列重要共识和成果。我们一致同意，面对错综复杂的世界政治经济形势，中德作为亚欧大陆两端最大的经济体，要继续巩固和提升政治互信，夯实中德全方位战略伙伴关系的基础，推动中德合作向更高水平迈进，以中德合作的确定性、维护多边主义的一贯性为世界和平与发展增添稳定性。

——我们要深挖贸易潜力。中德位居世界前三大贸易国之列，两国外贸总额占世界贸易的比重近 20%，但双边贸易额占两国外贸总额的比重不到 3%，有很大发展

---

① 李克强. 在第九届中德经济技术合作论坛上的致辞[EB/OL]. (2023-07-13)[2023-08-14]. https://www.gov.cn/gongbao/content/2018/content_5309419.htm.

空间。当前中国居民消费正在升级，产业转型步伐加快，德方企业扩大优势商品、高端装备、优质服务对华出口可以大有作为。希望德方进一步放宽对华高技术产品出口限制，提升高技术产品在双边贸易中的比重。中国货物出口世界第一，产品种类丰富、性价比高，深受各国消费者青睐和企业欢迎，也契合德国多样化的市场需求。我们要进一步提高贸易便利化水平，创新贸易方式，优化产品结构，大力发展跨境电子商务，推动双边贸易规模迈上新台阶。目前双方服务贸易规模不到300亿美元，要进一步扩大旅游、留学、运输等服务贸易，着力培育新的增长点，大力发展技术贸易、专业服务、维修维护、远程医疗、金融保险等现代服务贸易。今年11月，首届中国国际进口博览会在上海举行，德国将以主宾国身份参加，相信这会进一步提升德国产品和服务在中国市场的影响力。

——我们要扩大双向投资。中德双向投资虽然持续活跃，但目前德国对华投资占德国对外投资规模的比重不到7%，中国对德投资存量规模也较小。过去德国不少有远见的企业率先在中国投资兴业，获得了丰厚回报，实现了互利共赢。现在中国市场更加开放，营商环境不断改善，德国企业在华投资有新的更大机遇。欢迎德国企业继续扩大对华投资，特别是增加对发展潜力大的中国中西部、东北等地区投资，共办产业合作园区。中国企业对外投资遵循市场原则、商业规则，希望德方提供更加公平透明、可预期的投资环境，在安全审查、金融监管等方面给予公正待遇。尽早完成中欧投资协定谈判是业界的期盼，中方愿同德方一道推动协定谈判取得进展。

——我们要积极推动创新合作。中德在创新领域各有千秋，是创新合作的“黄金搭档”。德国在基础研究、原始创新、高端研发等领域全球领先，中国市场规模巨大、人力资源丰富、产业体系齐全、数字经济快速发展，在应用创新等方面具有独特优势。两国在全球创新链、价值链中处于不同位置，在创新发展中的合作面远大于竞争面，特别是新产业、新业态、新模式等领域的合作空间更广，双方互利共赢的基本态势没有改变。在新一轮科技革命和产业变革深入发展的背景下，双方利益交汇点更多、相互依存度更高。我们要加强创新发展战略对接，发挥好电动汽车、智能制造等合作平台的作用，加强在人工智能、新能源汽车、自动驾驶等新兴产业领域合作，共同做大产业、做大市场。此访期间，中德双方签署了自动网联驾驶领域合作联合意向声明，这是两国创新合作向更高水平迈进的一个重要标志。

——我们要拓展三方合作。三方合作是发挥中德各自优势、实现多赢共赢的重要途径，中德不少企业已开展了卓有成效的合作。比如，西门子公司与中国100多家企业携手，合作项目遍及60多个国家和地区，带动大量德国设备、技术和服务出口。共建“一带一路”为中德企业开展三方合作打开了新空间。中方愿与德方一道，引导双方企业在互联互通、工业建设、装备制造、节能环保、轨道交通等领域开展三方合作，探讨在中东欧、拉美、非洲等地区先行先试。双方可设立三方合作平台，为企业合作创造便利条件。

女士们,先生们!

中德经济技术合作离不开自由开放的贸易投资环境。以世贸组织为核心的多边贸易体制,是当今国际经贸秩序的基石,其权威和效力应得到尊重和维护。此次会晤中,中德双方再次重申,共同维护自由贸易,坚决反对一切形式的保护主义,维护以规则为基础的多边贸易体制。现行世贸组织的某些规则确有不完善之处,对此,可以适应形势适当予以调整和完善,但自由贸易等基本原则必须坚持,必须充分照顾包括发展中国家在内的各方利益诉求。否则,就会导致全球贸易失序、世界经济复苏进程受阻、南北发展差距拉大。中国是自由贸易和多边主义的坚定支持者、积极践行者,加入世贸组织后全面履行各项承诺,不断扩大对外开放,积极承担与自身实力相适应的国际义务。中方还多次自主降低进口关税,近三年降税产品超过2500项。今年7月1日起,又大幅降低汽车、部分日用品进口关税。刚刚修订发布的2018版外商投资准入负面清单,限制性条目压减24%,大幅放宽金融等服务业、汽车等制造业外资股比限制。中国对外开放的不断扩大,为促进全球经济贸易增长作出了重要贡献。2001—2017年,中国货物进口额累计20万亿美元,年均增长13.5%,高出全球平均水平6.9个百分点;服务进口额累计3.7万亿美元,年均增长16.7%,高出全球平均水平8.8个百分点;今年以来,中国货物和服务进口均增长20%左右。世贸组织成员特别是主要经济体,都应在开放举措上多做"加法",在市场壁垒上多做"减法",这符合各方的共同利益。

在国际经贸往来中,出现分歧甚至摩擦是难以避免的,也是正常的,关键是如何看待和解决。自由和公平是多边贸易体制的两大核心理念。中方倡导自由贸易,同样主张公平贸易,但公平贸易要体现包容性,考虑不同国家所处的不同发展阶段,反对以公平贸易之名行贸易保护主义之实。在经济全球化的今天,国际产业分工你中有我、我中有你,各国产业链、价值链"筋骨"相连。如果固守"零和博弈"的过时思维、任性挑起贸易争端,不仅损害当事双方利益,也伤及全球产业链上各方利益,不会有赢家。中国货物出口的40%、高科技产品出口的2/3都是在华外资企业实现的。我们反对单边主义、贸易保护主义,不仅是为了维护自身合法权益,也是为了维护外资企业利益,维护和促进世界自由贸易体制。

女士们,先生们!

今年以来,中国经济继续保持稳中向好发展势头。一季度国内生产总值同比增长6.8%,特别是就业形势稳定,5月份城镇调查失业率已经降到4.8%,处于多年来最低水平,这对于有着13亿多人口的发展中大国而言至关重要。用电量、货运量等实物量增长快于经济增速,企业利润、税收收入保持两位数增长,制造业PMI持续处于景气区间,说明企业生产经营活跃,经济基本面持续向好。

当前中国经济发展的良好势头，是近些年经济结构持续优化、内生发展动力增强的结果，具有较强的稳定性和可持续性。面对复杂严峻的国际环境和国内较大的经济下行压力，中国政府坚持稳中求进，坚持不搞强刺激，而是依靠改革创新推动转方式、调结构、破难题，深化供给侧结构性改革，着力培育壮大新动能。过去五年市场主体数量增加近80%，目前已超过1亿户，新兴产业迅猛成长，传统产业加快升级。这不仅创造了大量就业岗位，也带来新的消费和投资，推动经济发展形成良性循环。目前，中国的经济增长已从过度依赖投资、出口拉动转向主要依靠消费、服务业和内需支撑，最终消费和服务业对经济增长的贡献率均达到60%左右，经常项目顺差占国内生产总值的比例也已从2007年的约10%下降到去年的1.3%，新动能对经济增长的贡献率超过三分之一。中国经济发展潜力大、韧性强、后劲足，财政赤字率、政府负债率较低，商业银行资本充足率、拨备覆盖率较高，企业负债率趋于下降，宏观调控有足够的政策工具可以运用。我们有信心有能力应对各种风险挑战，长期保持经济持续健康发展。

过去40年，中国依靠改革开放取得了巨大成就，今后还要靠改革开放走向未来。我们要继续深化市场取向改革，加快完善社会主义市场经济体制，切实转变政府职能，深入推进简政放权，创新和加强事中事后监管，全面提升政府服务效能，打造国际一流、公平竞争的营商环境，进一步激发市场活力、释放内需潜力、增强内生动力。我们将继续扩大对外开放，加强与国际通行经贸规则对接，在资质许可、标准制定、政府采购、研发政策等方面对内外资企业一视同仁，保障企业基于自愿和商业规则开展技术合作，不允许强制转让技术。保护知识产权就是保护创新，我们将进一步加大知识产权保护力度，依法打击各类侵权假冒行为，大幅提高违法成本。这是为了维护内外资企业的合法权益，更是推动中国经济高质量发展的需要。

总之，我们将以继续主动扩大开放，对冲单边主义和保护主义；以公平、公开、透明的竞争，拓展多元化市场；以办好自己的事情，增强内生动力，保持经济持续增长、稳中向好的态势。

女士们，先生们！

中国先哲孔子说过，四十而不惑。意思是一个人年过四十，思想更加成熟，处事更加稳健，不为外界干扰因素所困，不为各种杂音所惑。我相信，中德务实合作经过40年的风雨洗礼，步伐一定会越来越稳健、合作之路一定会越走越宽广，更好造福两国、惠及世界！

谢谢大家。

4. 参考译文

Speech at the Ninth China-Germany Economic and Technological Cooperation Forum①

H. E. Li Keqiang
Late Premier of the State Council of
The People's Republic of China
Berlin, 9 July 2018

Honorable Chancellor Angela Merkel,

Ladies and gentlemen,

It gives me great pleasure to join Chancellor Merkel for the closing ceremony of the ninth China-Germany Economic and Technological Cooperation Forum. The Chancellor and I exchanged visits within six months of the formation of new governments in our countries. This speaks volumes about the strong ties between China and Germany, and our shared desire for closer cooperation. The theme of this forum "Maintaining and Enhancing Competitiveness and Innovation" is highly relevant as it is consistent with the trend of the times and reflects the needs of both sides. On behalf of the Chinese government, I wish to extend my warm congratulations on the success of the forum and pay high tribute to all those from both countries who have been committed to the China-Germany friendship and cooperation.

This year marks the 40th anniversary of China's reform and opening-up and the signing of the science and technology cooperation agreement between our governments. The past four decades have witnessed a sound momentum of steadily growing China-Germany cooperation, with expanded scale, widened scope and higher quality. In 2017, bilateral trade reached \$168.1 billion, 100 times that of 1978, accounting for nearly one third of the China-EU trade. China has been Germany's largest trading partner for two years in a row. Two-way investment has exceeded \$40 billion. Over 7,000 German companies and 2,000-plus Chinese firms run businesses in each other's countries.

Our cooperation has grown far beyond trade and spans a diverse range of areas, including finance, science and technology, environmental protection, energy and people-to-people exchange. And we have forged a strong partnership on innovation.

---

① 译之有道.李克强在第九届中德经济技术合作论坛上的致辞[中英][EB/OL].(2018-07-10)[2024-10-09]https://yizhiyoudao.kuaizhan.com/95/91/p5399918250a585.

With its steady performance, fruitful outcomes and enormous potential, China-Germany cooperation stands as an engine of China-EU cooperation and a fine example of cooperation between major countries. It has not only boosted the economies of both our countries but also injected vigor into global growth. We owe these achievements to our commitment to equality, mutual respect and mutual benefit and to our enterprising spirit. These achievements are also attributable to the hard work and contributions of the business communities of both countries.

During his successful visit to Germany last year, President Xi Jinping and German leaders jointly mapped out the development of our relations for the future. This morning, Chancellor Merkel and I co-chaired the fifth China-Germany inter-governmental consultation, which produced important agreements and deliverables in wide-ranging fields. We both believe that given the complicated political and economic environment in the world, China and Germany, as the largest economies at the two ends of the Eurasian continent, need to further consolidate and strengthen political trust, cement the foundations of our all-dimensional strategic partnership, and raise our cooperation to greater heights. We need to bring greater certainty to world peace and development with our unwavering commitment to China-Germany cooperation and to multilateralism.

—**We need to further unlock our trade potential**. China and Germany are among the top three trading nations in the world, and together we take up nearly 20% of global trade. However, trade between us only accounts for less than 3% of our total foreign trade, leaving huge space for further expansion.

With the upgrading of consumption structure and accelerating industrial transformation in China, German companies enjoy enormous opportunities in selling more competitive products, high-end equipment and quality services to Chinese consumers. We hope that Germany will further ease the restrictions on high-tech exports to China to raise the share of high-tech products in our bilateral trade. China is the largest exporter of goods, offering a rich variety of products at competitive prices. Chinese products are popular with their foreign customers, and can well meet Germany's diverse market demand. We need to advance trade facilitation, develop new forms of trade, improve the product mix and bolster cross-border e-commerce to take the scale of our trade to a new level.

Our trade in services currently stands at less than \$ 30 billion. We need to expand services trade in tourism, education and transportation, and nurture new growth areas in modern services such as technology trade, professional services, repair and maintenance, telemedicine, finance and insurance. In November this year, the first

China International Import Expo will be held in Shanghai. We believe Germany's participation as a Guest of Honor Country will further boost the popularity of German products and services in the Chinese market.

—**We need to expand two-way investment which has been active overall.** Yet, less than 7% of Germany's total outbound investment goes to China. And China's cumulative investment in Germany is modest at best. Many far-sighted German companies who took the lead in investing and doing business in China have reaped handsome returns and achieved win-win results. In an even more open Chinese market with an improving business environment, German investors enjoy new, greater opportunities. You are most welcome to expand your investment in China, especially in the central, western and northeastern regions where untapped potential lies and through setting up joint industrial parks.

In investing overseas, Chinese companies follow market principles and commercial rules. We hope that Germany will provide them with a fairer and more transparent and predictable investment environment, and see that they receive fair treatment in security reviews and financial regulation. The business communities of both countries hope to see early conclusion of the China-EU investment agreement. China is ready to work with Germany for new progress in the negotiations.

—**We need to promote innovation cooperation.** China and Germany have complementary strengths in innovation, which makes us "golden partners" for cooperation. Germany is a global leader in basic research, innovation and high-end R&D. China, on its part, has a big market, rich human resources, a sophisticated industrial system, and a booming digital economy, holding unique strengths in innovation application. Given our different places in the global innovation and value chains, there exist far more opportunities for cooperation than areas of competition. Cooperation in new industries, new forms of business and new models is particularly promising and will remain win-win in nature.

With the advance of a new round of technological revolution and industrial transformation, our two countries will only become more interdependent and our interests more closely intertwined. Against such a backdrop, we need to foster greater synergy between our innovation strategies, better leverage the platforms for cooperation on electric vehicles and smart manufacturing, and enhance cooperation in new industries, including artificial intelligence, new energy vehicles and autonomous driving. Through these efforts, we can make our industries stronger and foster bigger markets. During the visit, the two sides signed a letter of intent on cooperation in intelligent and connected vehicles (ICV), which marks a higher level of China-

Germany innovation cooperation.

—**We need to expand cooperation in third countries.** Such cooperation provides an important means for pursuing win-win and all-win cooperation based on our comparative strengths. This has been borne out by success stories among Chinese and German companies. For instance, Siemens has partnered with over 100 Chinese companies on projects in over 60 countries and regions, enabling expanded export of German equipment, technologies and services. The Belt and Road Initiative has provided new opportunities for our companies to continue such cooperation. China will work with Germany to encourage business-to-business cooperation in third countries in the areas of connectivity, industrial development, equipment manufacturing, energy conservation and environmental protection, and rail transport. We may roll out pilot programs starting with Central and Eastern Europe, Latin America and Africa. We may also set up special third-party cooperation platforms to facilitate such efforts by our businesses.

Ladies and gentlemen,

China-Germany economic and technological cooperation requires a free and open trade and investment environment. The multilateral trading regime with the WTO at its core is the cornerstone of the existing international trading order, and its authority and efficacy should be respected and upheld. During our meeting, we reaffirmed our shared commitment to upholding free trade, opposing protectionism in all its forms, and safeguarding the rules-based multilateral trading regime. It is true that some of the WTO rules are not perfect. They can be adjusted and improved in an appropriate way to reflect changing circumstances. However, free trade and other fundamental principles must be upheld, and the interests and demands of all countries, the developing countries in particular, must be fully accommodated. Failure to do so will lead to disruptions in global trade, stalling world recovery and a widening North-South gap.

China is a firm supporter and active practitioner of free trade and multilateralism. China has fully honored its commitments upon accession to the WTO to open up its markets and has undertaken international obligations commensurate with its capabilities. We have initiated several rounds of tariff cuts involving over 2,500 products in the last three years. On July the first this year, the import tariffs on automobiles and some daily consumer goods were significantly lowered. In the recently released 2018 version of the negative list on access to foreign investment, the number of restricted items was reduced by 24%. Foreign equity caps were substantially raised in the services and manufacturing sectors, financial services and automobiles in particular.

China's continued opening-up has contributed substantially to global economic and trade growth. Between 2001 and 2017, China imported a total of $ 20 trillion worth of goods, which comes down to an annual increase of 13.5%, 6.9 percentage points higher than the global average; during the same period, our total import in services reached $ 3.7 trillion, posting an annual growth of 16.7%, 8.8 percentage points higher than the global average. This year so far China's imports in goods and services have both increased by around 20%. WTO members, especially the major economies, all need to do more to open up their economies and less in raising market barriers, as this serves the interests of all parties.

Differences and even frictions are natural and unavoidable in global trade. The question is how to approach and address them. The notions of free and fair trade lie at the heart of the multilateral trading regime. China advocates free trade and also stands for fair trade. However, fair trade must be inclusive, considering the different development stages of countries. The practice of trade protectionism in the name of fair trade must be rejected. As we live in a globalized world, international division of labor has connected the industrial and value chains of different countries. The outdated zero-sum mentality and willful provocation of trade disputes will do harm not only to the interests of the parties directly concerned, but also to all in the global industrial chain. No one will emerge as a winner in such a situation.

Foreign-invested companies account for 40% of China's export in goods and two thirds of its high-tech exports. China's stance against unilateralism and trade protectionism, therefore, is meant not just to protect its own legitimate rights and interests, but also to safeguard the interests of foreign-invested companies and the global free trade system as a whole.

Ladies and gentlemen,

Let me now turn to the Chinese economy, which has kept a good momentum of steady growth this year. China's GDP grew by 6.8% year-on-year in the first quarter. It is particularly worth noting that employment has remained stable: the surveyed urban unemployment rate fell to 4.8% in May, the lowest level in several years, which is crucial to a major developing country with more than 1.3 billion people. Other economic indicators such as electricity consumption and freight volume have outpaced overall economic growth. Business profits and tax revenues have grown at double-digit rates. PMI reading has stayed above the threshold indicating economic expansion. All these point to robust business activities and sound economic fundamentals in China.

Such a good momentum is the result of our unrelenting efforts in recent years to upgrade the economic structure and foster stronger internal growth drive. Hence, we expect this growth to be solidly-based and sustainable in the long run. Despite a complex and challenging international environment and heavy downward pressure, the Chinese government sought to move forward while maintaining stable economic performance. Instead of resorting to massive stimulus, we relied on reform and innovation to shift the growth model and improve the economic structure and endeavored to overcome challenges by intensifying supply-side structural reforms and nurturing new growth drivers.

These efforts have paid off. The number of market entities surged by nearly 80% to more than 100 million in the past five years; emerging industries grow by leaps and bounds and traditional ones are upgrading at a faster pace; a big number of jobs have been added, and new consumption and investment generated, creating a virtuous circle of development. The economy has moved away from reliance on investment and export to being fueled mainly by consumption, the services sector and domestic demand. Final consumption and the services sector now both account for roughly 60% of China's growth; the current account surplus to GDP ratio dropped from around 10% in 2007 to 1.3% last year; and new drivers contribute over a third to economic growth.

Overall, the Chinese economy boasts vast potential, strong resilience and bright prospects. The combination of relatively low fiscal deficit ratio and government debt-to-GDP ratio, relatively high capital adequacy ratio and provision coverage ratio of commercial banks, and the paring down of corporate debt have handed us a host of policy tools for macro regulation. We have the confidence and capability to meet any potential risks or challenges and maintain sustained and sound development of the Chinese economy over the long run.

Reform and opening-up has been instrumental in China's remarkable achievements over the past 40 years; more reform and opening-up is the way to go for the future. China will continue to deepen market-oriented reforms and improve the socialist market economy by transforming government functions, streamlining administration, developing new and better ways of conducting compliance oversight, and increasing the efficiency of government services across the board. We aim to offer a world-class business environment with a level playing field that will unleash market vitality, unlock potential in domestic demand and boost internal drivers of growth.

China will open up further to the rest of the world and pursue greater convergence with established international business rules. We will treat all enterprises, Chinese or foreign, equally in license application, standards-setting, government procurement, and R&D policies. We will see to it that technological cooperation is conducted on a voluntary basis and according to commercial rules. No mandatory requirement for technology transfer will be allowed. Protecting intellectual property and encouraging innovation are the two sides of the same coin. We will step up IPR protection, crack down on all forms of infringements and counterfeiting according to law, and significantly raise the cost for offenders. Such efforts are not only aimed to protect the lawful rights and interests of Chinese and foreign companies alike, but more importantly, they are naturally required for achieving higher quality development of the Chinese economy.

All in all, we will continue to counteract the effects of unilateralism and protectionism with more proactive measures of opening-up, develop diversified markets through open, fair and transparent competition, and sustain the steady, upward momentum of the Chinese economy by managing our own affairs well and building up internal drivers of growth.

Ladies and gentlemen,

The great Chinese philosopher Confucius observed that a man at forty should be free from confusion. It means that upon turning forty, one would be more mature in outlook and more thoughtful in conduct, so he is no longer confused by distractions or noises of any sort. I believe, after all we have been through over the past forty years, China-Germany practical cooperation will move ahead with more steady steps, bear richer fruits, and bring greater benefits to our two countries and to the rest of the world.

Thank you.

5. 要点讲解

本篇讲话涉及较多有关中德贸易合作的数据，对译员的数字口译能力提出了较高要求。

大部分基数词都属于金额，如“1681 亿美元”（$168.1 billion）、“400 亿美元”（$40 billion）、“300 亿美元”（$30 billion）、“20 万亿美元”（$20 trillion）等，也涉及人口数量，如“13 亿多人口”（more than 1.3 billion people）和“1 亿”（100 million）。本篇演讲发表于政府间论坛，属于官方场合，译员必须准确无误地将这些数字翻译出来。

本文出现了两次序数词翻译：一处为“中国货物出口世界**第一**”，译员可译为“China is the **largest** exporter of goods”或者“China's goods export **ranks number one** in the world”；另一处为“中德位居世界**前三大**贸易国之列”，这句话可以译为“China and Germany are among the **top three trading nations** in the world”。

本文出现了多处小数翻译，如“6.9 个百分点”（6.9 percentage points）、“3.7 万亿美元”（$3.7 trillion）等。

本文出现了三处分数翻译：“2017 年双边贸易额达 1681 亿美元，占中欧贸易总额的**近三成**”中的“三成”就是 30%，所以这句话可译为“In 2017，bilateral trade reached US $168.1 billion，accounting for **nearly 30%** of the China-EU trade”，参考译文中采用的是 nearly one-third；“高科技产品出口的 **2/3** 都是在华外资企业实现的”可译为“Foreign-invested companies account for **two-thirds** of its high-tech exports”；“新动能对经济增长的贡献率超过**三分之一**”可译为“New drivers contribute over **a third** to economic growth”。

本文出现了多处百分数翻译，如“两国外贸总额占世界贸易的比重**近 20%**，但双边贸易额占两国外贸总额的比重**不到 3%**，有很大发展空间”可译为“China and Germany take up **nearly 20%** of global trade. However，trade between us only accounts for **less than 3%** of our total foreign trade，leaving huge space for further expansion”；“最终消费和服务业对经济增长的贡献率均达到 **60%左右**，经常项目顺差占国内生产总值的比例也已从 2007 年的**约 10%**下降到去年的 **1.3%**”可译为“Final consumption and the services sector now both account for **roughly 60%** of China's growth；the current account surplus to GDP ratio dropped from **around 10%** in 2007 to **1.3%** last year”。译员需格外注意“不到”“左右”“约”这些字眼，确保数字信息是完整无误的。

本文出现了一处倍数翻译：“2017 年双边贸易额达 1681 亿美元，是 1978 年的 **100 倍**”可译为“In 2017，bilateral trade reached US $168.1 billion，**100 times** that of 1978”。

整体而言，这篇演讲涉及较多数字翻译，在此类高规格的会议场合，数字翻译失误可能引发严重的后果，因此，译员需要尽力做到准确无误地转述数据。

二、实战训练篇

（一）欧洲强力支持太空发展，以此对抗气候危机

1. 词汇表

the 27th UN Climate Change Conference	第 27 届联合国气候变化大会
Josef Aschbacher	约瑟夫·阿施巴赫
European Space Agency (ESA)	欧洲航天局
Director General	局长
Pollsters Toluna and Harris Interactive	民意调查机构托伦纳和哈里斯互动
UN's Global Climate Observing System	联合国气候变化系统
ESA Council at Ministerial level	欧洲航天局成员国部长级会议
member states，associated states and cooperating states	成员国、联系国和合作国

2. 原　文

Strong European Support for Space to Combat Climate Crisis[①]

Europe should demonstrate responsibility, leadership and autonomy in space—and its highest priority should be to address climate change, according to a poll of European citizens.

Almost nine out of ten people questioned said that collecting insights on climate change and understanding what is happening on Earth should be the most important use of space.

"This survey shows that the citizens of Europe strongly support investing in space to improve life on Earth—and that there is an increased appetite for a greater ambition for space in Europe. As world leaders meet for the 27th UN Climate Change Conference (COP27), European citizens want to see space used even more to monitor and mitigate climate change. We must act now to increase European autonomy, leadership and responsibility in space," says Josef Aschbacher, Director General of ESA.

More than 21,000 people living in the 22 countries that are members of ESA responded to the survey, published today, which was conducted by pollsters Toluna and Harris Interactive between 25 September and 6 October.

Satellite data underpins more than half of the essential climate variables identified by the UN's Global Climate Observing System. Space helps scientists, policymakers and political leaders not only to monitor, understand, model and predict, but—crucially—to act on climate-induced and other crises.

The vast majority of people said that space is essential to further human knowledge of the Universe—as well as to improving life on Earth through always-on-everywhere communication connections and navigation signals.

However, in the future, most Europeans would like greater emphasis to be put on monitoring and mitigating climate change, while continuing to better understand the universe and to benefit from satellite-empowered connectivity and navigation.

More than 80% of respondents also said that Europe should pool its space activities to compete with other space-faring nations—and that European space activities should be independent of decisions made by other major space powers.

---

① The European Space Agency. Strong European support for space to combat climate crisis[EB/OL]. (2023-07-13)[2023-08-14]. https://www.esa.int/About_Us/Corporate_news/Strong_European_support_for_space_to_combat_climate_crisis.

Indeed the desire for the pooling and independence of European space has increased significantly since the survey was last undertaken in 2019.

Among respondents in France, Germany, Italy, Spain and the UK, some 84% of people favoured joining together to compete with the United States, Russia, China, India and Brazil (a rise of 14 percentage points) and 81% said that European space activities should be independent (a rise of 17 percentage points).

When it comes to venturing into outer space, most people thought that cleaning up space debris was the highest priority. Some 86% said it was important, ahead of organising a robotic exploration mission to Mars (77%), putting astronauts on the Moon (71%) and sending astronauts to Mars (70%).

However, the importance of space exploration has greatly increased in people's minds since the 2019 survey. Support for a robotic exploration of Mars rose by 18 percentage points among respondents in France, Germany, Italy, Spain and the UK, while backing for astronauts on the Moon and on Mars rose by 18 percentage points and 16 percentage points, respectively.

The poll was conducted ahead of the ESA Council at Ministerial level, which will be held in Paris on 22nd and 23rd November. At the meeting, ESA's member states, associated states and cooperating states will be invited to together strengthen Europe's space ambitions and ensure that space continues to serve European citizens.

(二) 专访工业和信息化部部长金壮龙

1. 词汇表

新能源汽车	new energy vehicle
大型客机	large passenger aircraft
光伏	photovoltaic
5G 基站	5G base station
千兆光网	gigabit optical network
"市市通千兆"	"gigabit coverage in every city"
"县县通 5G"	"5G coverage in every county"
"村村通宽带"	"broadband coverage in every village"
光纤光缆	optical fiber cable
关键工序数控化率	CNC rate of key processes

2. 原文

释放潜力　增强动力

——专访工业和信息化部部长金壮龙(节选)①

回顾刚刚过去的2022年,工业和信息化部部长金壮龙坦言,多种超预期因素冲击影响带来了巨大挑战。为了全力保持产业链供应链稳定畅通,工业和信息化部与各地区、各部门一道,及时出台振作工业经济系列举措,推动工业经济在较短时间内恢复回稳。

金壮龙用一系列统计数据总结了去年工业领域取得的成绩,2022年1—11月份,我国规模以上工业增加值同比增长3.8%,高技术制造业和装备制造业增加值同比分别增长8%和6.2%。制造业占GDP比重基本稳定,发挥了工业稳定宏观经济大盘的"压舱石"作用。与此同时,发展质量稳步提升,新能源汽车、大型客机、光伏等领域取得一批标志性成果。

今年是实施"十四五"规划承上启下的关键一年,2023年我国工业经济运行仍然面临不少困难和挑战,外部环境带来的影响在加深,需求收缩、供给冲击、预期转弱三重压力仍然较大。尽管如此,金壮龙对经济运行整体好转仍然充满信心,他的信心源自我国完备的产业体系和在一些重要领域形成的规模和技术优势。

先来看国家经济命脉——制造业。建设现代化产业体系,关键在于振兴制造业。经过多年发展,我国制造业规模已占全球比重约30%,连续13年位居世界首位。制造业是实体经济的基础,数字经济则是推动我国经济增长的另一个重要引擎,也是建设现代化产业体系的重要优势。我国数字经济经过几年的蓬勃发展,总体规模已稳居世界第二,数字经济赋能实体经济的作用也日趋凸显。

我国已经拥有了全球规模最大、技术领先的网络基础设施,建成开通5G基站228万个,5G应用案例累计超过5万个,千兆光网具备了覆盖5亿户家庭的能力,实现"市市通千兆""县县通5G""村村通宽带"。

坚实的信息通信基础设施为数字产业加快增长奠定了良好的基础,2022年1—11月,我国软件业务收入达到9.46万亿元,同比增长10.4%;移动支付年交易规模达到527万亿元,智能手机、光通信设备、光纤光缆等领域我国企业的全球份额达到50%。产业数字化全面提速,工业互联网广泛融入45个国民经济大类,有影响力的工业互联网平台达到248家,重点平台工业设备连接数超过8000万台(套),工业App数量近30万个,制造业企业数字化研发设计工具普及率达到76%,关键工序数控化率达到57.2%。

---

① 央视网.释放潜力 增强动力:专访工业和信息化部部长金壮龙[EB/OL].(2023-01-03)[2023-11-14]. https://news.cctv.com/2023/01/03/ARTINqdiodNdfiDvZWALXDPL230102.shtml.

(三) 王毅国务委员在2020年度"一带一路"国际合作高峰论坛咨询委员会会议上的开幕辞

1. 词汇表

"一带一路"国际合作高峰论坛咨询委员会	Advisory Council of the Belt and Road Forum for International Cooperation
单边主义	unilateralism
保护主义	protectionism
"一带一路"国际合作高级别视频会议	High-level Video Conference on Belt and Road International Cooperation
共商共建共享	extensive consultation, joint contribution and shared benefits
地缘政治工具	geopolitical tool
区域全面经济伙伴关系协定	Regional Comprehensive Economic Partnership (RCEP)
全面与进步跨太平洋伙伴关系协定	Comprehensive and Progressive Agreement for Trans-Pacific Partnership (CPTPP)
中欧班列	China-Europe Railway Express
新冠肺炎疫苗实施计划	COVAX (COVID-19 Vaccines Global Access) initiative
雅万高铁	the Jakarta-Bandung high speed railway
中老铁路	the China-Laos railway
匈塞铁路	the Budapest-Belgrade railway
中巴经济走廊	China-Pakistan Economic Corridor
二十国集团缓债倡议	G20's Debt Service Suspension Initiative (DSSI)
中国-东盟信息港	China-ASEAN Information Harbor
"一带一路"绿色发展国际联盟	BRI International Coalition for Green Development
绿色投资原则	BRI Green Investment Principles
绿色投资基金	BRI Green Investment Fund
《全球数据安全倡议》	*Global Initiative on Data Security*
森林蓄积量	forest stock volume
中国共产党第十九届五中全会	the Fifth Plenary Session of the 19th Central Committee of the Communist Party of China
二〇三五年远景目标	the Long-Range Goals for 2035
国内大循环	domestic circulation
国内国际双循环	domestic and international circulations
国内商品零售市场	domestic retail market

续表

数字交通走廊	digital transport corridor
跨境光缆信息通道	cross-border fiber optic cables for information transmission
碳中和	carbon neutrality
单位国内生产总值二氧化碳排放	CO_2 emissions per unit of GDP, carbon intensity
非化石能源占一次能源消费比重	the share of non-fossil fuels in primary energy consumption
风电、太阳能发电总装机容量	total installed capacity of wind and solar power

2. 原　文

在2020年度"一带一路"国际合作高峰论坛咨询委员会会议上的开幕辞①

国务委员　王毅

2020年12月18日，北京

各位委员：

大家晚上好。

很高兴同老朋友们见面。首先，我谨对"一带一路"国际合作高峰论坛咨询委员会2020年度会议的召开表示热烈祝贺！

今年以来，面对新冠肺炎疫情冲击，"一带一路"国际合作没有止步，而是展现出强大的韧性和活力，为各国抗疫情、稳经济、保民生发挥了重要作用，为落实联合国2030年可持续发展议程作出了积极贡献。

当前，世纪疫情和百年变局交织共振，各国人民生命安全和身体健康遭受威胁，全球产业链、供应链受到冲击，国际贸易和投资萎缩，世界经济陷入衰退，单边主义、保护主义抬头。但是，无论形势如何变化，国际社会对"一带一路"的需求没有改变，合作伙伴对"一带一路"的支持没有改变，中方推进"一带一路"国际合作的决心更没有改变。

各位委员，

习近平主席提出，要把"一带一路"打造成合作之路、健康之路、复苏之路、增长之路，为"一带一路"国际合作指明了方向。

打造"合作之路"，需要坚持开放合作、互利共赢。今年以来，"一带一路"合作伙伴克服疫情影响，在10多个"一带一路"多边合作平台下以视频形式举办了30多场

① 外交部.王毅国务委员在2020年度"一带一路"国际合作高峰论坛咨询委员会会议上的开幕辞[EB/OL].(2023-07-13)[2023-11-14]. https://www.fmprc.gov.cn/web/wjb_673085/zzjg_673183/gjjjs_674249/gjzzyhygk_674253/ydylfh_692140/zyjh_692150/202012/t20201220_10410189.shtml.

国际会议，在金融、税收、能源、绿色发展、反腐败等领域协调各方政策、推进务实合作。“一带一路”国际合作高级别视频会议发表联合声明，发出携手抗击疫情、加强互联互通、推动经济恢复的强烈信号，向世界传递了对“一带一路”国际合作的坚定信心。

“一带一路”是阳光倡议，秉持开放包容透明、共商共建共享的原则。“一带一路”是经济合作平台，从来不是地缘政治工具。“一带一路”是伙伴国家的合唱，不是中国一家的独奏。我们支持中外企业按照市场规则和公平原则，以公开透明方式参与具体合作项目，实现利益共享、风险共担。中方愿同各方一道，继续以双边、三方和多边形式推进各领域合作，推动建设开放型世界经济，维护全球产业链供应链稳定畅通。前不久，我们签署了区域全面经济伙伴关系协定(RCEP)，并宣布愿积极考虑加入全面与进步跨太平洋伙伴关系协定(CPTPP)，这表明了中方致力于推进自由贸易的坚定决心。

打造“健康之路”，需要持续推进抗疫国际合作。当前，疫情仍在全球蔓延，防控形势依然严峻。“一带一路”合作伙伴面临的最紧迫任务是团结合作战胜疫情。今年以来，中方和合作伙伴把抗击疫情作为重点，着力推进“健康丝绸之路”建设，开展疫情联防联控，分享信息和诊疗经验，相互提供急需的医疗物资，努力打通陆上、海上、空中运输物资的“生命之路”，共同筑起抗疫“防火墙”。中欧班列运送紧急医疗物资超过 800 万件，“空中丝绸之路”把超过 1700 吨的中国援助医疗物资运到世界各地。

疫苗是下阶段抗疫国际合作的重点。中国已加入世卫组织新冠肺炎疫苗实施计划。中国企业同俄罗斯、埃及、印度尼西亚、巴基斯坦、阿联酋等“一带一路”伙伴国家开展的疫苗三期试验已取得重要进展。中国疫苗研发成功后，将作为全球公共产品，为实现疫苗在发展中国家的可及性和可负担性作出贡献，“一带一路”合作伙伴必将从中受益。

打造“复苏之路”，需要统筹疫情防控和经济社会发展。中国已同很多国家特别是“一带一路”伙伴国家建立便利人员往来的“快捷通道”，同有需要的国家建立畅通货物流动的“绿色通道”。雅万高铁、中老铁路、匈塞铁路等项目进展良好，中巴经济走廊拉合尔城市轨道交通项目开通运行。很多“一带一路”项目在严格防控疫情的前提下不停工、不裁员，坚持运行，一批新项目启动实施。在世界经济衰退背景下，中方对“一带一路”整体投入没有减少，而是呈现出逆势增长的态势。今年前三季度，中国对“一带一路”共建国家非金融类直接投资同比增长近 30%。中国积极参与二十国集团缓债倡议，是二十国集团成员中缓债金额最大的国家，为多个共建“一带一路”国家抗击疫情、恢复经济提供了力所能及的支持和帮助。

当前，世界经济复苏仍面临诸多不确定因素，国际物流领域运力出现供需失衡。我们将继续同各方优化“快捷通道”和“绿色通道”合作机制，完善陆、海、空多式联运，进一步提升国际大通道运输效率和质量。

打造“增长之路”，需要挖掘世界经济新动能。“一带一路”合作伙伴正积极发展

"丝路电商",建设数字交通走廊、跨境光缆信息通道、中国—东盟信息港等。"一带一路"绿色发展国际联盟、绿色投资原则、绿色投资基金等多边平台不断走深走实。

数字丝绸之路是下阶段共建"一带一路"的重点领域之一。中国提出《全球数据安全倡议》,目的是在维护数据安全、促进数字经济发展与合作方面贡献力量,与各方共同推动构建和平、安全、开放、合作的网络空间。有的国家把高科技问题政治化,其实质是维护自身高科技垄断地位,这不是建设性做法。事实证明,共商共建共享才是解决全球数字治理的正路。

中国坚持用生态文明理念指导发展,积极参加气候变化国际合作,宣布将力争于2030年前达到二氧化碳排放峰值,2060年前实现碳中和。到2030年,中国单位国内生产总值二氧化碳排放将比2005年下降65%以上,非化石能源占一次能源消费比重将达到25%左右,森林蓄积量将比2005年增加60亿立方米,风电、太阳能发电总装机容量将达到12亿千瓦以上。这将为建设"绿色丝绸之路"提供新的动力。

各位委员,

今年以来,面对百年来全球发生的最严重的传染病大流行,中国迅速果断采取超常规的防控举措,统筹疫情防控和经济社会发展。前三季度中国经济增长已经由负转正。当前,中国生产需求持续回暖,就业形势总体稳定,财政金融平稳运行,市场预期不断改善。预计中国将成为今年全球唯一保持正增长的主要经济体。各方评估中国经济对全球经济增长的贡献率将会达到三分之一以上。

中国经济呈现回稳态势,对世界经济是利好消息。中国将继续坚持深化改革,扩大开放,持续释放中国经济对世界经济的拉动效应,为世界经济复苏发展作出更大贡献。

前不久,中国共产党第十九届五中全会审议通过了《关于制定第十四个五年规划和二〇三五年远景目标的建议》,擘画了未来一个时期中国发展宏伟蓝图。中国将科学把握新发展阶段,坚定贯彻新发展理念,积极构建以国内大循环为主体、国内国际双循环相互促进的新发展格局。

新发展格局是相互开放、相互促进的国内国际双循环。在新发展格局下,中国的市场潜力将会被充分激发,对外开放的大门将进一步敞开。中国有14亿人口,中等收入群体超过4亿,今年国内商品零售市场规模将超过6万亿美元,未来10年累计商品进口额有望超过22万亿美元。中国经济的高质量发展,必将为高质量共建"一带一路"提供更强劲动力和更广阔空间。

各位委员,

高峰论坛咨询委员会是"一带一路"多边合作的重要平台。各位委员都是经验丰富的政治活动家和专家学者。我期待各位在政策咨询、学术研究、国际合作等方面,为高质量共建"一带一路"建言献策,在国际上传播更多关于"一带一路"的真实故事,为"一带一路"国际合作贡献更多智慧和力量。

第五章 语言转换训练

在口译的信息处理过程中，译员对语言片段要经历“感知→识别→组合→转换”的过程。其中，语言转换是决定口译速度和质量的关键一环①。由于英汉两种语言的语言结构、思维方式和表达习惯不同，译员如果拘泥于源语的形式，口译输出质量会大打折扣，听众也不能轻松、准确地抓住源语要表达的意思。不同于同传，交传中译员有充分的时间梳理、记录信息，也可以采用灵活的方式进行语言转换，使输出符合目标语听众的语言习惯。本章重点介绍交传中译员常用的语言转换技巧，包括顺句驱动法、适当调整法、重组法、拆句法、合并法、转换法、增译法、减译法和简化法等方法。

第一节 语言转换技巧

口译实践中，译员可以运用不同语言转换技巧。从语序调整的角度来看，译员主要采用顺句驱动法、适当调整法和重组法；从长短句转换的角度来看，主要运用拆句法和合并法；从语言形式转化的角度来看，主要表现形式有语态转换和词性转换；从文本信息加工的角度来看，主要采用增译法、减译法和简化法。

一、顺句驱动法和适当调整法

顺句驱动法是指尊重源语，按照源语的语序进行翻译的方式；适当调整法则是指对源语语序进行一定程度的改动。译员常常交替使用这两种语言转换方式，既体现了翻译的灵活性，也能使译文更加符合目标语的表达习惯。

例 1：

原文：“一带一路”是经济合作平台，从来不是地缘政治工具。“一带一路”是伙伴国家的合唱，不是中国一家的独奏。

参考译文：The BRI is a platform for economic cooperation, not a geopolitical tool. It is a chorus by BRI partners, not a solo show by China.

不言而喻，这句话采用的是顺句驱动法，译员无需调整源语语序，就能以通顺自然的方式将源语的意思表达出来。

例 2：

原文：“一带一路”国际合作高级别视频会议发表联合声明，发出携手抗击疫情、

① 任文.英汉口译教程[M].北京：外语教学与研究出版社，2011：125.

加强互联互通、推动经济恢复的强烈信号，向世界传递了对“一带一路”国际合作的坚定信心。

参考译文：The joint statement issued at the High-level Video Conference on Belt and Road International Cooperation sent a strong message of collectively fighting the virus, strengthening connectivity and promoting economic recovery, and reaffirmed to the world the strong confidence in Belt and Road international cooperation.

译员在处理这句话时，使用的是适当调整法：译员将 statement 作为主语，并转换为被动语态，于是源语第一个分句的语序就发生了调整；后面在翻译两个平行的分句时，译员都将中心语 message 和 confidence 提前，将具体内容调后，用介词 of 和 in 来连接两者，这样的处理更符合英语的语法习惯。

二、重组法

重组法是指译员充分发挥自己的主观能动性，摆脱源语结构限制，将源语打破重组。

例 1：

原文：今年以来，“一带一路”合作伙伴克服疫情影响，在 10 多个“一带一路”多边合作平台下以视频形式举办了 30 多场国际会议，在金融、税收、能源、绿色发展、反腐败等领域协调各方政策、推进务实合作。

参考译文：Despite the onslaught of COVID-19 this year, BRI partners have convened 30-plus international video conferences under a dozen or so BRI multilateral platforms to coordinate policies and advance practical cooperation on finance, taxation, energy, green development, anti-corruption and other areas.

在处理这一长句时，译员首先将“克服疫情影响”前置，符合英语中让步状语常放在句首的表达习惯，“今年”很自然地被调整到第一个分句的末尾；然后，译员将“30 多场会议”和“10 多个……平台”顺序调换，符合英语“先局部后整体”的思维习惯；紧跟着的是 coordinate policies and advance practical cooperation，译员通过不定式 to 表目的，逻辑关系更加自然，最后再罗列合作的具体领域。打破重组后的语序通顺自然，整体上更加符合英文的表达习惯。

例 2：

原文：Middle income home buyers in the United States are finding little on the market to buy, even if they can qualify and afford a mortgage. These would-be buyers face the most severe housing shortage of any other income bracket, according to a new analysis from the National Association of Realtors and *realtor. com* that found the market is short more than 300,000 affordable homes for these buyers.

参考译文：美国中等收入购房者即使有资格贷款又能支付得起按揭，在市场上也难以找到合适的房。美国全国房地产经纪人协会及其官网 *realtor. com* 的最新分析表明，中等收入购房者面临着比其他收入阶层更严重的房源短缺问题，合适房源的缺口超过 30 万套。

译员首先将 even if 引导的让步状语从句提前，符合汉语的表达习惯；然后又将 according to 提前，因为汉语中习惯将引证来源放在前面以增强说服力；译员又整合句意，将 that 引导的定语从句和“These would-be buyers face…”放在了一起。

三、拆句法

顾名思义，拆句法是指把一个长句拆译成几个短句。拆句法常用于英译汉，译员可以在关系代词、关系副词等连接处断开，或者根据源语的意群切分。拆句法可以有效降低目标语听众的理解障碍。

例 1：

原文：As the AIIB pointed out in a statement, the bank is a diverse international team representing 65 different nationalities serving 106 members worldwide.

参考译文：正如亚投行在一份声明中所称，亚投行拥有一支多元化国际团队，成员来自 65 个国家和经济体，为全球的 106 个成员提供服务。

译员从非谓语动词 representing 和 serving 两处断句，将源语的第二个长分句拆分成了三个短分句。

例 2：

原文：The World Bank said that China saw a robust recovery in economic growth in early 2023 thanks to the release of pent-up demand during the pandemic.

参考译文：世界银行表示，中国经济在 2023 年初出现强劲复苏，这是因为疫情期间被压抑的需求得到释放。

译员从宾语从句引导词 that 和 thanks to 两处断开，将源语的一个长句分解成了三个分句。

四、合并法

和拆句法相反，合并法是将几个短句整合成一个长句，可以使信息更加凝练，常见于汉译英。

例 1：

原文：我们支持中外企业按照市场规则和公平原则，以公开透明方式参与具体合作项目，实现利益共享、风险共担。

参考译文：We support Chinese and foreign companies in participating in cooperation projects in an open and transparent manner following market rules and the principle of fairness and sharing benefits and risks.

在翻译这一句话时，译员通过使用介词短语 in a(n)… manner 和非谓语动词 following、sharing 的方式将源语的三个分句整合成了一个长句。

例 2:

原文:中方愿同各方一道,继续以双边、三方和多边形式推进各领域合作,推动建设开放型世界经济,维护全球产业链供应链稳定畅通。

参考译文:China is ready to continue advancing bilateral, trilateral and multilateral cooperation in various fields with all parties to build an open world economy and maintain the smooth functioning of global industrial and supply chains.

译员通过整合句意,使用动词不定式 to build 和连词 and 将源语的四个分句整合成了一个长句。

五、转换法

有时为了使译文更加符合目标语的表达习惯,译员需要对源语中的词性、句型和语态等进行转换。

例 1:

原文:"一带一路"国际合作为各国抗疫情、稳经济、保民生发挥了重要作用,为落实联合国 2030 年可持续发展议程作出了积极贡献。

参考译文:International cooperation on the Belt and Road Initiative has played an important role in helping countries fight the virus, stabilize the economy, and protect people's livelihoods. It has also contributed to the implementation of the UN 2030 Agenda for Sustainable Development.

源语中的"落实"为动词,译员将其翻译为名词 implementation,这种名词化的处理可以增强译文的正式性。"积极贡献"是名词,译员使用的是动词词组 contribute to,体现了口译的灵活性和多样性。

例 2:

原文:中欧班列运送紧急医疗物资超过 800 万件,"空中丝绸之路"把超过 1700 吨的中国援助医疗物资运到世界各地。

参考译文:Over eight million pieces of emergency medical supplies have been delivered by the China-Europe Railway Express, and over 1,700 tons of medical supplies from China have been delivered around the world through the Silk Road in the Air.

译员将汉语的主动语态转换为被动语态,更加符合英语的表达习惯。

六、增译法

由于英汉两种语言的思维方式和表达习惯不同,所以译员在翻译时常常需要增添一些单词、短句或句子来衔接语义,或者填补可能出现的语义短欠,以便更准确、完整地表达出源语所包含的含义,使之更符合目标语听众的表达习惯。

例 1:

原文:践行多边主义,要坚持以公平正义、合作共赢为宗旨。

参考译文：In pursuing multilateralism，we need to live up to the overarching principles of fairness，justice and win-win cooperation.

首先，译员在句首增加了介词 In 来表示“在践行多边主义的进程中”，以此来显化两个分句之间的逻辑关系。其次，译员补充了主语 we。汉语中无主句的现象非常普遍，这是由汉语的表达习惯所决定的。在这一译例中，无主句用来表倡议，表面上虽然没有主语，但源语听众非常明确其默认的主语就是“我们”。英语一般要遵循完整的语法结构，因此，在目标语语法规则的束缚下，译员自然补充出了主语 we，使之更符合英语的表达习惯。此外，因为汉语中“宗旨”的语义强度高于“原则”(principle)，所以译员增加了形容词 overarching 来强调这些 principle 的重要性。

例 2：

原文：Inadvertently or not，she highlighted the politicized motivation for the resignee's claim.

参考译文：无论是否有意，她都在声明中强调了辞职者所发声明的政治化动机。

译员在翻译时添加了连词“无论”，这样一来，源语中的语义逻辑更加清晰。就像上一个例子中的 In 一样，译员常常用增译的处理方式来显化逻辑关系。

七、减译法

在口译过程中，为了使译文更加简练通顺、更加符合译入语的表达习惯，译员可以删掉一些可有可无或者冗余累赘的词语，但必须要做到减词不减意。

例 1：

原文：人才是科技创新最关键的因素。

参考译文：Talent holds the key to innovation.

译员在翻译时省略了源语中的“因素”一词，用 key 一词简洁明了。

例 2：

原文：Domestically，Washington's trade war has caused sweeping losses borne by ordinary people.

参考译文：美国发起的贸易战在国内极大损害了普通民众利益。

如果直译源语的话，应该是“美国发起的贸易战在国内造成大范围的损失，而这些损失由普通民众所承受”。但译员省译了 borne 一词，直接译为“极大损害了普通民众(利益)”，表达简洁明了。

八、简化法

简化法是指使用目标语听众所周知的简称说法来达到省时省力的效果，比如使用一些缩略词等。

例 1：

原文：我们签署了区域全面经济伙伴关系协定，并宣布愿积极考虑加入全面与进步跨太平洋伙伴关系协定，这表明了中方致力于推进自由贸易的坚定决心。

参考译文: Not long ago, we signed the RCEP and announced our readiness to favorably consider joining the CPTPP. All this is testimony to China's firm commitment to free trade.

RCEP 和 CPTPP 分别是 the Regional Comprehensive Economic Partnership 和 the Comprehensive and Progressive Agreement for Trans-Pacific Partnership 的首字母缩略词。

例 2:

原文: 中海关系之所以实现跨越式发展,归根于深厚互信,中国和海合会国家始终相互支持彼此主权独立,尊重各自发展道路,坚持大小国家一律平等,坚定维护多边主义。

参考译文: The leapfrog growth of China-GCC relations is attributed to the profound mutual trust. China and GCC countries have all along supported each other's sovereignty and independence, respected each other's development paths, upheld equality between countries regardless of their size, and stood firm in defending multilateralism.

海合会是"海湾阿拉伯国家合作委员会"的缩略词,其英文全称为 Gulf Cooperation Council。海湾阿拉伯国家可用 GCC countries 这一缩略语表示。

第二节 篇章讲解与练习

一、重点解析篇

(一) 欧洲导航卫星系统管理局执行主任卡洛·德斯·多利德斯在"2019 年国际太空论坛——地中海分会"上的讲话

1. 词汇表

Global Navigation Satellite System (GNSS)	全球导航卫星系统
European Global Navigation Satellite Systems Galileo	欧洲"伽利略"全球导航卫星系统
European Geostationary Navigation Overlay Service (EGNOS)	欧洲地球静止导航重叠服务; 欧洲地球同步卫星导航增强系统
Copernicus	"哥白尼"计划
Earth Observation	地球观测
Differential GNSS, DGNSS	差分全球导航卫星系统
Satellite-Based Augmentation Systems (SBAS)	星基增强系统
Automatic Identification System and Long Range Identification and Tracking (AIS & LRIT)	自动识别系统和远程识别与跟踪

续表

situational awareness	态势感知
Global COSPAS / SARSAT service	国际卫星搜救系统服务
European GNSS Agency (GSA)	欧洲导航卫星系统管理局
International Space Forum 2019—The Mediterranean Chapter	2019 年国际太空论坛——地中海分会
EU Space Programme Regulation	《欧盟空间计划条例》
International Association of Marine Aids to Navigation and Lighthouse Authorities (IALA)	国际航标协会

2. 背景知识

(1) 全球导航卫星系统(GNSS)

GNSS是能在地球表面或近地空间的任何地点为用户提供全天候的三维坐标和速度以及时间信息的空基无线电导航定位系统。全球四大卫星导航系统供应商,包括中国的北斗卫星导航系统(BDS)、美国的全球定位系统(GPS)、俄罗斯的格洛纳斯卫星导航系统(GLONASS)和欧盟的伽利略卫星导航系统(GALILEO)①。

(2) 伽利略卫星导航系统(GALILEO)

伽利略卫星导航系统是由欧盟研制和建立的全球卫星导航定位系统,该计划于1999年2月由欧洲委员会公布,欧洲委员会和欧空局共同负责,为全球用户提供独立的导航、定位和授时信息。系统由轨道高度为23616千米的30颗卫星组成,其中27颗工作星,3颗备份星。总部设在捷克共和国的布拉格②。

(3) 欧洲地球同步卫星导航增强系统(EGNOS)

欧洲地球静止导航重叠服务是欧洲自主开发建设的星基导航增强系统,它通过增强GPS和GLONASS两个卫星导航系统的定位精度,来满足高安全用户的需求。它是欧洲GNSS计划的第一步,是欧洲开发的伽利略卫星导航系统计划的前奏③。

---

① 宁津生,姚宜斌,张小红. 全球导航卫星系统发展综述[J]. 导航定位学报,2013(1):3-8.

② 同①。

③ 北斗卫星导航系统. EGNOS[EB/OL]. (2014-02-11)[2023-07-13]. http://www.beidou.gov.cn/zy/kpyd/201710/t20171023_4775.html.

3. 原 文

Speech at "International Space Forum 2019—The Mediterranean Chapter"①

Carlo des Dorides

Executive Director, European GNSS Agency (GSA)

5 September 2019

Dear Heads of Delegations, Dear colleagues,

It is a pleasure for me as Executive Director of the European GNSS Agency, GSA, to be here today at this "International Space Forum 2019—The Mediterranean Chapter". The focus on "Space Technology and Applications meet Mediterranean Needs" could not be more relevant for GSA. The Agency is the EU user interface for business, institutions and user communities wanting to implement innovative and effective solutions—leveraging the European Global Navigation Satellite Systems (GNSS) Galileo and EGNOS—to respond to emerging economic, societal, and environmental challenges.

The core mission of GSA is to provide accurate positioning service worldwide with Galileo, and with EGNOS at European region level, augmenting GPS and specifically addressing the civil aviation world. Our current mission includes the safe and secure operation of these two complex satellite systems, to deliver high quality services accessible to all in the most cost-efficient manner. In the near future, according to the recently approved *EU Space Programme Regulation*, GSA will be endorsed also with responsibilities for the other European flagship space programme Copernicus, for Earth Observation, to maximise its use and create synergies among EU Space Programmes, for the benefits of business, institutions and society.

But before we move on, let me share a story.

For years I used to start my speeches explaining why Europe needed to invest in Space. The most frequently asked question was: "Why does Europe need Galileo?"

Fast forwarding to 2019, slowly everybody is understanding that we are already in the SPACE age. It is no longer a question of IF space applications—including satellite navigation—are going to change the way we live, but rather the question is HOW they will shape our societies. In this HOW lie uncertainties of course, but most importantly opportunities, opportunities for job creation, for innovation, for sustainable growth,

① International Astronautical Federation. Statement of GSA [EB/OL]. (2019-09-05) [2023-07-13]. https://www.iafastro.org/assets/files/events/isf/2019/isf2019-statement-of-gsa.pdf.

for resource efficiency, for social inclusion.

So today I am NOT here to tell you why we need space, or satellite navigation, but HOW space, and satellite navigation in particular, is contributing to the present and future of the delicate ecosystem we are surrounded by and how it is empowering the maritime segment with cost-effective solutions.

Space driven innovation is shaping all segments, from smart-mobility to smart agriculture, to emergency response. Today 90% of the apps running on our smartphones require location data, and 50% of them use positioning provided by satellites navigation systems.

Ladies and gentlemen,

One-third of the world's shipping passes through the Mediterranean.

Again, the question is not IF but HOW can Galileo and EGNOS help manage such a volume. WHAT applications are already in place or can be developed to manage such a fleet, using the provision of extremely precise Position, Navigation and Time, for navigation, for traffic management, for search and rescue etc.

And there is more, it is estimated that in the Mediterranean ships discharge between 100,000 and 150,000 tonnes of crude oil each year. What can the European Space Programme do to monitor and contain the damage, identify responsibilities and prevent eco-tragedies?

And I am not talking about the future, I am talking about right now! YES, we already have the capacity to act. To give you an example, it is estimated that by 2025 thanks to Galileo and EGNOS the EU will save 15 million tonnes of CO_2 emissions, 3.5 billion litres of fuel. With smart-farming thanks to EGNOS and Galileo we will be saving 1.5 million tonnes of fertilisers and 4,500 tonnes of pesticides. How can we transfer the lessons learned in aviation, automotive and other domains to the Mediterranean context, and develop smart solutions leveraging space? How can EU space programmes support the preservation of the Mediterranean, and secure growth and sustainability in one of the world's most populated and developed areas in the world.

To name but one of the roles that GNSS can take up in maritime, IALA, the International Association of Marine Aids to Navigation and Lighthouse Authorities, is providing guidance to maritime authorities in relation to the obsolescence problem of their Differential GNSS (DGNSS) infrastructure. One of the options preferred is the use of Satellite-Based Augmentation Systems (SBAS), which for Europe is EGNOS, to complement the DGNSS service.

The planned evolutions of this system include not only to fully benefit from Galileo on top of GPS, but also to provide a Maritime Safety of Life service.

In addition to the core contribution of GNSS to the maritime world and the blue economy, I would like to mention other less known usages:

- GNSS is the enabler of Automatic Identification System and Long Range Identification and Tracking (AIS & LRIT) position reports for Traffic management and situational awareness.
- GNSS (potentially with position authentication) is the enabler for position reporting (Blue box) in Vessel Monitoring System (VMS) used in Fisheries monitoring.
- The Galileo Search & Rescue (SAR) service is an important contributor to the Global COSPAS / SARSAT service. Galileo SAR reduces the detection time from 4 hours to 10 minutes and improves the position of the distress signal from a radius of 10 km down to 2 km, hence saving lives and reducing the exposure to risk. On average, six lives are saved every day with the assistance of Cospas-Sarsat.

Time does not allow me to give you a full picture of EU GNSS contributions to respond to the needs of the Mediterranean, but this speech would be incomplete if I were not to mention the other flagship European Space programmes, Copernicus.

Copernicus, "Europe's eyes on Earth" is the European Earth Observation programme that provides (free of charge, as is the case for GNSS) essential information for six main domains: Atmosphere monitoring, Marine environment monitoring, Land environment monitoring, Climate monitoring, Security services and Emergency services.

As you can see, I did not come here with any readymade answer, but rather with a toolbox of satellite-based services which I hope are relevant to the challenges that you and your communities are facing. These instruments are already available to you all, at no cost.

Ladies and gentlemen,

We are at the cusp of a revolution powered by space technologies and applications. Galileo, EGNOS and Copernicus are European assets that need to be translated into advantages, benefits, growth, and sustainability. We need your help in understanding how to accelerate progress in the convergence and deployment, and really get the biggest return on investment from European space infrastructures and services.

I look forward to making GNSS for marine and maritime a European success story and a business case that can be exported across the world.

Thank you for your attention.

4. 参考译文

在2019年国际太空论坛——地中海分会上的讲话[①]

欧洲导航卫星系统管理局执行主任 卡洛·德斯·多利德斯

2019年9月5日

尊敬的代表团团长,亲爱的同事们:

我很高兴代表欧洲导航卫星系统管理局,即GSA,出席"2019年国际太空论坛——地中海分会"。此次分会重点关注"满足地中海需求的空间技术和应用",这同GSA的工作高度契合。GSA如同欧盟的用户界面,让希望实施创新和有效解决方案的企业、机构和用户社区利用欧洲"伽利略"全球导航卫星系统(GNSS Galileo)和欧洲地球同步卫星导航增强系统,即EGNOS,应对新出现的经济、社会和环境挑战。

GSA的核心任务是借助伽利略系统和欧洲境内的EGNOS系统,提供全球范围内的精确定位服务,增强GPS功能,并着力解决全世界民用航空的定位问题。我们目前的任务包括保障上述两大复杂卫星系统的安全平稳运行,以最具成本效益的方式为各方提供高质量服务。根据最近批准的《欧盟空间计划条例》,在不久的将来,GSA也将负责另一个欧洲旗舰空间项目,即"哥白尼"地球观测计划,实现其最大程度的利用,并在欧盟各项空间计划之间形成合力,为企业、机构和社会带来利益。

在继续演讲之前,我想先分享一个故事。

多年来,我经常在演讲一开始就解释为什么欧洲需要投资发展太空。最常被问到的问题就是"为什么欧洲需要伽利略系统?"

时间快进到2019年,彼时,每个人都逐渐明白,我们已进入了太空时代。现在的问题已经不再是卫星导航等太空应用是否会改变我们的生活方式,而是将如何塑造我们的社会。在如何塑造的过程中当然存在不确定性,但更蕴藏着机会:创造就业、创新、可持续增长、资源效率和社会包容的机会。

因此,今天我并不是在这里告诉大家,我们为何需要太空或卫星导航,而是太空,尤其是卫星导航,如何为我们周围脆弱的生态系统及其未来做出贡献,以及如何为海事部门提供经济有效的解决方案。

太空探索驱动的创新正塑造着从智慧交通到智慧农业,再到应急响应的所有领域。如今,我们智能手机上运行的90%的应用程序都需要位置数据,其中50%使用卫星导航系统提供的定位。

女士们,先生们,

全球有三分之一的航运船舶会穿过地中海。

同样,问题不是"伽利略"系统和EGNOS"是否"能协助管理如此规模的吞吐量,

---

① 编者译。

而是“如何”协助管理。哪些已有或可开发的软件能够管理这些运输船舶，通过提供极其精准的定位、导航和授时数据，支持航运、交通管理、搜索救援等服务？

此外，据估计，船只每年在地中海排放10万至15万吨原油。欧洲空间计划又能做什么来监测和控制损害，认定责任和防止生态悲剧呢？

我说的不是未来，而是现在！是的，我们已经有能力采取行动。举个例子，据估计，到2025年，借助“伽利略”和EGNOS系统，欧盟将减少1500万吨二氧化碳排放量和35亿升燃料消耗量。两大系统支持下的智慧农业将为欧洲节省150万吨化肥和4500吨杀虫剂。我们该如何将航空、汽车和其他领域积累的经验教训应用到地中海地区，开发利用太空的智慧解决方案？地中海沿岸是世界上人口最多、经济最发达的地区之一，欧盟太空计划又该如何为保护地中海做出贡献，确保该地区的经济增长和可持续发展？

GNSS在海事中可发挥的作用颇多，在此仅举一例。国际航标协会，简称IALA，正在向海事管理机构提供差分GNSS(DGNSS)基础设施老化问题的指导。优先选择之一是星基增强系统，在欧洲也就是EGNOS系统，为差分GNSS基础设施服务提供补充。

该系统计划的演变不仅包括充分受益于建立在GPS之上的“伽利略”系统，还包括提供海上生命安全服务。

除了GNSS对海事和蓝色经济的重要贡献外，我还想提一下其他不太为人所知的用途：

- GNSS支持自动识别系统和远程识别与跟踪的位置报告，可用于交通管理和态势感知。
- 带有位置验证功能的GNSS可为渔业船舶监测系统(VMS)提供位置报告(蓝框)。
- 伽利略搜救(SAR)服务是国际卫星搜救系统服务的重要贡献者。伽利略SAR能将探测时间从4小时缩短到10分钟，并将遇险信号的位置半径从10公里降低到2公里，从而挽救了生命，降低了风险。在国际卫星搜救系统的协助下，平均每天挽救6条生命。

时间有限，我无法向大家全面介绍欧盟GNSS为响应地中海地区需要所作的贡献，但我还要提一下欧洲的另一个旗舰空间项目——“哥白尼”计划，否则演讲就不完整了。

“哥白尼”计划被称为“欧洲的地球之眼”，是欧洲的对地观测方案。该方案为六个主要领域提供基本信息，包括大气监测、海洋环境监测、陆地环境监测、气候监测、安全服务和应急服务。和GNSS一样，“哥白尼”计划供各方免费使用。

正如大家看到的，我并没有带来任何现成的答案，而是带来一个基于卫星的服务工具箱，希望能帮助您和您的社区解决面临的挑战。在座的各位都能免费获得这些工具。

女士们，先生们，

一场由空间技术和应用推动的革命已然来临，我们正处在这场革命的风口浪尖。“伽利略”、EGNOS和“哥白尼”三大系统是欧洲的资产，需要转化为我们的优势、效益，助力经济增长和可持续发展。了解如何加速各个系统融合和部署，并真正从欧洲空间基础设施和服务的投资中获得最大回报，我们需要在座各位的帮助。

我希望海洋和海事GNSS能够在欧洲取得巨大成功，越来越多的业务能够由此延伸至世界各地。

谢谢大家！

5. 要点讲解

本篇演讲稿中专业术语俯拾皆是，因为默认的目标语听众对于“卫星导航系统”这一话题有基本的认识，所以其译文最直观的语言转换技巧是简化法，通过使用缩略词的形式来实现“省时省力”的效果，比如GNSS、GSA、EGNOS和GPS等。

英译汉过程中，译员经常采取增译的手法，以使译文更加符合汉语的语言思维。比如将addressing the civil aviation world译为着力解决全世界民用航空的定位**问题**，将Our current mission includes the safe and secure operation of译为“我们目前的任务包括**保障**……的安全平稳运行”，将for navigation, for traffic management译为“提供航运……等**服务**”，将Copernicus, Europe's eyes on Earth译为“‘哥白尼’观测系统**被称为**‘欧洲的地球之眼’”，将Galileo, EGNOS and Copernicus are European assets译为“‘伽利略’、EGNOS和‘哥白尼’**三大系统**是欧洲的资产”，这些字眼使译文更加通畅、完整，更符合汉语的语言惯性。另外，译员还会补充完整句意，避免歧义，比如将free of charge, as is the case for GNSS译为“和GNSS一样，‘哥白尼’系统免费供各方使用”。译员偶尔也会采用减译法，比如将“With smart-farming thanks to EGNOS and Galileo we will… ”一句中的we省去不译，而是直接将smart-farming作为主语。

译员也做到了对语态进行灵活转换，最常用的手段是化被动为主动，比如将six lives are saved every day译为“每天挽救6条生命”，将are already available to you all译为“在座的所有人都能免费获得”，将that need to be translated into advantages译为“需要转化为优势”，而非“被转化为”。此外，将complement the DGNSS service译为“为……提供补充”，化动词为名词，也是转换法的应用。

面对源语中的长难句，译员多采用拆句法，将其拆译成汉语分句，比如将“How can EU space programmes… world's most populated and developed areas in the world”一句拆成两句，先译出“地中海沿岸是世界上人口最多、经济最发达的地区之一”的部分，再译出前面的问题，逻辑过渡自然；将“To name but one of the roles… infrastructure”一句拆成两句，先概述GNSS“作用颇多”，再以IALA“仅举一例”，译文逻辑清楚明白；将“Copernicus… six main domains”从关系代词that处切断，拆成两句。

本篇译文中，译员也常常采取适当调整法来使行文更加自然流畅：①定语性成分前移：比如将 wanting to implement innovative and effective solutions、we are surrounded by、that you and your communities are facing、powered by space technologies and applications 等结构前置；②方式、方法前置：比如将 with Galileo, and with EGNOS at European region level、in the most cost-efficient manner、with the assistance of Cospas-Sarsat 等结构提前；③举例提前：比如将"Space driven innovation is shaping all segments, from smart-mobility to smart agriculture, to emergency response."译为"太空探索驱动的创新正塑造着从智能移动到智能农业，再到应急响应的所有领域。"通过这些语序上的调整，译文更加符合汉语的表达习惯，因此更加地道。

（二）时任国务院总理李克强在国家科学技术奖励大会上的讲话

1. 词汇表

神舟十一号载人飞船	Shenzhou-11 manned spacecraft
天宫二号空间试验室	Tiangong-2 space lab
中期驻留	a medium-term stay in space
单口径射电望远镜	aperture spherical telescope
量子科学实验卫星"墨子号"	quantum science experimental satellite "Mozi"
中国自主研发芯片的超级计算机"神威·太湖之光"	supercomputer "Sunway TaihuLight" powered by a home-made chip
全国科技创新大会	National Science, Technology and Innovation Conference
创新驱动发展战略	innovation-driven development strategy
创新领军人才	leading professionals of innovation
普惠性创新政策	extensive policy incentives to encourage innovation
费用加计扣除	additional deduction of R&D spending in taxable income
固定资产加速折旧	accelerated depreciation of fixed assets
"小核心、大协作"的"双创"模式	the innovation approach based on core-group innovation enabled by wider circles of collaboration
"两个一百年"奋斗目标	Two Centenary Goals
中华民族伟大复兴的中国梦	the Chinese dream of the great rejuvenation of the Chinese nation

2. 背景知识

(1) 国家科学技术奖励大会(National Science and Technology Award Conference)

国家科学技术奖励大会是根据1999年5月国务院发布的《国家科学技术奖励条例》所设立的。根据《国家科学技术奖励条例》,国务院设立了五项国家科学技术奖:国家最高科学技术奖,国家自然科学奖,国家技术发明奖,国家科学技术进步奖和国际科学技术合作奖①。

(2) 国家最高科学技术奖(State Preeminent Science and Technology Award)

国家最高科学技术奖由国家科学技术奖励工作办公室负责,每年评选一次,是中国五个国家科学技术奖中等级最高、分量最重的奖项。该奖项是为表彰在基础科学研究领域取得重大突破、带来深远影响的科学工作者,以及创造了巨大经济效益、促进了社会进步的杰出科学家而设立的②。

(3)《国家创新驱动发展战略纲要》(*Outline of the National Strategy on Innovation-driven Development*)

2016年1月18日,中共中央、国务院印发《国家创新驱动发展战略纲要》,明确我国科技事业发展的目标是:到2020年时使我国进入创新型国家行列,到2030年时使我国进入创新型国家前列,到新中国成立100年时使我国成为世界科技创新强国③。

3. 原 文

在国家科学技术奖励大会上的讲话④

时任国务院总理　李克强

2017年1月9日,北京

同志们,朋友们:

今天,我们隆重召开国家科学技术奖励大会,表彰为我国科技事业和现代化建设

---

① 国家科学技术奖励工作办公室. 国家科学技术奖励条例[EB/OL]. (2020-12-01)[2023-07-13]. https://www.nosta.gov.cn/pc/zh/gjkjjl/zcfg/38.html.

② 同①。

③ 新华社. 中共中央 国务院印发《国家创新驱动发展战略纲要》[EB/OL]. (2016-05-20)[2023-07-13]. https://news.cnr.cn/native/gd/20160520/t20160520_522193821.shtml.

④ 中国政府网. 李克强:在国家科学技术奖励大会上的讲话[EB/OL]. (2017-01-09)[2023-07-13]. https://www.gov.cn/xinwen/2017-01/09/content_5158191.htm.

作出突出贡献的科技工作者。刚才,习近平总书记等党和国家领导同志,向获得国家最高科学技术奖的赵忠贤院士、屠呦呦研究员和其他获奖代表颁了奖。在此,我代表党中央、国务院,向全体获奖人员表示热烈祝贺!向全国广大科技工作者致以崇高敬意和诚挚问候!向参与和支持中国科技事业的外国专家表示衷心感谢!

刚刚过去的一年,面对复杂严峻的国内外环境,在以习近平同志为核心的党中央坚强领导下,我国经济社会发展取得了显著成就,科技战线大事喜事多、创新成果多。党中央、国务院召开全国科技创新大会,明确提出要建设世界科技强国。创新驱动发展战略深入实施,《国家创新驱动发展战略纲要》颁布施行,面向2030年的科技创新重大项目部署启动,科技体制改革和管理方式创新加快推进,以增加知识价值为导向的分配政策制定实施,有效调动了广大科技人员的积极性。一批具有标志性意义的重大科技成果涌现,不少达到国际先进水平。神舟十一号载人飞船与天宫二号空间试验室成功交会对接,航天员实现中期驻留,世界最大单口径射电望远镜建成使用,世界首颗量子科学实验卫星"墨子号"发射升空,使用中国自主研发芯片的超级计算机"神威·太湖之光"再次刷新世界纪录。科技创新成果加速转化,大众创业、万众创新蓬勃兴起,创新作为引领发展的第一动力作用更加显现。中国创新令世界瞩目、让人民自豪。中华大地在创新中展现出勃勃生机与活力。

当前,世界新一轮科技革命和产业变革孕育兴起,抢占未来制高点的国际竞争日趋激烈。我国经济结构深度调整、新旧动能接续转换,已到了只有依靠创新驱动才能持续发展的新阶段,比以往任何时候都更加需要强大的科技创新力量。必须认真学习贯彻习近平总书记系列重要讲话精神,把创新摆在国家发展全局的核心位置,以新发展理念为引领,以供给侧结构性改革为主线,深入实施创新驱动发展战略,加快培育壮大新动能、改造提升传统动能,推动经济保持中高速增长、迈向中高端水平。

我们要全面提高科技创新能力,筑牢国家核心竞争力的基石。瞄准世界科技前沿,紧扣经济社会需求,在战略必争领域前瞻部署、超前研究,推进国家科技重大项目、重大工程和重大基础设施建设,夯实科技创新的基础支撑。要大力加强基础研究和原始创新,充分发挥科研院所和高校的主力军作用,建立长期稳定的支持机制,鼓励从事基础研究和原始创新的科研人员潜心研究,可以十年不鸣,争取一鸣惊人。要建立以企业为主体、以市场为导向的技术创新机制,引导社会各方面力量投入创新领域。推动开放式科技创新,深化国际科技合作,利用互联网等新平台新模式,加强产学研协同,集聚优化创新要素,提高科技创新和成果转化效率。

我们要深化科技体制改革,充分调动科技人员积极性。人才是科技创新最关键的因素。必须充分尊重科技人才,保障科技人才权益,最大限度激发科技人才的创造活力。要深入推进科技领域简政放权、放管结合、优化服务改革，推行科研管理清单制度，实施更加方便简约有效的规则，赋予科研院所和高校更大的科研自主权，赋予创新领军人才更大的人财物支配权。要加大成果处置、收益分配、股权激励、人才流动、兼职兼薪等政策落实力度，使创新者得到应有的荣誉和回报，增强科技人员的持久创造动力。

我们要推动大众创业、万众创新，着力激发全社会创新潜能。人民群众是历史的创造者，也是推动创新的根本力量。我们有1.7亿多受过高等教育或拥有专业技能的人才，蕴藏着巨大的创新潜能，这是我国发展用之不竭的最大"富矿"。要不拘一格用好各方面创新人才，集众智、汇众力，提高社会创新效率。既要支持专业人才在创新上不断突破，也要激发普通民众的创造潜力；既要支持本土人才勇攀高峰，也要吸引海归人才、外国人才来华创业创新。我们要以海纳百川、求贤若渴的气度，为各类创新人才施展才华提供更大空间、更广阔的舞台。

我们要全面提高创新供给能力，推动科技创新成果向各行业各领域覆盖融合，加快新旧动能转换。新动能既来自新兴产业成长，也来自于传统产业的改造提升。在科技创新的推动下，我国新兴产业快速成长，数字经济、分享经济、平台经济等新业态方兴未艾，对这些产业要审慎监管，使之健康发展。同时，要促进新技术、新业态、新模式加快与一二三产业融合发展，推动实体经济升级，使传统产业焕发新的生机与活力。要实施普惠性创新政策，落实和完善研发费用加计扣除、固定资产加速折旧等措施，支持企业与高校、科研院所、创客合作建立协同创新平台，推广"小核心、大协作"的"双创"模式，促进源头创新、成果转化、市场开发齐头并进，重点围绕提升产业竞争力、满足多层次消费需求和助力破解医疗、环保等领域民生难题，大力研发新品、多出优品、打造精品，着力提升"中国制造"的品质和"中国创造"的影响力。

我们要加强知识产权保护，打造良好创新生态环境。保护知识产权就是保护和激励创新。要展开知识产权综合管理改革试点，构建知识产权创造、保护、运用体系，严厉打击侵权假冒行为，使创新者的合法权益得到切实有力的保护，使知识产权更多转化为现实生产力。要努力营造支持创新、追求卓越的社会氛围，让尊重劳动、尊重知识、尊重人才、尊重创造蔚然成风，让人人皆可创新、处处是创新之地，促进科学创新精神与企业家精神、工匠精神相结合，形成推动创新发展的强大动力。

同志们，朋友们，科技改变世界，创新决定未来。让我们更加紧密地团结在以习近平同志为核心的党中央周围，倍加珍惜荣誉，切实担当使命，奋力创造辉煌，推动科技事业更好更快发展，以优异成绩迎接党的十九大胜利召开，为实现"两个一百年"奋斗目标和中华民族伟大复兴的中国梦、建设富强民主文明和谐的社会主义现代化国家，作出新的更大贡献。

4. 参考译文

Speech at the National Science and Technology Award Conference①

H. E. Li Keqiang

Then Premier of the State Council of

The People's Republic of China

Beijing, 9 January 2017

Comrades,

Friends,

Today, we are gathered for the important occasion of the National Science and Technology Award Conference to honor professionals in science and technology who have made outstanding contribution to advances in their own fields and to the modernization of China overall. Just now, General Secretary Xi Jinping and other Party and state leaders presented the Highest National Award of Science and Technology to academician Zhao Zhongxian and researcher Tu Youyou and conferred other awards on their winners. Here, on behalf of the CPC Central Committee and the State Council, I wish to extend warm congratulations to all the award-winners, and convey cordial greetings and high regard to all the science and technology professionals in China. My deep appreciation also goes to foreign experts who have participated in and supported the development of science and technology in China.

Last year, thanks to the strong leadership of the CPC Central Committee with Comrade Xi Jinping as the core, China made notable strides in economic and social development despite complex and grim internal and external environment. Major milestones were recorded in fields of science and technology and numerous outcomes in innovation were achieved. The CPC Central Committee and State Council convened the National Science, Technology and Innovation Conference, issuing a clarion call to build China into a science and technology giant. Vigorous efforts were made to implement the innovation-driven development strategy, such as the promulgation of *the Outline of the National Strategy on Innovation-driven Development* and the launch of programs for major science, technology and innovation projects toward 2030. Reform of science and technology institutions and innovation in management were accelerated,

---

① The State Council of the People's Republic of China. Speech by Premier Li Keqiang at national science award conference[EB/OL]. (2017-01-10)[2023-07-13]. https://english.www.gov.cn/premier/speeches/2017/01/10/content_281475538340580.htm.(文字有校改)

and a distribution policy that puts premium on the value of knowledge was introduced, which effectively incentivized science researchers. Significant breakthroughs were made in science and technology, many of which reached an internationally advanced level. Let me list a few. The Shenzhou-11 manned spacecraft successfully docked with Tiangong-2 space lab and astronauts managed a medium-term stay in space. The world's largest single aperture spherical telescope went into operation. The world's first quantum science experimental satellite "Mozi" was launched. The supercomputer "Sunway TaihuLight" powered by a home-made chip held the current world record. Outcomes of science, technology and innovation are now commercialized at a faster pace. The initiative of mass entrepreneurship and innovation gained a strong momentum. Innovation as the primary driving force is exerting a notably positive impact on growth. "Innovated in China" won international recognition, made the Chinese people proud and unleashed enormous energy that invigorated the whole nation.

A new round of technological revolution and industrial transformation is in the making, further sharpening international competition to stay ahead of the curve. China has entered a new phase of economic development featuring deep structural adjustments and renewal of driving forces. Only through innovation can China sustain the strong momentum of development. More than ever the powerful force of science, technology and innovation is called for. We must conscientiously study and act on the thinking laid out in the important speeches of General Secretary Xi Jinping, put innovation at the center of development and follow the new vision on development. Specifically we will focus on supply-side structural reform, deepen the implementation of innovation-drivendevelopment, and transform traditional drivers of growth while fostering new ones, with a view to maintaining a medium-high rate of growth and reaching a medium-high level of development.

We will build up the capability of scientific and technological innovation across the board to lay a solid foundation for our country's core competitiveness. We should closely follow frontier trends in science and technology, and make early and visionary planning for research and application in focal areas in light of the needs of our economy and society. National scientific and technological programs and projects and key infrastructure must be implemented and built to support innovations in science and technology. We must strengthen basic research and original innovation, give full play to the central role of scientific research institutes and universities, and develop long-term and stable supporting mechanisms for researchers engaged in these areas to delve deep into their areas of endeavor. It is by no means easy for them to work quietly and thanklessly for long years. Yet their efforts will be adequately rewarded once the breakthrough comes. To encourage social input into innovation, a mechanism needs to be put in place to encourage companies to play a leading role based on market rules. We should also encourage greater openness in scientific and technological innovation by enhancing international cooperation and leveraging the Internet and other new platforms and models to develop synergy among industries, universities and research institutes. We should pool and upgrade factors of innovation and raise the efficiency of commercialization of innovation and research outcomes.

We will deepen institutional reform for scientific and technological development and fully motivate professionals working in these fields. Talent holds the key to innovation. We must give full respect to scientists, engineers and technicians, guarantee their rights and interests, and tap their potential to the full. The government must further streamline science and technology administration, delegate powers and provide better services. A checklist-based management approach should be taken to strengthen and simplify rules, whereby greater autonomy will be given to institutes and universities in their research activities, and greater say to leading professionals of innovation in allocating research personnel, funding and resources. The government will vigorously implement policies regarding the usage of research outcomes, distribution of research yields, equity incentives, personnel mobility and permission for researchers to take part-time roles and be remunerated for that. All these will help to duly recognize and reward innovators, thus providing long-term motivation for scientific and technological research and innovation.

We will encourage mass entrepreneurship, and bring out the society's potential for innovation. Our people are makers of history and also the fundamental driving force for innovation. We have over 170 million people with higher education or professional skills, who are endowed with great potential for innovation and form the biggest treasure house for our country's development. We must enable professionals to apply their potential, pool their talent and strength, and raise the efficiency of the innovation. We should help both professionals to make new innovations and the common people to explore their creativity. We should both support domestically educated talent to scale the heights of their professions, and encourage overseas professionals to pursue start-ups and innovation in China. The Chinese government welcomes talent with an open mind and open arms. We will provide greater space and broader platforms for all innovators to fulfill their potential.

We will comprehensively improve our supply capacity for innovation and facilitate the application and integration of scientific achievements into all sectors and fields to speed up the shift to new growth drivers, which are derived both from the emerging industries and the upgrading of traditional sectors. Driven by scientific innovations, the emerging industries in China are thriving, spearheaded by new forms of business, like the digital economy, sharing economy and platform economy. Proper regulation must be exercised to ensure healthy development of these new sectors. Meanwhile, the new technologies, new forms of business and new models must be integrated into the first, second and tertiary industries at a faster pace, to infuse traditional industries with new vitality. We need to introduce extensive policy incentives to encourage innovation. Such measures as additional deduction of R&D spending in taxable income and accelerated depreciation of fixed assets should be better implemented. Greater support should be given to companies in setting up collaborative innovation platforms with universities, research institutes and makers. We should popularize the innovation approach based on core-group innovation enabled by wider circles of collaboration. We need to strive for parallel progress of innovation at the source, commercialization of R&D results and market development, and aim to develop more new, high-quality products that boost the competitiveness of our industries, meet the diverse consumer demand and help tackle thorny issues such as in medical care and environmental protection, with a view to raising the quality of Chinese manufacturing and influence of products "Created in China".

We need to enhance protection of intellectual property rights and foster an innovation-friendly environment. Protecting intellectual property is protecting and incentivizing innovation. We willpilot integrated management of IPR, establish a system for IPR creation, protection and application, and crack down on IPR infringement and counterfeiting to safeguard the legal rights and interests of innovators and turn more intellectual property into real productivity. We need to foster a social environment that embraces innovation and excellence, and respects hard work, knowledge, talent and creativity. We should enable each and every one to innovate wherever possible, and celebrate the spirit of innovation together with entrepreneurship and workmanship to generate strong impetus for innovative development.

Comrades,

Friends,

As science and technology changes the world, innovation shapes the future. Let us rally more closely around the CPC Central Committee with Comrade Xi Jinping as the core, cherish the honor the country has conferred, conscientiously fulfill our mission and strive for greater achievements. Attaining better and faster progress of our scientific and technological endeavors would be the best way for you to greet the opening of the

19th CPC National Congress. By joining hands we will make fresh and greater contribution to the realization of the "two centenary goals" and the Chinese dream of the great national renewal and to the building of a prosperous, strong, democratic, culturally advanced, harmonious and modern socialist country.

5. 要点讲解

本篇讲话稿的翻译综合运用了多种语言转换技巧,从而使译文更加符合译入语的表达习惯,便于目标语听众的接收和理解。

从语序调整的角度来看,本篇译文主要采用顺句驱动法,这一点无须赘述,而在微观层面上,译员也进行了适当调整,主要是将源语中较长的修饰语后置,以符合译入语的表达习惯。

今天,我们隆重召开国家科学技术奖励大会,表彰为我国科技事业和现代化建设作出突出贡献的科技工作者。

Today, we are gathered for the important occasion of the National Science and Technology Award Conference to honor professionals in science and technology who have made outstanding contribution to advances in their own fields and to the modernization of China overall.

将"为……突出贡献的科技工作者"译为定语从句"who have made outstanding contribution… "(这种处理最为常见)。

刚刚过去的一年,面对复杂严峻的国内外环境,**在以习近平同志为核心的党中央坚强领导下**,我国经济社会发展取得了显著成就,科技战线大事喜事多、创新成果多。

Last year, **thanks to the strong leadership of the CPC Central Committee with Comrade Xi Jinping as the core**, China made notable strides in economic and social development despite complex and grim internal and external environment. Major milestones were recorded in fields of science and technology and numerous outcomes in innovation were achieved.

将"以习近平同志为核心的党中央"译为复合结构"with Comrade Xi Jinping as the core"。

使用中国自主研发芯片的超级计算机"神威·太湖之光"再次刷新世界纪录。

The supercomputer "Sunway TaihuLight" **powered by a home-made chip** held the current world record.

将"使用中国自主研发芯片的超级计算机"译为过去分词作定语"powered by a home-made chip"。

同时,译员也适当采用了重组法,比如将"我国经济结构深度调整……新阶段"这句话打破重组,将各层逻辑娓娓道来,展现出译员娴熟的语言组织能力。

China has entered a new phase of economic development featuring deep structural adjustments and renewal of driving forces. Only through innovation can China sustain the strong momentum of development.

从长短句转换的角度来看，本篇译文综合运用了拆句法和合并法。拆句法多从逗号处切断，另起一句，避免因句子过长而增加目标语听众的理解成本，比如将“科技战线大事喜事多”拆分出来，可以突出重点；将“使创新者得到应有荣誉和回报”拆成一句，译为“All these will help to… ”，可以作为上一段话的总结；将“以供给侧结构性改革为主线”译为“Specifically we will… ”，突出递进关系。

合并法主要是将两个具有递进、平行、从属等关系的短句合为一句，使译文更加紧凑，比如将“表彰……科技工作者”译为“to honor professionals… ”表目的，将“鼓励……潜心研究”译为“for researchers… ”表递进，将“新动能既来自……”转化为非限制性定语从句衔接上一句。

从语言形式转化的角度来看，转换法是本篇翻译的一大常用技巧，主要有两种表现形式：

① 转换为被动语态：比如将“我们隆重召开”译为 we are gathered，将“大事喜事多”译为 milestones were recorded，将“推行科研管理清单制度”译为 should be taken 等，可以增强译文语气的客观性和句式的多样性；

② 词性转换：比如将“明确提出”译为 issue a clarion call（动词变名词，也是最常见的处理），“更加方便简约有效的规则”译为 strengthen and simplify rules（形容词变动词），“得到切实有力的保护”译为 safeguard（名词变动词）等，用词富于变化，避免同一词性的堆砌。

从文本信息加工的角度来看，本篇翻译综合运用了增译和减译两种技巧。增译法的运用主要表现在：

① 补充主语：讲话稿中表倡议、建议的无主句比比皆是，但在英语中，主语一般是不可或缺的语法成分，因此译员通过补充默认主语 we 或 the government 来使输出符合英语的语法结构。

② 增加单词、词组或短句，显化逻辑关系：比如在“筑牢国家核心竞争力的基石”“加快新旧动能转换”“加强产学研协同”等多处分句前增加 to 表目的；在“深化国际科技合作”前增加 by 表方式，在“赋予科研院所和高校更大的科研自主权”前增加 whereby 表递进；在“创新驱动发展战略深入实施”后增加 such as 表列举；在“科技改变世界”前增加 as 表类比；在“不少达到国际先进水平”后增加 Let me list a few 表列举等。这样的处理可以使源语中隐含的语义关系明晰化，有效衔接分句，增强译文的连贯性。

③ 补充说明性成分：比如在“国家科学技术奖励大会”前插入 the important occasion 来说明大会性质，在“科研自主权”后添加 in their research activities 来限制范围，将“兼职兼薪”译为 permission for researchers to take part-time roles and be remunerated for that 来说明政策具体内容。

④ 译出隐含义：比如将意蕴丰富的"可以十年不鸣，争取一鸣惊人"译为"It is by no means easy for them to work quietly and thanklessly for long years. Yet their efforts will be adequately rewarded once the breakthrough comes"，将源语内涵进行了充分诠释。

同时，译员也采取了减译法，在不影响句意传达的前提下删掉一些冗余的表述，比如对"广大""一批""大力""领域""因素""蔚然成风"等修饰性成分不进行翻译，再比如将"多出优品、打造精品"的同义对仗结构译为 develop more high-quality products 等。此外，使用 CPC、R&D、IPR 等缩写属于简化法。

二、实战训练篇

（一）国际民航组织秘书长胡安·卡洛斯·萨拉萨尔先生在联合国世界旅游组织第 24 届全体大会上发表的主旨演讲

1. 词汇表

Visible Digital Seal (VDS)	可视数字印章
decarbonizing	去碳化
Sustainable Aviation Fuels (SAF)	可持续航空燃料
International Civil Aviation Organization (ICAO)	国际民航组织
World Tourism Organization (UNWTO)	联合国世界旅游组织
short haul international route	国际短途航线
Cancun	坎昆
Airports Council International Assembly	国际机场理事会大会
Agenda 2030	《2030 年可持续发展议程》
Common Agenda	《共同议程》
Memorandum of Understanding (MOU)	谅解备忘录
International Civil Aviation Day	国际民航日
Ministerial Declaration	《部长宣言》
26th UN Climate Change Conference of the Parties (COP26)	第 26 届联合国气候变化大会
Group of Seven (G7)	七国集团
Group of Twenty (G20)	二十国集团
Net-zero 2050 commitments	2050 年净零承诺
Carbon Offsetting and Reduction Scheme for International Aviation (CORSIA)	《国际航空碳抵消和减排计划》
Council's Aviation Recovery Task Force (CART)	航空恢复工作组

2. 原 文

Keynote Speech to the UNWTO 24th General Assembly①

Mr. Juan Carlos Salazar
ICAO Secretary General
Madrid, 30 November 2021

It is a great honour to address you today, and to be able to help set the stage for your discussions with these keynote remarks.

I wish to extend my deepest thanks to UNWTO Secretary General Pololikashvili for his kind invitation to address you, and at such an important crossroads for air transport and global tourism.

I would like to begin today by highlighting, that despite some substantial volatility over 2021, during the summer some noticeable improvements have been observed in international travel.

This was especially evident on short haul international routes like those you're very familiar within Europe, but which also characterize international air transport in the Asia Pacific and the Americas and Caribbean.

I was just in Cancun a few days ago in fact, for an Airports Council International Assembly, and I remarked there on how seat capacity in that region has jumped by more than 14% in just the past month.

Similar signs are being seen all over the world, and the latest passenger trends globally are also pointing to a further increase in both domestic and international reservations during the end-of-year holiday period.

As these early signs of recovery begin to expand in their scope and significance, I think it's clear to all of us that the innovation and agility of the travel and tourism sectors will be put to an important test in terms of how we pivot together and build back better.

I would highlight here as well, that these very early stages of recovery are an especially critical moment for cooperation and multilateralism to play their role—

① International Civil Aviation Organization. Keynote speech by ICAO Secretary General Mr. Juan Carlos Salazar to the UNWTO 24th General Assembly[EB/OL]. (2021-11-30)[2023-07-13]. https://www.icao.int/secretariat/SecretaryGeneral/Documents/Addresses%20and%20Messages/20211230_SG-SPEECH_UNWTO_GA24.pdf.

whether to amplify and accelerate the momentum being seen among the diverse partners involve, or to forge consensus among them on where we need to go and how we're going to get there.

I think this clearly underscores how the role of UN agencies in these times will be absolutely fundamental to us achieving collective, global outcomes consistent with *Agenda* 2030 and the *Common Agenda*.

This explains in part why the UNWTO and ICAO are now working more closely than ever.

We've formalized a new MOU recently addressing a comprehensive range of collaboration priorities, and which will guide us in specific recovery areas such as the modernization of the passenger experience, and the greening of travel and tourism on a comprehensive end-to-end basis.

We'll also be issuing a joint statement soon for International Civil Aviation Day on 7 December, the first joint issuance for this occasion in ICAO's history. And that day will also mark the start of a new communications campaign we're conducting on the theme of "Reconnecting the World".

The Ministerial Declaration adopted at ICAO's High-level Conference on COVID-19 one month ago clearly demonstrated the strong political commitment and resolve of States, and of all industry stakeholders, to restore aviation air connectivity safely and efficiently, and to build its resilience and sustainability.

To accomplish this ICAO needs to closely oversee the safe recovery of aviation, managing a complex path back to full standards compliance and engaging with industry operators and national regulators facing unparalleled financing and resource challenges.

As just underscored, we also need to assure a modernized recovery for every stage of the passenger experience, addressing both aviation security and traveler health priorities.

Post-pandemic travelers will be arriving at airports with clear expectations for an air travel experience which is secure, contactless, and health prioritized. We will need to work with States and suppliers to leverage new screening approaches and technologies to realize this.

Fortunately, the increasing digitization of these processes should also lead to greater passenger convenience and throughput at border entry and security and customs checkpoints. But they also raise cybersecurity concerns which will need to be carefully tended to.

The objective here is both greater public confidence in air travel and tourism, and a more efficient and comfortable overall experience for passengers, even as they are being more vigilantly protected.

A recent and important contribution ICAO has made to enhance passenger confidence in contactless and secure aviation processes are the new specifications for the ICAO Visible Digital Seal for encoding personal health proofs.

The VDS provides for mutually-recognized global trust and acceptance in these proofs and certifications, safeguarding data and privacy and greatly improving passenger confidence in international travel options.

Its current global roll-out is also highlighting how our passenger security, processing, and health screening objectives will become far more integrated and synergistic moving forward, and a key aspect of aviation's post-pandemic new normal.

Underscoring all of our recovery efforts, of course, is the imperative for our sectors to decrease and ultimately eliminate our carbon footprints.

The decarbonizing of aviation has been called for very strongly of late, and by many countries—whether at the recent COP/26, G7, or G20 events, or ICAO's own High-level Conference on COVID-19.

This reflects the fact that the climate crisis represents one of our gravest threats. From natural disasters to economic downturn, from disease outbreaks to conflicts, global warming and its ripple effects impact everyone, everywhere, including aviation and tourism.

Bold action must therefore be taken now if we are to mitigate the most severe impacts, and it is clear that business as usual is not good enough.

I am very much supportive of Secretary General Guterres' recent calls for our sectors to be rapidly moving toward carbon neutrality. I am optimistic in reaching that goal based on the detailed Net-zero 2050 commitments now emerging from all over the air transport industry.

Considerable progress is already being realized by more fuel-efficient aircraft technologies, better optimized operations, Sustainable Aviation Fuels (SAF), and the transparent and accountable carbon offsetting being established under ICAO's CORSIA initiative.

ICAO has been taking stock of new innovations in aviation emissions reduction at events designed for that purpose this year and last. Based on what we're seeing, it's clear that innovation will be as critical to meeting aviation's climate objectives, as it is to our efforts to re-imagine the aircraft, or to modernize the passenger experience.

ICAO's key role is to understand and enable all of this innovation through effective global standardization, and to assure that all new technologies and operations are safely, securely and sustainably integrated into our global network.

In light of the incredible pace at which things are now advancing, we're also faced with the challenge of transforming ourselves from the ground up to make ICAO more agile, efficient, and results-based in the face of this incredible new era of imagination and invention.

Virtually everyone in this room today for example can one day look forward to an autonomous air taxi carrying you from an airport to a hotel or other travel destination.

But the path from the videos we see on social media of these amazing new aircraft, to you sitting in that seat, is one paved with detailed and inclusive research, and diligent standardization and certification procedures.

I'd like to highlight that vaccination is making air travel and tourism possible again. However, for countries to begin reaping the benefits of revitalized travel and tourism markets, border restrictions for doubly-vaccinated passengers need to be quickly reviewed and lifted wherever conditions permit.

Prior to the pandemic, some 1.4 billion tourists crossed borders every year, well over half of whom arrived at their destination by air.

These air travelers represent an important economic lifeline for many countries, but most especially those characterized by Least Developed, Landlocked Developing, or Small Island Developing status.

Together we therefore share a solemn duty to bring travelers back to their destinations and reconnect our world, so that our common targets for global sustainability can be put back on track, and so that millions of people can have their livelihoods restored.

In concluding today, I wish to express ICAO's deepest gratitude to the UNWTO for being such an effective partner and contributor to aviation's response and recovery efforts, including through the ICAO Council's Aviation Recovery Task Force (CART), and at our recent High-level conference.

I think our Organizations have demonstrated an excellent level of solidarity and efficiency throughout these challenge. And we now have a very solid foundation and some excellent motivation to guide our future collaboration.

I look forward to seeing just how much progress we can achieve together in the important months ahead.

Thank you.

（二）中国商飞董事长贺东风的致辞

1. 词汇表

中国商飞公司	Commercial Aircraft Corporation of China，Ltd.（COMAC）
客机	commercial aircraft
大飞机	trunk liner
喷气式商用飞机	business jet
习近平新时代中国特色社会主义思想	Xi Jinping Thought on Socialism with Chinese Characteristics for a New Era
创新驱动发展战略	the strategy of driving development by innovation

2. 原　文

中国商飞董事长贺东风的致辞①

衷心感谢各级领导、社会各界和国际友人一直以来对中国商飞公司的关心、支持和帮助！

2008年5月11日，中国商飞公司在上海这片改革开放的热土上组建成立，吹响了我国自主研制大型客机的嘹亮号角，扬起了我国大飞机创新发展的希望之帆。十年来，特别是党的十八大以来，我们始终以习近平新时代中国特色社会主义思想为指引，深入学习贯彻习近平总书记关于大飞机事业重要指示精神，高举科技创新的大旗，团结协作、攻坚克难，实现了我国喷气式商用飞机从无到有的历史性跨越，开启了我国大飞机产业从弱到强的新征程。

回顾过去，我们只争朝夕、奋力拼搏，迈出了大飞机事业万里长征的第一步。展望未来，公司发展进入新阶段、迈向新征程。我们要树雄心壮志，坚决落实公司“三步走”总体安排，努力实现“三个一”阶段目标和“两个建成”奋斗目标，矢志打造“四个世界级”航空强企，为建设社会主义现代化强国贡献力量。

我们将始终坚持创新驱动发展战略，坚定不移推进管理创新、技术创新、产品创新、商业模式创新，努力实现关键核心技术自主可控，不断提升主制造商核心能力，全力打造具有国际影响力的自主品牌。

我们将始终坚持开放合作，发挥产业龙头作用，推动大飞机产业体系建设，与全球伙伴精诚合作，共同构建“共享成就、共享荣耀、共享利益”的“事业共同体”，“相互理解、相互支持、合作共赢、共担风险”的“生命共同体”，和“关心大飞机、支持大飞机、促进大飞机”的“梦想共同体”。

① 中国商飞.董事长致辞[EB/OL].(2017-07-27)[2023-07-13].http://www.comac.cc/gywm/dsz/.

我们将始终坚持以客户为中心，秉承"精湛设计、精细制造、精诚服务、精益求精"的质量方针，为客户提供更加安全、经济、舒适、环保的商用飞机。

我们将坚定不移全面从严治党，坚持"严党建、强支部、大监督、聚群团"，弘扬航空强国、"四个长期"、永不放弃的大飞机创业精神，弘扬劳模精神、工匠精神，一以贯之，锲而不舍，扎扎实实，脚踏实地，坚持"勤俭研制大飞机"，打造"廉洁商飞"，为大飞机事业营造良好政治生态。

让我们紧密团结在以习近平同志为核心的党中央周围，深入学习贯彻习近平新时代中国特色社会主义思想和党的二十大精神，不忘初心，牢记使命，为实现中华民族伟大复兴的中国梦作出新的更大贡献！

（三）北京航空航天大学校长王云鹏院士在2022年新生开学典礼上的讲话

1. 词汇表

总师	chief designer
轻型客机"北京一号"	light passenger plane "Beijing No. 1"
探空火箭"北京二号"	sounding rocket "Beijing No. 2"
电磁兼容	electromagnetic compatibility
高温涂层材料	high-temperature coating materials
型号火箭总指挥	commander-in-chief of the model rocket
长征三号甲系列	Long March 3A series
探月工程	lunar exploration project
北斗组网	BeiDou network
"神舟"飞船	Shenzhou spacecraft
东方红系列卫星	Dongfanghong series satellites
两弹一星	Two Bombs, One Satellite
载人航天	manned spaceflight
歼15	J-15 fighter jet
2022级新生开学典礼	2022 Opening Convocation
沈元学院	Shenyuan College
北京冬奥会	Beijing Winter Olympics
全面建设社会主义现代化国家	build a modern socialist country in all respects
第二个百年奋斗目标	the second centenary goal
抗美援朝	War to Resist U. S. Aggression and Aid Korea
德才兼备、知行合一	Integrating Virtue with Brilliance and Combining Knowledge with Practice

续 表

爱国奉献、敢为人先、团结拼搏、担当实干	patriotic dedication, daring to be the first, unity and hard work, taking responsibility and being practical
集成电路学院	School of Integrated Circuit Science and Engineering
“雪飞燕”	National Alpine Ski Centre

2. 原 文

勇担青春使命　逐梦空天报国

——北京航空航天大学校长王云鹏院士
在 2022 年新生开学典礼上的讲话①
2022 年 9 月 10 日，北京

亲爱的同学们、家长们、老师们：

大家上午好！

今天我们在两校区隆重举行 2022 级新生开学典礼，共同迎接 3871 名本科生、6650 名研究生加入北航这个大家庭，开启你们的人生新篇章。在这里，我代表赵长禄书记和全校师生员工，对你们的到来表示热烈的欢迎！向多年来辛勤培育你们的家长、师长致以衷心的感谢！

我注意到，你们当中年龄最小的本科生是来自沈元学院的罗维高同学，只有 15 岁；年龄最小的研究生是来自计算机学院的朱琛同学，只有 18 岁。重名最多的是“张瀚文”和“杨帆”，各有 6 名同学。名字带有“航”“天”元素的同学有 108 名。航空学院的赵剑辉、士谔书院的辛卓妍等 37 名同学今天过生日，让我们向他们表示生日的祝福。同学们来自天南海北，从今天开始，你们将有一个共同的名字——北航人！

2022 年，注定是极不平凡的一年。我们共同见证了北京冬奥会上那些“无与伦比”的美好瞬间，向全世界展示了大国的自信和风范；我们即将迎来党的二十大，全国各族人民正在阔步迈向全面建设社会主义现代化国家新征程、向第二个百年奋斗目标进军；我们也在共同经历百年变局和世纪疫情的相互交织，国际形势的错综复杂带来了多种风险挑战。你们克服疫情影响，居家学习、相约网课，在艰苦环境中磨练成长，终得圆梦北航，你们每个人都了不起！我要为你们点一个大大的赞！

2022 年，也是北航建校 70 周年。北航诞生于抗美援朝的烽火硝烟之际，成长于改革开放和社会主义现代化建设的火热浪潮之中，奋进在实现中国特色社会主义新时代的伟大复兴之时。70 年来，一代代北航人承续红色基因，秉持“德才兼备、知行

---

① 北京航空航天大学.勇担青春使命，逐梦空天报国！北航校长王云鹏院士寄语 2022 级新生[EB/OL].(2022-09-10)[2023-07-13]. https://mp. weixin. qq. com/s? —biz＝MzA3NTY1MTUzOQ%3D%3D&mid＝2651414962&idx＝1&sn＝0335778cce1deaad6437f0c341589d79&scene＝45#wechat_redirect.

合一”的校训，培养一流人才、做出一流贡献，在建设中国特色世界一流大学的征程中涵养了“爱国奉献、敢为人先、团结拼搏、担当实干”的“空天报国”精神。

爱国奉献，是北航人忠贞不渝的使命追求。建校初期，以沈元、屠守锷、林士谔、陆士嘉等为代表的一大批杰出科学家，心怀报国志从海外学成归来，组建北航，创业兴学。70年栉风沐雨，学校已培养优秀人才23万余名，一大批院士、总师和行业精英从这里启航，把事业书写在祖国大地上。

敢为人先，是北航人与生俱来的基因标识。从中国第一架轻型客机“北京一号”、亚洲第一枚探空火箭“北京二号”、中国第一架无人机“北京五号”横空出世，到3D打印、电磁兼容、高温涂层材料等一大批核心技术直接应用于国家重点型号。70年追求卓越，新世纪以来15项国家级科技奖励一等奖的成绩是对北航创新最“硬核”的诠释。

团结拼搏，是北航人深入血脉的奋斗底色。从建校初期全校师生大干一百天，送“北京一号”上天，到创新团队200余名师生全年集智攻关，创造连续可控飞行时间的世界纪录。70年斗转星移，北航人的团队文化和团队精神一脉相承，凭团结成功、靠拼搏取胜的优良传统更加历久弥新。

担当实干，是北航人始终如一的价值传承。刚才发言的型号火箭总指挥岑拯学长，三十多年如一日，为祖国航天事业踏实工作，带领团队研制的长征三号甲系列，被誉为中国的“金牌火箭”，有力保障了探月工程、北斗组网等重大任务。70年砥砺前行，北航人干一行、爱一行，钻一行、精一行，“国家的需要就是我们的选择”。

习近平总书记强调，“实现中国梦是一场历史接力赛，当代青年要在实现民族复兴的赛道上奋勇争先”。同学们是与新时代共同前进的一代，你们的成长期和奋斗期与国家实现第二个百年奋斗目标的历程高度重合，“天将降大任于斯人也”，勇担青春使命，逐梦空天报国，舍我其谁！在即将接过空天报国的“接力棒”之际，我有三点期望与大家共勉。

第一，坚定理想信念，把爱党报国作为毕生志向。古人云：“志不立，天下无可成之事。”一个人能成就多大事业，首先要看志向是不是远大、理想信念是不是坚定。古往今来凡成大事者，无一不是将报效祖国、服务人民作为毕生追求。

我国空间技术专家、“神舟”飞船总设计师戚发轫院士，1950年在大连读高中时，正值抗美援朝战争爆发，祖国领土接连遭到美军轰炸机扫射的欺侮，使他受到极大的触动，下定决心今后一定要学航空，造飞机保家卫国。高考后填报志愿，他全都选择了航空专业，最终如愿进入北航。他一生中主持了多型东方红系列卫星的研制和“神舟”号飞船总体方案设计，为我国“两弹一星”和载人航天事业做出了重大贡献。

心有所信，方能行远。同学们今后的理想追求和职业选择各不相同，面临各种社会思潮的影响，难免有疑惑彷徨的时候。希望你们坚定理想信念，把人生理想融入党和人民的事业中，在扎根祖国、服务人民的过程中成就一番事业。

第二，练就过硬本领，勇担民族复兴的历史使命。对于在座的本研新生来说，你们即将迎来人生中学习成长的黄金阶段。信息化时代加速了知识的产生和更迭，打牢学识根基、练就过硬本领，对培养适应时代要求的能力素质尤为重要。

在博学强识中夯实基础。唯勤学可广才，积小流成江海。“航空工业英模”、歼15总指挥罗阳学长，求学期间刻苦用功、品学兼优，理论力学和数学考试均是满分，他的本科毕设图纸已基本达到工程实际应用的水准，到现在仍被导师王浚院士大加赞许。

在矢志创新中提升能力。创新一直都是北航的靓丽名片。集成电路学院朱道乾同学，博士期间不断追求科技创新，经过缜密研究，他提出了一种新型方案，为支撑国家重点研发计划提供了重要技术路径，用实际行动诠释了创新不问年龄、英雄不问出处。

在实践实干中增长本领。“道虽迩，不行不至；事虽小，不为不成”，本事和能力不会自己长出来，祖国大地幅员辽阔，处处留心皆有学问，书本知识只有与实践行动相结合，才能增益其所不能，迸发出巨大能量。

学如弓弩，才如箭簇。成绩的取得绝不会一蹴而就。希望同学们珍惜时光，把学习作为首要任务，“苟日新，日日新，又日新”，如饥似渴、孜孜不倦地学习，努力成为堪当大任、能做大事的优秀人才。

第三，锻造刚健勇毅，涵养奉献向上的时代品格。人生之路有坦途也有崎岖，从来都不是一马平川。同学们正处于人生中重要的“拔节孕穗期”，多经历一些摔打挫折和困难考验，才能成为人生的强者。

人生的强者总是甘于无私奉献、永葆乐观向上。冬奥会期间，北航428名学生志愿者领命出征，驻守“雪飞燕”，他们决战高山、笑对严寒，以满腔热忱谱写了一曲动人的奉献者之歌。还有你们当中的研究生新生刘鑫同学，24年来不曾屈服命运的安排，坚持与病魔斗争，本科期间一直致力于发明创造，四年中获得多个国家级奖项，拥有十余项专利，以优异的成绩保送北航。在此我提议，让我们以热烈的掌声向他表示祝贺和敬意！

艰难困苦，玉汝于成。无论什么时候，顽强拼搏、永不言弃的精神不能丢，甘于奉献、团结互助的品质不能垮。希望同学们学会正确看待成败得失，从困难挫折中启迪心智、磨砺意志，使人生获得升华和超越。

奋斗是青春最亮丽的底色，行动是青年最有效的磨砺。同学们，属于你们的大学时光即将启程，希望你们不务虚功、不负韶华，让青春在祖国和人民最需要的地方绽放绚丽之花！

今天是教师节，也是中秋节。在这里祝全体老师教师节快乐，祝全体师生和家长们中秋佳节快乐、阖家幸福！谢谢大家！

第六章　跨文化交流训练

要做好口译工作，译员不仅需要熟练掌握源语和目标语，还须具备对中外文化的深入理解。语言不仅仅是信息传递的工具，更承载着文化的丰富内涵。译员在翻译过程中，除了确保信息的准确传达外，还肩负着充当文化桥梁的重任。这意味着译员必须深入了解各国和各地区的政治、历史、风俗习惯等背景知识，从而能够在传译时敏锐地意识到潜在的文化差异，避免因文化误解导致的沟通障碍。同时，跨文化理解不仅能帮助译员在传译过程中准确传达原文信息，还能使他们在面对复杂的交流环境时游刃有余，确保各方沟通顺畅。通过这种方式，译员不仅在促进国际交流中发挥了重要作用，还为增进不同文化间的理解与友谊做出了贡献。因此，译员的职责远不止语言转换，更在于通过深入的文化认知，实现真正的跨文化沟通。

第一节　跨文化交流技巧

本节从言语和非言语两个维度介绍跨文化交流技巧。希望同学们通过本节的学习，能够把握中英双语在语义和思维方式上的不同，恰当处理口译中涉及的跨文化交流难点，并掌握基本的非言语交流技巧，了解各国习俗与禁忌。

一、言语交际技巧

（一）语义差别

语义差别是影响口译中言语交际的主要因素，中英文的语义差别包括多种情况。一是，一种语言的有些词在另一种语言中没有对应词，例如汉语中的“风水”“气”“阴阳”等概念很难让外国人理解；二是，在两种语言里，某个词表面含义大致相同，但在内涵意义上却有较大的差异①。

1. 没有对应词

由于文化背景不同，中文和英文都有独一无二的词汇，在目标语中找不到对应词。这种情况可以采用音译或者“音译＋释义”的方法。

例 1：

原文：Before long, her posts went viral on **Twitter** with thousands of **retweets**,

① 任文. 英汉口译教程[M]. 北京：外语教学与研究出版社，2011：197.

likes and replies.

参考译文:很快,她的文章也在**推特**“走红”,收获了数千**转发**、点赞和评论。

例 1 中提到了国外社交媒体 Twitter,以及 retweet,即在该社交平台转发他人发布的消息。面对这类词汇,“音译+释义”是翻译中常用的方法,如将 Twitter 译为“国外社交平台推特”,将 retweet 译为“转推”。但由于口译场合中对译文的简洁性提出了较高要求,且 Twitter 在中国已具备一定的影响力,直接译为“推特”和“转发”也是可取的。

例 2:

原文:He is one of the top influencers on Instagram, gaining millions of followers every year.

参考译文:他是 Instagram 上的顶级网红,每年吸引数百万粉丝。

例 2 中 Instagram 作为全球知名的社交平台,直接保留源语。而 influencer 对应到中文语境可以翻译为“网红”,体现了这个词汇在社交媒体上的特殊含义。

例 3:

原文:欢迎海方参与中国**嫦娥**和**天问**等航天任务的搭载合作,研究成立中海联合月球和深空探测中心。

参考译文:China welcomes GCC countries' participation in payloads cooperation in its aerospace missions including **Chang'e** and **Tianwen**, and will consider establishing a China-GCC joint center for lunar and deep space exploration.

例 3 中航天任务的命名独具中国特色,例如,以神话人物“嫦娥”命名绕月探测工程,以屈原的长诗《天问》命名火星探测器,体现了中华民族从古至今对自然的探索。这些独特的中国文化负载词可能对目标语听众的理解造成了一定挑战。但对于航空航天类专有名词,音译法基本能够满足文化传播的要求,因为在跨文化信息交流中,名字背后的含义常常不是翻译的重点。

2. 表面对应但意思不同

中英文有些词表面上对应,但实际含义不同。这种情况可酌情采用增译或减译等方法。

例 1:

原文:这种团队精神值得我们学习。

参考译文:This sense of teamwork is worth learning from.

中文的“精神”不仅仅指心理或意识层面的东西,也可以指特定的价值观或态度。而在英语中,若直译为 spirit,可能显得模糊,甚至容易引起误解。因此,译者将“团队精神”译为 sense of teamwork,使其更符合英语表达习惯。

例 2:

原文:……并就坚定实施创新驱动发展战略,加快建设创新型国家,做出了重大战略部署。

参考译文:**Arrangements** were made to further the implementation of the innovation-driven development strategy and accelerate the building of an innovative country.

此例中"战略部署"对应的英文表达是 strategic deployment,但是 strategic deployment 有很强的军事色彩,为了避免外交上的误解,弱化语气,采用减译法,译为 Arrangements。

(二) 思维方式差异

除了语义上的差异外,中英文表达的另外一个不同点在于,中文重意合,英文重形合。汉语句子之间常靠内部的逻辑关系联系在一起,像一串珠子被一根逻辑线串起来,"形散而神聚";而英文句子之间往往靠各种语言形式紧密结合,像一根枝繁叶茂的大树。因此,英译汉时,译员要采用意合法,不必贴近英文的语法结构,将核心内容翻出即可;汉译英时,译员要使用形合法,在适当的时候补充语法成分,符合英语形式完整的要求。

例 1:

原文: She first wrote in Chinese, and then attached an accurate translation of the lines in both Italian and English **to** help more people understand.

参考译文: 她先用中文打出这段话,随后附上了意大利语与英语翻译,帮助更多网友理解其中的含义。

例 2:

原文: 遥感卫星地面系统进一步完善,基本具备卫星遥感数据全球接收、快速处理与业务化服务能力。

参考译文: **With** further improvements to the ground system of its remote-sensing satellites, **China** is now able to provide remote-sensing satellite data receiving and quick processing services across the world.

例 3:

原文: 深化载人登月方案论证,组织开展关键技术攻关,研制新一代载人飞船,夯实载人探索开发地月空间基础。

参考译文: Continue studies and research on the plan for a human lunar landing, develop new-generation manned spacecraft, **and** research key technologies **to** lay a foundation for exploring and developing cislunar space.

上述例 1 中英文以动词不定式引出目的状语 to help more people understand,译员在翻译过程中没有译出 to 所体现的目的关系,却很符合中文表达简明扼要的特点。例 2 和例 3 是中文典型的无主句罗列,例 2 译员增加了主语,并将"遥感卫星地面系统进一步完善"译为介词 with 引导的伴随状语,符合英文语言形式完整、结构复杂的特点。例3表面上看是四个并列的无主句，但译员分析出它们之间的逻辑关系，前三个单句是并列关系，在第四个单句前增加 to，表示目的。

（三）跨文化言语交际难点

跨文化交往过程中，译员要始终保持清醒头脑，在忠实传达发言人意思的同时，谨记文化差异，充分考虑目标语听众所在国家的政治制度、社会规范、风俗习惯和宗教信仰，注意禁忌事项，既要让对方听得懂，也要听得舒服。以下介绍跨文化言语交际的难点，包括称谓、迎来送往和习语翻译。

1. 称谓（头衔、职位、职称）

对外交往中，译员最先开口说的就是对方的称谓，译员要事前做好充分的功课，通过互联网搜索、打电话或者当面询问的方式，了解外方主要代表的姓名和职位。姓名要发音正确，职位要译得准确。

对来自君主制国家的外宾要多用敬称，例如称瑞典国王“卡尔十六世·古斯塔夫国王陛下”为 Your Majesty King Carl XVI Gustaf，称英国国王“查尔斯三世国王陛下”为 Your Majesty King Charles III。对国际组织和各国副部级以上高级官员用 Your Excellency 尊称，例如“尊敬的古特雷斯秘书长”可译为 Your Excellency Secretary-General António Guterres，“尊敬的部长阁下”称为 Your Excellency Minister，“尊敬的大使阁下”可译为 Your Excellency Ambassador。但对美国外宾则少用 Your Excellency 等敬称，可以直接称呼 Mr. President 和 Madam Secretary 等。

勿用职业名称直呼人，不要称“王老师”为 Teacher Wang，可以采用职称、学位、女士、先生等词加姓氏的方式，例如：Professor Wang，Dr. Wang，Miss Wang，Mr. Wang 等。

中国人特别重视家庭，尊重长者，也将这种理念扩展到社会层面，对陌生人也会尊称爷爷奶奶，叔叔阿姨，常见的有：“解放军叔叔”“警察叔叔”“邻居爷爷奶奶”等，但在国际交往中不能将其直接译为 Uncle PLA man、Uncle Policeman 和 Neighbor grandpa and grandma。

2. 迎来送往

中国人比较含蓄，西方人比较直接，迎来送往的时候各自的表达习惯不同，如果按照中文的习惯，直接译成英文，可能会产生误解。

中方主人欢迎外宾时，常常表示关心，“您不远万里来到中国，时差十几个小时，一定很辛苦。”但如果直译可能会让外宾感到不安——担心自己给人留下疲惫的印象。

3. 习语的翻译

英汉两种语言发展历史悠久,形成了大量的习语。习语一方面体现了中西方的独特文化,另一方面也为翻译和跨文化交流设置了障碍。在翻译习语时,译员应当理解其深层含义,通过借用、直译、意译等多种方式进行翻译。

(1) 借用法

例 1:

原文:岁月不居,时节如流。

参考译文: Time keeps rolling on.

例 2:

原文:春耕夏耘,万物生长。

参考译文: All crops grow through cultivation.

(2) 直译法

例 1:

原文:新中国成立以来,几代人**逢山开路、遇水架桥**,建成了交通大国,正在加快建设交通强国。

参考译文: Since the founding of New China, generation after generation of the Chinese people have worked **in the spirit of opening roads through mountains and putting bridges over rivers**, and turned China into a country with vast transport infrastructure.

例 2:

原文:小河有水大河满,大河无水小河干。

参考译文: A big river is full when its tributaries are filled with water; and tributaries must be dry when there is no water in the big river.

(3) 意译法

例 1:

原文: There is an American Indian quotation that says— **"Listen or thy tongue will keep thee deaf."**

参考译文:美洲印第安人有句名言:**"只说不听与聋人无异。"**

例 2:

原文:地方政府要当**"铁公鸡",不该花的钱一分钱也不能花,该给市场主体的钱一分都不能少,多一分那是添光彩。**

参考译文: All local governments must **count every cent and make every cent count** to ensure that all funding support for businesses will be fully delivered, and it would be even better if more could be done.

例 3:

原文:这些年不管在哪里,始终难以忘怀的是追逐航天梦想的**激情燃烧岁月。**

参考译文: No matter where I was working in these years, I could never forget **the golden days** of passionate pursuit of space dreams.

4. 诗词的翻译

诗词是中华文化博大精深的又一体现,可为文章增光添彩。直译能够让目标语受众对中华文化有进一步的了解。

例 1:

原文:毛泽东同志在诗词中更是抒发了**"可上九天揽月,可下五洋捉鳖"**的豪迈情怀和**"欲与天公试比高"**的雄心壮志。

参考译文: Chairman Mao further expressed the heroic feeling of **"we can bring down the moon from the ninth heaven, or catch the giant turtles in the sea"** and the ambition of **"all trying to match the sky in height"** in his poems, adhering to a long-established "Quest to the Heaven" spirit.

同样,如果诗词在文章中保留的意义不大,而且直译会造成冗余,引起听众的误解,那么意译不失为一种很好的翻译策略。

例 2:

原文:要**咬定青山不放松**,一张蓝图绘到底。

参考译文: We will **persevere in** carrying out the set blueprint until it becomes reality.

二、非言语交流注意事项

在跨文化交往过程中,译员除了要掌握必须的言语交际技巧外,还需要学习国际通行的外事礼仪,尊重对方的风俗习惯。

1. 打招呼

国际交往中最通行的打招呼方式是握手。握手的时候,右手握住对方的右手,看着对方的眼睛,上下轻轻摇动右手,不要大幅度摇晃。在西方常见的打招呼方式还有拥抱礼、贴面礼和吻手礼等。此外,不同的国家和地区还有独具特色的打招呼方式。2014 年 11 月,国家主席习近平和夫人彭丽媛访问新西兰,新西兰政府举行了隆重的毛利族传统欢迎仪式,毛利族男女长者分别向习近平主席和夫人彭丽媛行"碰鼻礼"。

2. 入乡随俗

外事场合我们遵循客随主便和主随客便的原则。到访一个国家,要尊重当地的文化习俗。比如到曼谷大皇宫,女士不能穿短裙或短裤,男士不能穿短裤,所有人都要穿有袖的衣服;参观印度泰姬陵,在跨上寝宫台阶前必须脱鞋;访问沙特,女士需要裹头巾。

作为主人,我们要展现礼仪之邦的风范,尊重外方的习俗,例如为素食的外宾安排专门的饮食。

3. 禁 忌

在进行跨文化交流的时候,还要注意不同文化的禁忌。与外方交谈时,通常不应涉及年龄、宗教、婚姻和收入等话题。在接待基督教国家的客人时,尽量避开数字"13";和伊斯兰国家的人士打交道时,不要用左手拍对方的肩膀,宴请时不安排猪肉;在接待信奉印度教的人士时,不要使用牛制品。

第二节 篇章讲解与练习

一、重点解析篇

(一) 意大利航天员在太空展现中国古文之美

1. 词汇表

manned spaceflight	载人航天
Astronaut Center of China (ACC)	中国航天员中心
Italian Air Force	意大利空军
astronaut corps	宇航员队伍
Preface to Poems Composed at the Orchid Pavilion	《兰亭集序》
Bohai Bay	渤海湾
the Yellow Sea	黄海

2. 背景知识

(1) 中欧航天合作

中国和欧洲在航天领域的合作始于 20 世纪 80 年代。2001 年 7 月,中欧签署"双星计划"合作协议,这是双方开展的重大空间科学探测合作,树立了中欧航天合作典范。近年来,双方不断扩大航天合作领域,实现互利共赢:2019 年,中国嫦娥四号探测器实现人类探测器首次月背软着陆,搭载了来自德国、荷兰等欧洲国家的仪器和设备;2021 年,中国"天问一号"与欧洲航天局"火星快车"任务团队合作,开展了"祝融号"火星车与"火星快车"轨道器在轨中继通信试验,取得了圆满成功。

此外,早在 2015 年中欧就签署了载人航天领域合作相关协议,中国航天员叶光富曾赴意大利撒丁岛开展训练。2017 年 8 月,中国航天员与 2 名欧洲宇航员在山东烟台附近海域,完成了为期 17 天的海上救生训练,包括海上自主出舱、海上生存、海上搜救船救援及海上直升机悬吊营救等科目[①]。

(2) 萨曼莎·克里斯托弗雷蒂

萨曼莎·克里斯托弗雷蒂(Samantha Cristoforetti),1977 年 4 月 26 日出生于意大利米兰。2009 年 5 月,32 岁的萨曼莎成为欧空局新一批航天员选拔的首位女性人选,她也同时成为意大利的第一位女性航天员、欧空局的第三位女性航天员。

萨曼莎·克里斯托福雷蒂会说意大利语、英语、德语、法语、俄语和中文,她还是欧空局内一个中国联络组的成员,欧空局官网显示该工作组的任务是与中国同行保持联系,以确定和实施双方航天员领域的合作计划。2017 年,萨曼莎·克里斯托福雷蒂还曾赴华,与中国航天员一同参加了在中国黄海组织的海上生存演习。这也是中国航天员和国外宇航员首次在中国开展联合训练[②]。

3. 原 文

Italian Astronaut Shows Beauty of Ancient Chinese Composition in Space[③]

Besides "Wow" and "Great", what else would you say when flying over the Earth in space? Samantha Cristoforetti, the first Italian female astronaut in space, posted several lines of a famous ancient Chinese composition on Twitter to share her joy.

She first wrote in Chinese, and then attached an accurate translation of the lines in both Italian and English to help more people understand.

"Looking up, I see the immensity of the cosmos; bowing my head, I look at the multitude of the world. The gaze flies, the heart expands, the joy of the senses can reach its peak, and indeed, this is true happiness," it said.

The text she quoted is from the *Preface to Poems Composed at the Orchid Pavilion*,

① 郑彬."欧中航天合作潜力巨大"[EB/OL].(2022-05-15)[2023-07-17]. https://www.yidaiyilu.gov.cn/p/243335.html.

② 中国妇女报.意大利女航天员引用中国古文,外交部:点赞![EB/OL].(2022-10-13)[2023-07-17]. https://m.thepaper.cn/baijiahao_20288676.

③ Xinhua. Italian astronaut shows beauty of ancient Chinese composition in space[EB/OL].(2022-10-16)[2023-07-17]. https://global.chinadaily.com.cn/a/202210/16/WS634b4f1da310fd2b29e7cb18.html.

a Chinese calligraphy masterpiece by Chinese calligrapher Wang Xizhi (303—361) of the Eastern Jin Dynasty (317—420).

She also attached three photos taken in space of China's Bohai Bay and the day-night view of Beijing, the country's capital.

Two days before she landed on Earth, the beauty of space and Chinese literature resonated on her Twitter account. Before long, her posts went viral on Twitter with thousands of retweets, likes and replies.

The poem and views are both magnificent; if everyone could enjoy the same sight, a lot of pain in the world would be dissipated, one user commented.

A European astronaut in space recites Chinese ancient poetry, how romantic, said another replier.

As the first Italian female astronaut in space, Cristoforetti carried out her first space mission on Nov. 23, 2014. She started the second mission in April and returned to Earth on Friday night after a 127-day long stay in orbit.

When not traveling for work, she is an avid reader and has a great passion for language learning.

"Always looking to the future, Samantha (Cristoforetti) is learning Chinese in her spare time," the European Space Agency (ESA) said on its website.

Meanwhile, Cristoforetti has been active in Sino-European cooperation in the field of manned spaceflight. She was a working group member tasked with liaising with Chinese counterparts to define and implement cooperation in the field of astronaut operations. In 2017, she took part in a sea survival exercise organized by the Astronaut Center of China in the Yellow Sea, which was the first joint training of Chinese and foreign astronauts in China.

An agreement was signed in 2015 between China and Europe on cooperation in the field of manned spaceflight, specifying the participation of China and Europe in each other's astronaut training activities. Speaking clear Chinese, Cristoforetti once told the media that the teams from China and Europe were working very well together in the field of astronaut training.

She joined the Italian Air Force in 2001 and had accumulated more than 500 hours of flight experience. In May 2009, Cristoforetti was one of the six chosen from 8,000 European applicants to join the ESA astronaut corps.

4. 参考译文

意大利航天员在太空展现中国古文之美①

如果你飞出地球，遨游太空，除了发出“哇塞”“太棒了”之类的感叹，你还会想到什么？正在太空执行任务的首位意大利女宇航员萨曼莎·克里斯托福雷蒂日前在推特发文，引用了中国古文名篇来分享自己的内心喜悦。

她先用中文打出这段话，随后附上了意大利语与英语翻译，帮助更多网友理解其中的含义。

文中写道：“仰观宇宙之大，俯察品类之盛，所以游目骋怀，足以极视听之娱，信可乐也。”

她所引用的古文来自东晋书法家王羲之的书法名作《兰亭集序》。

她同时还配上了三张照片，分别是从太空拍摄的中国渤海湾以及首都北京的白天和夜景照片。

在她返回地球的前两天，太空之美以及中国文学在她的推特账户下引起热烈反响。很快，她的文章也在推特“走红”，收获了数千转发、点赞和评论。

一名网友评论说，词句意境绝美，照片蔚为壮观；如果每个人都有这样的眼界，很多痛苦都会烟消云散。

另一名网友则表示，欧洲宇航员在太空吟诵中国古文，画面实属浪漫！

萨曼莎·克里斯托福雷蒂是意大利首位进入太空的女宇航员，欧洲航天局此前曾在官网发布过她的照片。作为首位进入太空的意大利女宇航员，克里斯托福雷蒂于 2014 年 11 月 23 日首次执行航天任务。她于 2022 年 4 月开始执行第二次航天任务，而后在轨飞行 127 天，并于周五返回地球。

在不需要外出工作时，她酷爱读书，并对语言学习保有极大热情。

欧航局网站的介绍中写到，克里斯托福雷蒂总是着眼未来，正在利用业余时间学习中文。

同时，克里斯托福雷蒂还活跃在中欧载人航天领域的相关合作中。她曾是联络工作组成员，与中国同行共同制定和实施航天员培养合作计划。2017 年，她参加了中国航天员中心在黄海组织的海上生存演习，这是中外航天员在中国的首次联合训练。

2015 年，中欧就载人航天领域合作签署协议，明确中欧参与对方的航天员训练活动。克里斯托福雷蒂曾在接受媒体采访时用清晰的中文说道，双方团队在航天员训练领域的合作非常顺利。

---

① 编者译。

她于2001年加入意大利空军,积累了超过500小时的飞行经验。2009年5月,克里斯托福雷蒂从约8000名欧洲籍申请人中脱颖而出,入选成为欧洲航天局宇航员。

5. 要点讲解

本篇英译汉材料中,欧洲女宇航员萨曼莎·克里斯托福雷蒂在社交平台分享中国古文名篇,表达了遨游太空的喜悦之情。译员在口译时应当注意中英文在语义和思维方式上和差异,通过多种跨文化言语交际技巧妥善处理译文。

英文的句子重心在前,中文的句子重心在后,为了使译文更加符合汉语的表达习惯,译员运用了词序调整法,对句子中的信息进行重组,如在翻译"what else would you say when flying over the Earth in space"这句话时,译员将 when 引导的条件状语从句提前,将重要的提问信息置后,翻译为"如果你飞出地球,遨游太空,……你还会想到说什么"。这样的处理方式既使得译文更加通顺,又更能使汉语读者更快地找出该句的重点信息。

基于中英文的表达差异,译员还善用转换法,在翻译时对原文的部分句子的语态、词性或定语的位置进行转换。例如在翻译"Photo of Samantha Cristoforetti, … published on the official website of the European Space Agency."这句话时,由于中文中主动句居多,译员把 the European Space Agency 当成主语,将译文从原文的被动语态转换为主动语态,使该句译文表达更加地道。对于"When not traveling for work, she is an avid reader"这句话,考虑到英文中多用名词、形容词,而中文里多用动词,译员便将原文中的名词 reader 转换为动词"读书",这种处理方式比起直接译成"热爱读书的人"更为简洁明了。除此之外,原文中后置定语和插入语较多,然而中文中更加偏向使用前置定语,于是在翻译过程中,译员也有注重调整译文定语的位置。如在翻译 a sea survival exercise organized by the Astronaut Center of China in the Yellow Sea 时,译员将该句的后置定语前置,译为了"中国航天员中心在黄海组织的海上生存演习",使中文译文的可读性更强。

为了更好地让英文读者理解文章,了解中国古代的文化常识,原文补充了一些相关的背景知识,如在介绍《兰亭集序》的作者王羲之时,原文补充了 Chinese calligrapher 和东晋的时间等信息。在把这个句子译成中文时,由于中文读者对以上文化常识都有所了解,为了避免赘述,译员使用了减译法,删去了 Chinese 和东晋的时间等信息。

（二）合肥市委常委袁飞在航天日新闻发布会上的讲话

1. 词汇表

综合性国家科学中心	comprehensive national science center
量子国家实验室	quantum national laboratory
“芯屏汽合、急终生智”	chips, displays, smart EVs, integration of AI with manufacturing, safety emergency devices, smart terminals, life science and AI
集成电路	integrated circuits
新型显示	new displays
国家战略新产业集群	national strategic emerging industry clusters
智能语音	intelligent voice
国家先进制造业集群	national advanced manufacturing industry cluster
空天信息产业	aerospace information industry
深空探测实验室	deep space exploration laboratory
北斗导航技术安徽省重点实验室	Anhui Provincial Key Laboratory of BeiDou Navigation Technology
国家高分辨率对地观测系统安徽数据与应用中心	Anhui Data and Application Center of the National High-Resolution Earth Observation System
国际先进技术应用推进中心	International Advanced Technology Application Promotion Center
“新华地球”	“Xinhua Earth”
巢湖一号	Chaohu No. 1
墨子号	Mozi No. 1
天链二号	Tianlian No. 2
高分十号	Gaofen No. 10
中国地理信息产业大会	China Geographic Information Industry Conference
空天信息技术与应用创新论坛	Aerospace Information Technology and Application Innovation Forum
数字地球峰会	Digital Earth Summit
《空天信息产业高质量发展若干政策》	*Several Policies for High-quality Development of Aerospace Information Industry*
“中国星城”	“Chinese Star City”
“中国航天日”	“China Aerospace Day”
深空探测实验室	the Deep Space Exploration Laboratory
三大科创引领高地	“Three Major Scientific and Technological Innovation Clusters”
深空探测	deep space exploration
量子信息	quantum information
聚变能源	fusion energy

2. 背景知识

(1) 中国首个深空探测实验室

2022 年 2 月,国家航天局、安徽省人民政府、中国科学技术大学三方共建的中国首个深空探测实验室(天都实验室)总部落户安徽合肥。实验室面向世界航天科技前沿和国家航天强国战略需求,围绕深空探测领域国家重大科技工程和国际大科学计划,开展战略性、前瞻性、基础性研究。作为实现科学、技术、工程融合发展的新型科技研发机构,该实验室志在打造代表国家水平的大型综合性研究基地、具有国际重要影响力的人才中心和创新高地。

自成立以来,实验室紧紧围绕深空探测领域国家重大科技工程,在全国范围内吸纳尖端人才和一流青年人才,支撑探月工程与行星探测工程的实施,在月球、行星及小行星、太阳系边际探测等领域的研究不断深入,引领未来深空探测事业高质量发展。目前,实验室数字化仿真、基础研究等能力持续提升,两地七址科研办公条件建设稳步推进,为后续发展形成了较好的基础条件支撑①。

(2) "芯屏汽合、急终生智"

"芯屏汽合、急终生智"作为安徽省合肥市战略新兴产业的发展口号,早已深深烙印在合肥产业发展的血液中。芯,即集成电路产业;屏,即新型显示产业;汽,即新能源汽车和智能网联汽车产业;合,即人工智能赋能制造业融合发展。急,即城市应急安全产业;终,即智能终端产业;生,即生物医药和大健康产业;智,即智能语音及人工智能产业。

伴随着以产业链为纽带的转型升级,这八个字有过两次"微雕"。2021 年 1 月,合肥市委十一届十二次全会决议上,把"芯屏器合"调为了"芯屏汽合",从"器"到"汽",建设"新能源汽车之都"的目标锚定。同年 11 月,"急终生智"取代"集终生智"登上文稿,至此,围绕芯片、新型显示、新能源与智能网联汽车、人工智能和制造业融合、公共安全和应急产业、家电及智能终端、生物医药和大健康、智能语音及人工智能的大产业格局完成构建②。

---

① 国家航天局.安徽聚力发展航天和空天信息产业 加快科技强省建设纪实[EB/OL].(2023-05-05)[2023-07-17].https://www.cnsa.gov.cn/n6758823/n6758838/c10013024/content.html.

② 合肥日报.揭秘:合肥是如何从"芯屏器合"到"芯屏汽合"的[EB/OL].(2023-04-20)[2023-07-17].https://www.zhihu.com/question/488647887/answer/3000064237.

3. 原　文

合肥市委常委袁飞在2023年“中国航天日”新闻发布会上的讲话①

女士们，先生们，朋友们：

大家好！

我是合肥市委常委袁飞，很高兴同大家见面。

合肥，居皖之中，承东启西，左右逢源，是长三角城市群副中心城市。党的十八大以来，习近平总书记先后两次考察安徽、亲临合肥，点赞合肥是“养人的地方”“创新的天地”。我们谨记总书记的教诲，在创新的道路上勇毅前行，全国第二个获批建设综合性国家科学中心，量子国家实验室全国首个挂牌，形成“芯屏汽合、急终生智”的战略性新兴产业格局，集成电路、新型显示、人工智能入选首批国家战新产业集群，智能语音入选国家先进制造业集群，“大力培育发展战新产业”连续5年获国务院督查激励。

近年来，合肥市积极依托科技和产业创新优势，聚力打造空天信息产业。去年，全国首个深空探测实验室在合肥组建运行。目前，合肥市已汇聚空天信息产业各类主体近70家，初步形成了航天产品设计、制造、集成、测试、试验及信息服务的全产业链。拥有北斗导航技术安徽省重点实验室、国家高分辨率对地观测系统安徽数据与应用中心等创新平台，国际先进技术应用推进中心、新华社融媒体“新华地球”等国家级平台也相继落户合肥；主导和参与了巢湖一号、墨子号、天链二号、高分十号等多颗卫星的研发和发射工作；中国地理信息产业大会、空天信息技术与应用创新论坛、数字地球峰会等一系列行业高端会议在合肥成功举办。此外，我们还出台了《空天信息产业高质量发展若干政策》，持续加大高层次专业人才的招引和扶持力度，同时建立空天信息产业基金。我们看到，一座“中国星城”正加速向我们走来。

本次“中国航天日”活动由工业和信息化部、国家航天局、安徽省人民政府主办，合肥市人民政府、中国科学技术大学、深空探测实验室承办，合肥成为第八个“中国航天日”主场活动举办地，充分体现了工信部、国家航天局对合肥的重视和支持。在此，我们致以衷心感谢。为承办好本次主场活动，合肥市高度重视、精心组织、认真落实，在各方面都做好了充分准备。比如，在展览展示方面，专门设计“三大科创引领高地”展区，重点展现合肥市在强化国家战略科技力量方面，围绕深空探测、量子信息、聚变能源等领域，和国家重大科技工程、国际大科学计划中，取得的战略性、前瞻性研究成果。合肥敞开大门，迎接八方来宾，航天日期间，组织全市重点商业综合体、特色商业

① 国家航天局. 2023年“中国航天日”新闻发布会[EB/OL]. (2023-04-18)[2023-07-17]. https://www.cnsa.gov.cn/n6758967/n6758969/n10008374/index.html.

街区开展航天主题宣传活动，全市所有收费A级景区对持有"会议电子票"的来宾免费开放。在航天日活动期间，将开展航天日主题大型灯光秀，在全市营造浓厚的节日氛围。

草木蔓发，春山可望。在这个充满生机的季节里，合肥将以一座科创名城、产业名城、巢湖名城、活力之城、幸福之城的美好状态，热情期待大家的到来。

最后，预祝2023年"中国航天日"活动取得圆满成功。谢谢大家。

4. 参考译文

Speech at the 2023 "Space Day of China" Press Conference①

Yuan Fei

Member of the Standing Committee of the Hefei Municipal Party Committee

Ladies and Gentlemen, Friends,

Good afternoon.

I am Yuan Fei, a member of the Standing Committee of the CPC Hefei Municipal Committee. It is nice to meet you all.

Hefei, located in central Anhui Province, is a subcenter of the Yangtze River Delta region. General Secretary Xi Jinping took Hefei as one of his main stops during his two visits to Anhui Province after the 18th National Congress of the Communist Party of China. He praised Hefei as a place of talent and innovation. Bearing in mind the General Secretary's encouragement, we have forged ahead in pursuing innovation. Hefei is the second comprehensive national science center approved by the government, and boasts China's first national laboratory on quantum. It has put in place a pattern for strategic emerging industries, covering integrated circuit, new displays, new energy vehicles and V2X, AI in manufacturing, emergency response, smart terminals, life science, intelligent speech and AI. Integrated circuit, new displays, and AI have been included in the first batch of national strategic emerging industry clusters, and intelligent speech as a national advanced manufacturing industry cluster. For five consecutive years, the State Council has rolled out incentive measures for Hefei based on its exceptional performance in fostering and developing the strategic emerging industries.

In recent years, Hefei has leveraged its strengths in technological and industrial innovation to develop the aerospace information industry. Last year, the first deep

① 编者译。

space exploration laboratory in the country started operation in Hefei. At present, Hefei has brought together nearly 70 entities in aerospace information industry, and has initially formed a complete industrial chain from aerospace product design, manufacturing, integration, testing, experimentation to information services. It is home to the Anhui Provincial Key Laboratory of BeiDou Navigation Technology, the Anhui Data and Application Center of the National High-Resolution Earth Observation System and other innovation platforms. Hefei also hosts multiple national-level platforms such as the International Advanced Technology Application Promotion Center and Xinhua News Agency's integrated media "Xinhua Earth". It has led and participated in the research and development and launch of satellites such as Chaohu No. 1, Mozi No. 1, Tianlian No. 2, and Gaofen No. 10. A series of top industry conferences such as the China Geographic Information Industry Conference, Aerospace Information Technology and Application Innovation Forum, and Digital Earth Summit have been successfully held in Hefei. We have also issued *Policies for High-quality Development of Aerospace Information Industry*, intensifying efforts to attract and support top professionals. We have established an aerospace information industry fund. We see a "Star City" emerging.

The "China Aerospace Day" event is hosted by the Ministry of Industry and Information Technology(MIIT), the National Space Administration, and the People's Government of Anhui Province. It is organized by the People's Government of Hefei, the University of Science and Technology of China, and the Deep Space Exploration Laboratory. Hefei has become the eighth host city of "China Space Day". Here, we express our sincere gratitude to the MIIT and the National Space Administration for the support given to Hefei. Taking this event very seriously, we have made full preparations to make it a success. For example, in terms of exhibition, a pavilion on "Three Major Scientific and Technological Innovation Clusters" is designed to focus on Hefei's efforts to boost national strength in strategic science and technology and its major cutting-edge achievements in deep space exploration, quantum information, fusion energy and other fields, as well as in major national sci-tech projects and international mega-science programs. Hefei opens its door to guests from across the world. During the Space Day, key commercial complexes and commercial districts throughout the city will organize aerospace-themed activities. All A-level tourist attractions in the city will be free to guests with a Conference E-Ticket. During the Space Day event, a large-scale light show with an aerospace theme will be held to create a festive atmosphere throughout the city.

The grass and trees are flourishing, and the view of spring mountains are there for you to enjoy. In this season full of vitality, Hefei will present its best as a city of innovation, industrial development, vitality and happiness along the Chaohu Lake, and embrace your arrival with open arms.

Finally, I wish the 2023 China Space Day a complete success. Thank you.

5. 要点讲解

本篇材料选自安徽省合肥市委常委袁飞 2023 年 4 月在"中国航天日"主场活动上的讲话,演讲主要介绍了合肥市相关情况及空天信息产业发展概况。材料体现了中英文在言语及非言语交际中的多方面差别,值得译员的关注和思考。例如:

目前,合肥市已汇聚空天信息产业各类主体近 70 家,初步形成了航天产品设计、制造、集成、测试、试验及信息服务的全产业链。

译文: At present, Hefei has brought together nearly 70 entities in aerospace information industry, and has initially formed a complete industrial chain from aerospace product design, manufacturing, integration, testing, experimentation to information services.

东西方文化的差异还体现在思维方式的不同,中国人习惯于从个别推导出一般的归纳式思维方式,而西方人则习惯于从一般去认识个别的演绎式论证,使得中英文的句子语义中心不同,汉语的语言重心后置,而英语则刚好相反。例句中,"全产业链"作为总结性词汇,在汉语中被置于最后,而在翻译的过程中,译员应将 complete industrial chain 前置,这样译文才更加流畅自然,符合目标读者的语言习惯。

除此之外,口译过程中译员还应当注意中英文文化中社会习俗的不同,如打招呼、迎来送往、宴请招待等。本篇演讲材料中,发言人在开头和结尾常使用一些四字成语来提升文采、构成对仗,这一演讲策略体现了中国社会习俗在中国人欢迎、告别、祝贺等礼仪场景中产生的影响,译员应当有意识地注意到这类问题,理解成语的内涵,并在翻译过程中采用多种技巧,转换为符合英语国家读者习惯的译文。

二、实战训练篇

(一) 美国联邦航空管理局前任局长斯蒂芬·M·迪克森在部落航空研讨会开幕式和闭幕式上的致辞

1. 词汇表

Tribal Aviation Symposium	部落航空研讨会
Federal Aviation Administration (FAA)	美国联邦航空管理局
Bipartisan Infrastructure Law	两党基础设施法

续表

the Choctaw Nation of Oklahoma	俄克拉荷马州乔克托族
the Great Plains	大平原地区
Williston, North Dakota	北达科他州威利斯顿

2. 原 文

Tribal Aviation Symposium Opening and Closing Remarks①

FAA Former Administrator Stephen M. Dickson

14 March, 2022

Opening Remarks

Thanks, Arlando. I'm happy to be here.

This is an exciting time in aviation. There are more opportunities than ever before to start a career, grow a business, and be part of the aerospace community. At the FAA, it's our responsibility to make sure everyone knows about these opportunities and how to get involved. That's what today is all about. We want to make aviation safe, efficient, environmentally friendly, and accessible to everyone.

This week's symposium is a chance for us to connect with Tribal Nations. We want to hear directly from you about your concerns and needs regarding aviation. And we want you to know what the FAA is doing, and how you can benefit from it.

Today, we'll talk about several FAA grant programs, like the Airport Improvement Program, through which we provide grants to airports across America. We'll also discuss the Bipartisan Infrastructure Law—a historic, once-in-a-generation opportunity to invest $20 billion in airport and air traffic control infrastructure.

We want you to know what grants are available, and how to navigate the application process.

Tomorrow, we'll promote aerospace education and career opportunities for Tribal youths and people living in Tribal Nations.

And we'll also discuss how drones and Advanced Air Mobility can benefit Native Nations as well. For instance, the FAA and the Choctaw Nation of Oklahoma have teamed up to test how drones can deliver cargo, including parcels, at low altitudes. As I said at the top, this is a really exciting time in aviation.

---

① Stephen M Dickson. Tribal Aviation Symposium Opening and Closing Remarks[EB/OL]. (2022-03-14)[2023-07-17]. https://www.faa.gov/speeches/tribal-aviation-symposium-opening-and-closing-remarks.

Our intention is to create a trusted community between the FAA and Tribal Nations, so we can communicate openly and regularly about all areas of aviation, including opportunities to be part of it.

For this to happen, we have to acknowledge that Native Nations have been mistreated by the Federal Government. We must be mindful of how this history shapes the challenges and conditions that affect Tribes today.

We strongly support President Biden's Executive Order calling for the Federal Government to advance equity for ALL people, including those who have been historically underserved.

The President has also reaffirmed that Tribes are sovereign Nations. We must accord you with the respect we provide all nations. And we must consult with Tribes in areas of aviation policy that affect them, just as we do with other nations.

There is an American Indian quotation that says—"Listen or thy tongue will keep thee deaf."

It's important that we listen to you—really listen—and respect your needs and concerns. And when we listen, we will do so without pre-existing bias, recognizing that each Tribe has a unique perspective.

Ultimately, we must earn and maintain your trust.

The more we engage with each other, the better we can develop solutions, and extend the benefits of aviation to everyone. Let me give you some examples of where we have achieved positive engagement in aviation.

The FAA has initiated consultations with more than 130 Native American Tribes to ensure they have a meaningful opportunity to express concerns regarding air tours over U. S. national parks, many of which encompass sacred sites and other areas of importance to Tribes.

We also consulted with 11 Tribes in the Great Plains region to protect sacred sites during construction of a new airport project in Williston, North Dakota. This effort included having 41 Traditional Cultural Specialists present as monitors during construction.

And when the Native Village of Kwinhagak spoke to us about the need to rehab the runway attheir airport in southwest Alaska, we worked to establish funding for this ongoing project.

These are just a few examples of collaboration between the FAA and Tribal Nations. We continue to learn from these efforts, so we can connect with Tribes better in the future, and help you take advantage of the opportunities that are out there.

This week's Symposium will help us do more of that. Thank you; I look forward to an engaging session today.

Closing Remarks

Thanks Arlando.

As I noted in my earlier remarks, we want to establish a trusted community between the FAA and Tribal Nations to address all issues related to aviation.

We covered some important ground today.

Through the Bipartisan Infrastructure Law, along with the FAA's Airport Improvement Program, we have a tremendous opportunity over the next several years to strengthen aviation infrastructure across America. These projects can help improve airports that serve Tribal Nations.

We want Native authorities to feel confident navigating the grant application process for Tribal-owned airports. And we're eager to help you do that.

Aside from our airport grant process, Native American businesses can also contact the agency's Small Business Office, and look at their website—*sbo. faa. gov*—for information and resources about FAA contracting opportunities.

And check out the FAA's YouTube site where we've posted a playlist containing videos about the Bipartisan Infrastructure Law, which includes webinars about contract opportunities for small businesses. I encourage you to share these resources on your Facebook accounts and other social media platforms.

Tomorrow, the FAA's Deputy Administrator and other speakers will discuss education and career opportunities in the aerospace field, and how these efforts can benefit young people in Tribal Nations.

We'll also cover emerging innovation like drones, Advanced Air Mobility, and commercial space transportation, and what these activities can mean for Tribes.

And we'll delve into more detail about flights over sacred sites, and how we can better collaborate with each other to address those concerns.

As some of you may know, I'll be leaving the agency at the end of this month. After more than 43 years as an aviation professional, it's time for me to go home. In due time, the agency will name an acting Administrator to guide things along, until a permanent Administrator is named.

Many times, changes in leadership make people wonder how the agency's priorities are going to change. Please know that the FAA will remain steadfast in our commitment to collaborate with Tribal Nations. This commitment is an agency-wide priority.

It's imperative that we establish and maintain trust with each other, so that even in times of changing leadership, we'll always be able to maintain strong ties going forward. I am confident that strong foundation already exists.

Future generations of Tribal Nations should know that they will always have a partner in the FAA.

I'm reminded of another Native American quotation attributed to Howard Rainer of the Taos Pueblo-Creek. It says "Children learn from what they see. We need to set an example of truth and action."

When we set a positive example of collaboration, we'll maintain that example through future generations of leadership in the Tribal Nations.

Thank you for engaging with us today. And we look forward to tomorrow's session, where we'll connect again.

(二) 习近平在意大利媒体发表的署名文章

1. 词汇表

卫星	satellite
载人航天	manned space exploration
文艺复兴	the Renaissance
"大秦"	"Da Qin"
"丝绸之国"	Seres, the land of silk
《马可·波罗游记》	*The Travels of Marco Polo*
"中国热"	"China fever"
人类命运共同体	a global community of shared future
警务联合巡逻	joint police patrols
《神曲》	*Divine Comedy*
《意大利与中国》	*Italy and China*
"汉学热"	a long-running boom of China studies
全面战略伙伴关系	comprehensive strategic partnership
"北方港口建设计划"	Italy's plan to develop its northern ports
"投资意大利计划"	Invest Italian program
中国国际进口博览会	China International Import Expo
联合国教科文组织世界遗产地	UNESCO world heritage sites
二十国集团	Group of 20 (G20)
亚欧会议	Asia-Europe Meeting
世界贸易组织	World Trade Organization (WTO)

2. 原 文

习近平在意大利媒体发表的署名文章
东西交往传佳话 中意友谊续新篇[①]

中华人民共和国主席 习近平

2019 年 3 月 20 日

很高兴在万象更新的时节应马塔雷拉总统邀请，对意大利共和国进行国事访问。2011 年我曾在罗马出席“意大利统一 150 周年”庆典活动，2016 年又过境撒丁岛。意大利将古老和现代、经典和创新相结合的生活方式和工业理念，给我留下了深刻印象。即将再次踏上这个美丽国度，见到热情的意大利朋友，我感到十分亲切。

中国和意大利是东西方文明的杰出代表，在人类文明发展史上留下浓墨重彩的篇章。作为古罗马文明的发源地和文艺复兴的摇篮，意大利雄壮华美的历史古迹、文学艺术巨匠的恢宏杰作在中国广为人知。中国和意大利两个伟大文明的友好交往源远流长。早在两千多年前，古老的丝绸之路就让远隔万里的中国和古罗马联系在一起。汉朝曾派使者甘英寻找“大秦”。古罗马诗人维吉尔和地理学家庞波尼乌斯多次提到“丝绸之国”。一部《马可·波罗游记》在西方掀起了历史上第一次“中国热”。马可·波罗成为东西方文化交流的先行者，为一代代友好使者所追随。

进入当代，沿着古人友好交往的足迹，中意关系不断焕发出新的勃勃生机。1970 年中华人民共和国同意大利共和国建立外交关系。2020 年两国将迎来建交 50 周年。建交以来，无论国际风云如何变幻，两国始终相互信任、密切合作，树立了不同社会制度、文化背景、发展阶段国家互利共赢的典范。中意两国牢固的传统友谊历久弥新，成为双边关系快速稳定发展的坚实支柱。

——中意友谊扎根在深厚的历史积淀之中。两千多年交往史为中意两国培育了互尊互鉴、互信互谅的共通理念，成为两国传统友谊长续永存、不断巩固的保障。面对当今世界的变革和挑战，两国从历史沧桑中汲取宝贵经验，共同畅想构建相互尊重、公平正义、合作共赢的新型国际关系，构建人类命运共同体的美好愿景。

——中意友谊凝结在深厚的战略互信之中。两国领导人坚持从战略高度和长远角度看待和发展双边关系。2004 年两国建立全面战略伙伴关系以来，双方发挥高层交往的引领和推动作用，在事关彼此核心利益和重大关切问题上相互理解、坚定支持，成为确保中意关系长期稳定发展的坚实后盾。

——中意友谊体现在丰富的务实合作之中。中意互为重要贸易和投资伙伴，两

---

① 外交部. 习近平在意大利媒体发表署名文章[EB/OL]. (2019-03-20)[2023-07-17]. https://www.mfa.gov.cn/web/gjhdq_676201/gj_676203/oz_678770/1206_679882/1209_679892/201903/t20190320_9352358.shtml.

国利益深度交融。2018 年,双边贸易额突破 500 亿美元,双向投资累计超过 200 亿美元。“意大利制造”是高品质产品的代名词,意大利时装、家具广受中国消费者青睐,比萨饼、提拉米苏为青少年所喜爱。两国在卫星、载人航天等领域合作喜报频传,中意科技创新合作周、警务联合巡逻、足球培训等活动受到两国人民热烈欢迎。

——中意友谊传承于密切的文化交流之中。中意两国人民对研习对方文化抱有浓厚兴趣。中国一位教授在古稀之年开始翻译但丁的《神曲》,几易其稿,历时 18 载,在临终病榻上最终完成。意大利汉学家层出不穷,为中欧交往架起桥梁。从编写西方第一部中文语法书的卫匡国,到撰写《意大利与中国》的白佐良和马西尼,助力亚平宁半岛上的“汉学热”长盛不衰。

意大利著名作家莫拉维亚写道:“友谊不是偶然的选择,而是志同道合的结果。”当今世界正面临百年不遇之大变局。把中意关系提高到新的更高水平,共同维护世界和平稳定和发展繁荣,是历史赋予我们的责任。我愿通过这次访问,同意大利领导人一道擘画中意关系蓝图,引领中意关系进入新时代。

我们愿同意方提升全面战略伙伴关系,密切高层交往,加强政府、议会、政党、地方各级别合作,强化政策沟通,增进战略互信和战略对接,继续相互理解和支持彼此核心利益和重大关切,夯实双边关系政治基础。

我们愿同意方共建“一带一路”,发挥两国“一带一路”合作的历史、文化、区位等优势,把“一带一路”互联互通建设同意大利“北方港口建设计划”“投资意大利计划”等对接,在海上、陆地、航空、航天、文化等多个维度打造新时期的“一带一路”。

我们愿同意方拓宽务实合作领域。中国将扩大对外开放,通过每年举办中国国际进口博览会等方式,同包括意大利在内的世界各国分享中国市场机遇。双方可以深入挖掘在港口物流、船舶运输、能源、电信、医药等领域合作潜力,鼓励两国企业开展第三方市场合作,实现互利多赢。

我们愿同意方密切人文交流。作为两个拥有最多联合国教科文组织世界遗产地的国家,中意拥有丰富的文化和旅游资源。双方要加强两国世界遗产地结好,鼓励两国文化机构和个人互办高水平文物和艺术展,联合拍摄影视作品,加强语言教学,促进人员往来,为世界文明多样性和不同文化交流互鉴作出新贡献。

我们愿同意方加强国际事务和多边组织内的协调。中方愿在联合国、二十国集团、亚欧会议、世界贸易组织等框架内,同意方加强在全球治理、气候变化、联合国改革、世界贸易组织改革等重大问题上的沟通和配合,维护共同利益,促进多边主义和自由贸易,维护世界和平稳定和发展繁荣。

回首 50 年,中意关系深耕厚植、硕果累累。展望新时期,中意合作欣欣向荣、前景广阔。中国人民期待着同友好的意大利人民携手努力,为两国关系发展培育更加艳丽的花朵,让中意友谊不断焕发新的生机活力。

（三）习近平主席在亚洲文明对话大会开幕式上的主旨演讲

1. 词汇表

亚洲命运共同体	an Asian Community with a Shared Future
亚洲文明对话大会	Conference on Dialogue of Asian Civilizations
底格里斯河-幼发拉底河	the Tigris and the Euphrates
印度河-恒河流域	the Indus and the Ganges
黄河-长江流域	the Yellow River and the Yangtze
《诗经》	*The Book of Songs*
《论语》	*The Analects of Confucius*
《塔木德》	*The Talmud*
《一千零一夜》	*One Thousand and One Nights*
《梨俱吠陀》	*The Rigveda*
《源氏物语》	*Genji Monogatari*
楔形文字	cuneiform script
阿拉伯数字	Arabic numerals
麦加大清真寺	Great Mosque of Mecca
泰姬陵	Taj Mahal
吴哥窟	Angkor Wat
丝绸之路	Silk Road
茶叶之路	Tea Road
香料之路	Spice Road
“一带一路”	Belt and Road Initiative
“两廊一圈”	Two Corridors and One Belt
“欧亚经济联盟”	Eurasian Economic Union
中亚的古城撒马尔罕	Central Asian city of Samarkand
埃及的卢克索神庙	Luxor Temple in Egypt
新加坡的圣淘沙	Sentosa in Singapore
泰国的曼谷玉佛寺	Wat Phra Kaew in Bangkok
希腊的雅典卫城	Acropolis in Athens
天文历法	China's astronomical knowledge, calendar system
民本理念	the people-centered doctrine
佛教东传	the introduction of Buddhism

续表

"伊儒会通"	the confluence of Islam and Confucianism
"西学东渐"	the introduction of Western learning
新文化运动	New Culture Movement

2. 原 文

深化文明交流互鉴 共建亚洲命运共同体

——在亚洲文明对话大会开幕式上的主旨演讲①

中华人民共和国主席 习近平

2019 年 5 月 15 日，北京

尊敬的各位国家元首、政府首脑、国际组织负责人，

尊敬的各位嘉宾，

女士们，先生们，朋友们：

在这个草木生长的美好季节，来自亚洲 47 个国家和五大洲的各方嘉宾，为深化文明交流互鉴共聚一堂，共襄盛举。首先，我谨代表中国政府和中国人民，并以我个人的名义，对亚洲文明对话大会的召开，表示诚挚的祝贺！对各位嘉宾的到来，表示热烈的欢迎！

当前，世界多极化、经济全球化、文化多样化、社会信息化深入发展，人类社会充满希望。同时，国际形势的不稳定性不确定性更加突出，人类面临的全球性挑战更加严峻，需要世界各国齐心协力、共同应对。

应对共同挑战、迈向美好未来，既需要经济科技力量，也需要文化文明力量。亚洲文明对话大会，为促进亚洲及世界各国文明开展平等对话、交流互鉴、相互启迪提供了一个新的平台。

女士们、先生们、朋友们！

亚洲是人类最早的定居地之一，也是人类文明的重要发祥地。亚洲地大物博、山河秀美，在世界三分之一的陆地上居住着全球三分之二的人口，47 个国家、1000 多个民族星罗棋布。从公元前数千年起，生活在底格里斯河—幼发拉底河、印度河—恒河、黄河—长江等流域的人们，开始耕耘灌溉、铸器造皿、建设家园。一代又一代亚洲先民历经岁月洗礼，把生产生活实践镌刻成悠久历史、积淀成深厚文明。广袤富饶的平原，碧波荡漾的水乡，辽阔壮美的草原，浩瀚无垠的沙漠，奔腾不息的江海，巍峨挺拔的山脉，承载和滋润了多彩的亚洲文明。

① 习近平. 深化文明交流互鉴 共建亚洲命运共同体[EB/OL]. (2019-05-15)[2023-07-17]. https://www.mfa.gov.cn/web/zyxw/201905/t20190515_346483.shtml. 文字有校改

在数千年发展历程中，亚洲人民创造了辉煌的文明成果。《诗经》《论语》《塔木德》《一千零一夜》《梨俱吠陀》《源氏物语》等名篇经典，楔形文字、地图、玻璃、阿拉伯数字、造纸术、印刷术等发明创造，长城、麦加大清真寺、泰姬陵、吴哥窟等恢宏建筑……都是人类文明的宝贵财富。各种文明在这片土地上交相辉映，谱写了亚洲文明发展史诗。

亚洲先人们早就开始了文明交流互鉴。丝绸之路、茶叶之路、香料之路等古老商路，助推丝绸、茶叶、陶瓷、香料、绘画雕塑等风靡亚洲各国，记录着亚洲先人们交往交流、互通有无的文明对话。现在，"一带一路""两廊一圈""欧亚经济联盟"等拓展了文明交流互鉴的途径，各国在科技、教育、文化、卫生、民间交往等领域的合作蓬勃开展，亚洲文明也在自身内部及同世界文明的交流互鉴中发展壮大。

璀璨的亚洲文明，为世界文明发展史书写了浓墨重彩的篇章，人类文明因亚洲而更加绚烂多姿。从宗教到哲学、从道德到法律、从文学到绘画、从戏剧到音乐、从城市到乡村，亚洲形成了覆盖广泛的世俗礼仪、写下了传承千年的不朽巨著、留下了精湛深邃的艺术瑰宝、形成了种类多样的制度成果，为世界提供了丰富的文明选择。

回顾历史、展望世界，我们应该增强文明自信，在先辈们铸就的光辉成就的基础上，坚持同世界其他文明交流互鉴，努力续写亚洲文明新辉煌。

女士们、先生们、朋友们！

亚洲各国山水相连、人文相亲，有着相似的历史境遇、相同的梦想追求。面向未来，我们应该把握大势、顺应潮流，努力把亚洲人民对美好生活的向往变成现实。

——亚洲人民期待一个和平安宁的亚洲。维护和平是每个国家都应该肩负起来的责任。没有和平，冲突不断甚至战火纷飞，经济增长、民生改善、社会稳定、人民往来等都会沦为空谈。亚洲各国人民希望远离恐惧，实现安居乐业、普遍安全，希望各国互尊互信、和睦相处，广泛开展跨国界、跨时空、跨文明的交往活动，共同维护比金子还珍贵的和平时光。

——亚洲人民期待一个共同繁荣的亚洲。经济发展是文明存续的有力支撑，繁荣富强是国家进步的重要基石。亚洲一些民众特别是妇女儿童正忍受着贫困、饥饿、疾病的折磨，这样的局面必须改变。亚洲各国人民希望远离贫困、富足安康，希望各国合力推进开放、包容、普惠、平衡、共赢的经济全球化，共同消除一些国家民众依然面临的贫穷落后，共同为孩子们创造衣食无忧的生活，让幸福和欢乐走进每一个家庭。

——亚洲人民期待一个开放融通的亚洲。亚洲近几十年快速发展，一条十分重要的经验就是敞开大门，主动融入世界经济发展潮流。如果各国重新回到一个个自我封闭的孤岛，人类文明就将因老死不相往来而丧失生机活力。亚洲各国人民希望远离封闭、融会通达，希望各国秉持开放精神，推进政策沟通、设施联通、贸易畅通、资金融通、民心相通，共同构建亚洲命运共同体、人类命运共同体。

女士们、先生们、朋友们！

文明因多样而交流，因交流而互鉴，因互鉴而发展。我们要加强世界上不同国家、不同民族、不同文化的交流互鉴，夯实共建亚洲命运共同体、人类命运共同体的人文基础。为此，我愿提出4点主张。

第一，坚持相互尊重、平等相待。每一种文明都扎根于自己的生存土壤，凝聚着一个国家、一个民族的非凡智慧和精神追求，都有自己存在的价值。人类只有肤色语言之别，文明只有姹紫嫣红之别，但绝无高低优劣之分。认为自己的人种和文明高人一等，执意改造甚至取代其他文明，在认识上是愚蠢的，在做法上是灾难性的！如果人类文明变得只有一个色调、一个模式了，那这个世界就太单调了，也太无趣了！我们应该秉持平等和尊重，摒弃傲慢和偏见，加深对自身文明和其他文明差异性的认知，推动不同文明交流对话、和谐共生。

我访问过世界上许多地方，最吸引我的就是韵味不同的文明，如中亚的古城撒马尔罕、埃及的卢克索神庙、新加坡的圣淘沙、泰国的曼谷玉佛寺、希腊的雅典卫城等。中国愿同各国开展亚洲文化遗产保护行动，为更好传承文明提供必要支撑。

第二，坚持美人之美、美美与共。每一种文明都是美的结晶，都彰显着创造之美。一切美好的事物都是相通的。人们对美好事物的向往，是任何力量都无法阻挡的！各种文明本没有冲突，只是要有欣赏所有文明之美的眼睛。我们既要让本国文明充满勃勃生机，又要为他国文明发展创造条件，让世界文明百花园群芳竞艳。

文明之美集中体现在哲学、社会科学等经典著作和文学、音乐、影视剧等文艺作品之中。现在，大量外国优秀文化产品进入中国，许多中国优秀文化产品走向世界。中国愿同有关国家一道，实施亚洲经典著作互译计划和亚洲影视交流合作计划，帮助人们加深对彼此文化的理解和欣赏，为展示和传播文明之美打造交流互鉴平台。

第三，坚持开放包容、互学互鉴。一切生命有机体都需要新陈代谢，否则生命就会停止。文明也是一样，如果长期自我封闭，必将走向衰落。交流互鉴是文明发展的本质要求。只有同其他文明交流互鉴、取长补短，才能保持旺盛生命活力。文明交流互鉴应该是对等的、平等的，应该是多元的、多向的，而不应该是强制的、强迫的，不应该是单一的、单向的。我们应该以海纳百川的宽广胸怀打破文化交往的壁垒，以兼收并蓄的态度汲取其他文明的养分，促进亚洲文明在交流互鉴中共同前进。

人是文明交流互鉴最好的载体。深化人文交流互鉴是消除隔阂和误解、促进民心相知相通的重要途径。这些年来，中国同各国一道，在教育、文化、体育、卫生等领域搭建了众多合作平台，开辟了广泛合作渠道。中国愿同各国加强青少年、民间团体、地方、媒体等各界交流，打造智库交流合作网络，创新合作模式，推动各种形式的合作走深走实，为推动文明交流互鉴创造条件。

第四，坚持与时俱进、创新发展。文明永续发展，既需要薪火相传、代代守护，更需要顺时应势、推陈出新。世界文明历史揭示了一个规律：任何一种文明都要与时偕行，不断吸纳时代精华。我们应该用创新增添文明发展动力、激活文明进步的源头活水，不断创造出跨越时空、富有永恒魅力的文明成果。

激发人们创新创造活力，最直接的方法莫过于走入不同文明，发现别人的优长，启发自己的思维。2018 年，中国国内居民出境超过 1.6 亿人次，入境游客超过 1.4 亿人次，这是促进中外文明交流互鉴的重要力量。中国愿同各国实施亚洲旅游促进计划，为促进亚洲经济发展、增进亚洲人民友谊贡献更大力量。

女士们、先生们、朋友们！

中华文明是亚洲文明的重要组成部分。自古以来，中华文明在继承创新中不断发展，在应时处变中不断升华，积淀着中华民族最深沉的精神追求，是中华民族生生不息、发展壮大的丰厚滋养。中国的造纸术、火药、印刷术、指南针、天文历法、哲学思想、民本理念等在世界上影响深远，有力推动了人类文明发展进程。

中华文明是在同其他文明不断交流互鉴中形成的开放体系。从历史上的佛教东传、“伊儒会通”，到近代以来的“西学东渐”、新文化运动、马克思主义和社会主义思想传入中国，再到改革开放以来全方位对外开放，中华文明始终在兼收并蓄中历久弥新。亲仁善邻、协和万邦是中华文明一贯的处世之道，惠民利民、安民富民是中华文明鲜明的价值导向，革故鼎新、与时俱进是中华文明永恒的精神气质，道法自然、天人合一是中华文明内在的生存理念。

今日之中国，不仅是中国之中国，而且是亚洲之中国、世界之中国。未来之中国，必将以更加开放的姿态拥抱世界、以更有活力的文明成就贡献世界。

女士们、先生们、朋友们！

这次亚洲文明对话大会议题广泛、内容丰富，希望大家集思广益、畅所欲言，提出真知灼见，共同创造亚洲文明和世界文明的美好未来！

最后，预祝亚洲文明对话大会圆满成功！

谢谢大家。

第七章 译前准备

要顺利完成一项口译任务，译员不仅需要积累语言知识和专题背景，还必须在每次任务前完成充分且有针对性的准备工作。这种准备工作包括但不限于深入了解会议的背景信息、议程安排、参会人员的详细名单、翻译的具体形式、会议的地点和时间安排、交通方式和着装要求等细节。如此繁杂的准备工作要求译员具备高效的时间管理能力和强大的全局把控能力，确保在面对任何突发情况时能够从容应对。本章将围绕三条线索，系统性地梳理译前准备的各项事宜，帮助译员在任务前有条不紊地完成准备工作。

第一节 译前准备事项

俗话说“台上一分钟，台下十年功”，口译工作亦是如此。译员需要长时间不断地积累语言知识和专题背景知识，磨练翻译技能，锻炼心理素质。除此之外，当译员接到一项新的口译任务时，还需要积极地、有针对性地做好短期相关准备。

1. “粗线条”准备工作

首先，译员需要明确活动的翻译形式，是会议翻译还是参观陪同翻译，是同声传译还是交替传译，是双向翻译还是单向翻译。

其次，译员需要了解活动的基本情况，包括名称、主题、主办单位和承办单位等。

再次，译员需要主动了解背景情况，包括行业发展现状和国别情况。以国际民航组织第三届航空和可替代燃料大会为例，译员需要事前了解会议的背景、参会国别、往届大会情况和产业发展挑战和机遇等。如果会议的主题是气候变化，那么译员需要准备相关词汇和背景知识，如《京都议定书》、IPCC、《巴黎协定》、“双碳”目标、碳税、中美气变合作和 COP28 等。

2. “细线条”准备工作

首先，译员需请主办方提供参会人员名单，最好是中英文对照的名单，包括机构名称、人名和职位，因为直译或者网上查到的译文可能跟实际情况有所出入。例如，有些行业协会的英文名称用 association，但也有的协会称 council 或者 society 等。有些中国企业的英文名称是汉语的音译，但也有一些中国企业的英文名称跟其中文发音没有任何关系。所以机构的专有名词应力求准确，不能想当然。外方人名的发音也要提前确认清楚，特别是母语为非英语人士的姓名发音，例如法国前总统奥朗德的姓 Hollande 中的 H 不发音。有些外国人有中文名，但不能用汉语拼音的方式称呼，例如，不能称呼联合国教科文组织驻华代表处代表夏泽翰教授为 Professor Xia，而应称其为 Professor Shahbaz Khan。如果借助互联网和工具书也查不到人名读音，译员可在会前电话联系或在会场直接询问外方工作人员，以准确称呼外方代表名字。

其次，译员需和主办方沟通，拿到中英文的议程，熟悉议题顺序，做到心中有数。如果收到文字稿或者 PPT 等文件，需按议程先后顺序整理，便于现场快速准确地找到发言材料。译员会前应仔细研读会议资料，遇到不明白的术语或者概念，应该查阅工具书、上网搜索或请教专业人士，并整理词汇表。

如果确实无法获取发言人的演讲稿等相关文件，译员则可以上网搜索该发言人在其他场合的视频或者报道，事先了解其对这一话题的立场，熟悉其讲话口音、语速和风格等。

3. 后勤准备工作

除了会议内容的准备外，还有很多细节需要译员注意。

首先是会议时间。一般来说，译员需比会议开始时间提前至少半个小时到达会场，熟悉嘉宾座次和译员位置。在交替传译场景下，译员一般站在或坐在发言人旁边。有些情况下，会场很大，译员需要照顾到很多发言人，如果不使用话筒，译员和坐在远端的参会人员之间可能互相听不清，这时候一定要跟主办方协商，安装话筒，并提前调试，熟悉话筒的开关按键，确认是否啸叫，保证能从扩音器里听到清晰的声音。所以译员务必要提前到达会场，及时发现问题、解决问题。

其次，会议地点也需仔细确认。例如，北京有若干个会议中心，名字很相似，一定要弄清楚会议举办的具体地点，千万不要搞错了，导致不能按时执行口译任务。译员还需提前规划好出行路线，留出充足的时间，以防堵车耽误行程。

最后，着装也是译员需要提前考虑的因素。一般来说要大方得体，既不要 underdress，也不要 overdress，要看具体的会议场合和地点。如果是工厂里的技术交流活动，不建议女译员穿裙子，不要穿露脚趾的鞋或高跟鞋。如果是正式的公务或商务活动，着装一般较为正式。总的原则是跟参会的其他人员保持一致。当然，也可以向活动主办方询问着装要求。

除上述细节外，译员还要提前准备好文具和设备。记笔记的小册子建议选择前后有硬纸板、可上下翻页的记录本，因为译员有时候需要站着记笔记，有硬纸板支撑的小册子会更舒服。随身携带至少两支笔，以防一支笔出现意外。打印好会前准备好的词汇表，尽量做到简洁、有序，急需的时候可以瞄上几眼。保证电脑或者手机联网，需要时可以查询电子词典或者网页。

总而言之，无论是翻译新手还是翻译经验丰富的译员，准备工作都必不可少。准备得越充分，现场发挥便越得心应手。如果怠于准备，翻译经验丰富的译员也可能“翻车”。

第二节　篇章讲解与练习

一、重点解析篇

（一）美国副总统哈里斯和法国总统马克龙在加强太空合作会议上的讲话

1. 词汇表

Artemis I	阿尔忒弥斯一号飞行器
Webb Telescope	韦伯太空望远镜
Space for Climate Observatory (SCO)	空间气候观测平台
Comprehensive Dialogue on Space	太空全面对话
European Space Agency (ESA)	欧洲航天局
International Space Station (ISS)	国际空间站
Artemis Accords	《阿尔忒弥斯协定》
direct-ascent anti-satellite missile	直升式反卫星导弹
National Space Council	美国国家空间委员会
Argos advanced Researchand Global Observation Satellite (Argos)	先进研究和全球观测卫星
Surface Water and Ocean Topography (SWOT)	地表水和海洋地形观测卫星
Atmosphere Observatory System	大气观测系统
mock-up	实体模型
French Guyana	法属圭亚那

2. 背景知识

(1) 詹姆斯·韦伯空间望远镜(James Webb Space Telescope)

詹姆斯·韦伯空间望远镜是美国航空航天局、欧洲航天局和加拿大航空航天局联合研发的红外线观测用太空望远镜，为哈勃空间望远镜的继任者，于2021年12月25日发射升空，于2022年1月24日顺利进入围绕日地系统第二拉格朗日点的运行轨道①。

(2) 国际空间站(International Space Station)

国际空间站是在轨运行最大的空间平台,是一个拥有现代化科研设备、可开展大规模、多学科基础和应用科学研究的空间实验室,为在微重力环境下开展科学实验研究提供了大量实验载荷和资源,支持航天员在地球轨道长期驻留。国际空间站项目由16个国家共同建造、运行和使用,是有史以来规模最大、耗时最长且涉及国家最多的空间国际合作项目。自1998年正式建站以来,经过十多年的建设,于2010年完成建造任务转入全面使用阶段。国际空间站主要由美国国家航空航天局、俄罗斯联邦航天局、欧洲航天局、日本宇宙航空研究开发机构、加拿大国家航天局共同运营②。

(3)《阿尔忒弥斯协定》(*Artemis Accords*)

《阿尔忒弥斯协定》是美国国家航空航天局与澳大利亚、加拿大、意大利、日本、卢森堡、阿联酋和英国航天机构签署的有关探索月球的协定,于2020年10月13日签定。该协定还概述了未来探索月球及其他领域的行为③。

3. 原 文

Remarks by Vice President Harris and President Emmanuel Macron of France in Meeting on Space Cooperation④

Washington, D.C., 30 November 2022

VICE PRESIDENT HARRIS: Good morning to everyone. President Macron, it is my pleasure — my great pleasure to welcome you to Washington, D.C., and to NASA headquarters.

France is a vital ally to the United States. And this visit demonstrates the strength of our partnership, of our friendship, and our cooperation, and truly all as — as the background to an enduring relationship between the United States and France, and one that is based on shared democratic principles and values.

---

① 央视新闻.韦布空间望远镜拍摄到太阳系外行星直接图像[EB/OL].(2022-09-03)[2023-07-17]. https://content-static.cctvnews.cctv.com/snow-book/index.html? item_id=3699881094338345725&toc_style_id=feeds_default&share_to=qq&track_id=026efb1e-415f-4b42-9c69-f0dc57bda9ad.

② 中国载人航天.国际空间站各舱段命名由来[EB/OL].(2011-04-23)[2023-07-17]. https://www.cmse.gov.cn/art/2011/4/23/art_313_3625.html.

③ 环球网.NASA公布登月最新安排 2024年将宇航员送上月球[EB/OL].(2020-10-15)[2023-07-17]. https://world.huanqiu.com/article/40IK2MmCXxi.

④ The White House. Remarks by Vice President Harris and President Emmanuel Macron of France in Meeting on Space Cooperation [EB/OL]. (2022-11-30)[2023-07-17]. https://www.whitehouse.gov/briefing-room/speeches-remarks/2022/11/30/remarks-by-vice-president-harris-and-president-emmanuel-macron-of-france-in-meeting-on-space-cooperation/.(文字有校改)

I want to thank NASA Administrator Bill Nelson for hosting us today and for your very impressive year of leadership of NASA.

During your leadership, we redirected the trajectory of an asteroid. We successfully launched Artemis I, the mission, which is orbiting the Moon as we speak, in partnership with our French and European allies. In fact, the Administrator was able to show the President a mock-up of Artemis I, and it truly is quite impressive.

And, of course, we showed the world incredible images from the Webb Telescope, which was launched from French Guyana with a French launch vehicle.

So, Mr. President, today we have the occasion to celebrate the cooperation of our two nations in space. You and I have discussed this topic many times, including my visit in November in France, in Paris. And as a result of that meeting, a number of things happened.

One, the United States agreed to join the Space for Climate Observatory as a result of that meeting that you and I had. You and I created the Comprehensive Dialogue on Space. And we agreed to advance specific rules and norms in space — international rules and norms.

And the United States, of course, welcomes and is elated that France and the European Space Agency will continue operations at the International Space Station through 2030.

Today, we build on the progress that we have made. And today we will also identify additional areas of collaboration and cooperation.

Regarding the climate crisis, something you and I have discussed at length and which we agree is an existential threat, satellite technology and the data that it provides are critical to our efforts to combat and address the climate crisis.

In Paris, I made a commitment that the United States would join the French-led Space for Climate Observatory, which we formally did in June. Through this work, we will develop tools for decision-makers to better prepare for and respond to climate disasters such as wildfires, floods, and changing agricultural patterns.

And in two weeks, the United States and France will launch a satellite from Vandenberg, in my home state of California, which will conduct the first-ever global survey of the Earth's water surface — the surface on the Earth of water.

It's very exciting because this innovation will actually help us to improve water management and inform — and better inform agricultural decisions. And it will also help coastal communities around the world prepare for rising seas.

During my visit to Paris, we also agreed to create a Comprehensive Dialogue on Space. And this brings leading experts together from France and the United States to strengthen our commercial, our civil, and our national security cooperation. And they had a very productive meeting, actually, already this month.

Regarding rules and norms, we are working together to develop norms for the responsible and peaceful uses of outer space. And this is critically important work. To that end, Mr. President, I thank you for joining the Artemis Accords in response to my invitation during my visit with you. These accords are a set of principles that guide civil use in space. They contribute to a safe and transparent environment for space exploration, science, and commercial activities.

More specifically, regarding rules and norms and international rules and norms: In April, I made a commitment on behalf of the United States not to conduct destructive, direct-ascent anti-satellite missile testing. And I called on other nations to join us. I thank you, Mr. President, for your commitment, which was announced yesterday, to adopt this norm.

And finally, I believe space remains a place of undiscovered and unrealized opportunity. And for that reason, there is so much potential in terms of the work that nations can do there — and, in particular, when we work together based on shared principles and values.

To seize these opportunities, as the head of the National Space Council in the United States, I have focused on three priorities in particular, which is to expand our STEM workforce, to address the climate crisis in every way that is about building on the innovation and also creating a workforce to do that, and establish rules and norms to govern space activities in close partnership with the private sector.

In this mission, we are so very proud to work with France. The United States and France have partnered on space exploration for more than 60 years. In this time, we have made great strides. And yet, in so many ways, we are beginning a new journey together.

And so, with that, again, I welcome you, Mr. President, and I look forward to our ongoing work in this matter, including the work that we will do today. Welcome.

PRESIDENT MACRON: Thank you. Thank you very much, Ms. Vice President, dear Kamala. Thank you for hosting us and having this meeting all together.

And dear ministers, ambassadors, and administrators, I want to thank as well. And, NASA Administrator, to welcome us in this NASA headquarters. Thank you so much. And thank you for the great commitment in our cooperation.

I mean, you said everything, and I don't want to repeat, because I want us to be engaged in a common discussion. But speaking about space is obviously speaking about both science and having this journey you mentioned. But this is, as well, the story of a great cooperation between our two countries. And we did a lot during the past decades together, and I think we can do a lot. And we have a lot of projects together.

And I want to thank you personally because, indeed, when you came one year ago in Paris, and we had this very long discussion, we decided together to strengthen this cooperation and to launch a strategic dialogue on space. And we did it, and we already delivered concrete results, based on the past experience between our research leaders, our academics, our astronauts, as well. And some of them are around this table, and I want to thank them for this commitment and their achievement.

Having said that, we want to work now very closely together indeed on climate and environment as one of the key verticals where we do believe, together, we have a common objective. Argos, launched more than 40 years ago, is a—45 years ago now already—was the best evidence of this strong cooperation between our research agencies.

We have, and you mentioned them, on top of that, the Surface Water and Ocean Topography, with this launching the 12th of December, which will be the symbol of a new type of cooperation.

We will have the Space for Climate Observatory. And thank you for joining it. And I think this initiative was launched three years ago, but now our cooperation will accelerate at this point.

And we have, as well, the Atmosphere Observatory System.

And all this initiative, we can launch together.

On all these issues, I want to thank our institutions and researchers, and I think our commitment is great.

Second, we have, indeed, exploration. James Webb Space Telescope is obviously one of the best examples of this cooperation and how we can team up and how science made on our side can help in working in common missions with you.

And you mentioned Artemis Ⅰ, Ⅱ, Ⅲ. And we are very keen in participating, and we have here the French team. And Thomas is very keen on—and very excited—to be part of it.

And it's very important for us, as long as you can propose the French leader to fly to the Moon. Quite happily—(laughter) we are quite happy with that.

And Sophie is also now selected for participating to this international mission and follow up this common effort.

But I saw myself—with your presence, with our discussion with NASA teams, and so on how strong is the partnership and how we can do more for the years to come.

And I think, from a scientific point of view and a human point of view, all these upcoming changes regarding exploration—the Moon, Mars, and so on—are extremely important for us.

And I want to thank you for your leadership and to tell you our commitment to work closely together.

My last point is about norms, as you mentioned. Thank you for having launched these—these very important items. And I do believe, indeed, space is a new place for conflictuality.

4. 参考译文

美国副总统哈里斯和法国总统马克龙在加强太空合作会议上的讲话①

2022年11月30日,华盛顿

美国副总统哈里斯:大家上午好。马克龙总统,很荣幸与您见面,欢迎来到华盛顿特区参观美国国家航空航天局总部。

法国一直是美国的重要盟友。此次访问展现了我们合作与友谊的力量,这源于美国与法国之间的长久伙伴关系,以及我们共同的民主原则和价值观。

我要感谢美国国家航空航天局局长比尔·纳尔逊(Bill Nelson)的热情款待,感谢您这一年中所展现出的优秀领导力。

在您的带领下,我们改变了一颗小行星的运行轨道。我们还与法国和欧洲盟友合作,成功发射了阿尔忒弥斯一号飞行器,此时它正在执行绕月飞行任务。此外,纳尔逊局长还向总统展示了阿尔忒弥斯一号的实体模型,十分精彩。

我们通过法国运载火箭从法属圭亚那发射了韦伯太空望远镜,向全世界展示了韦伯太空望远镜拍摄的瑰丽图像。

总统先生,我想借此机会祝贺两国在太空领域的合作。在我11月访问巴黎等多个场合时,我们都讨论过这一话题,并取得了一系列成果。

首先,美国决定加入空间气候观测平台。我们还举行了太空全面对话。此外,我们还一致同意推进太空领域具体国际规则和准则的制定。

当然,对法国和欧洲航天局将继续在国际空间站运行至2030年,美国热烈欢迎,并倍感鼓舞。

今天,就让我们共同努力,再接再厉。今天,我们还将明确可开展合作的其他领域。

我们充分讨论了气候危机,并一致认为气候威胁关乎我们生存,而卫星技术及数据对应对气候危机至关重要。

---

① 编者译。

之前在巴黎，我承诺美国将加入法国领导的空间气候观测平台，并于6月正式加入。通过这项工作，我们将为决策者开发工具，更好应对山火、洪水等气候灾害以及日新月异的农业生产方式。

未来两周，在我的家乡加利福尼亚州范登堡(Vandenberg)，美国和法国将联合发射一颗卫星，首次对地表水进行全球调查。

这一计划令人振奋，因为它能够有效帮助我们改善水资源管理，更好地为农业决策提供支撑，并帮助全球沿海地区做好应对海平面上升的准备。

访问巴黎期间，我们还同意就太空问题开展全面对话，齐聚法国和美国的顶尖专家，加强两国商业、民用及国家安全合作。就在本月，两国专家开展了一次富有成效的对话。

在规则和准则方面，我们正在共同制定负责任的、和平利用外太空的相关准则。这项工作至关重要。总统先生，十分感谢您在我访问期间应邀加入《阿尔忒弥斯协定》。这项协定为民用太空活动制定了一系列原则，有助于为空间探索活动、科学活动和商业活动创造安全、透明的环境。

具体来说，关于规则和准则，以及国际规则和准则：今年4月，我代表美国承诺不再进行破坏性的直升式反卫星导弹试验。我强烈呼吁其他国家也加入我们。总统先生，您昨天承诺法国将采纳这一准则，我对您表示由衷的感谢。

最后，我认为太空仍然充满机遇，等待着人类去发现，去把握。因此，各国在太空开展的工作有巨大的潜力可以挖掘。如果我们能在共同的原则和价值观基础上携手合作，这一潜力空间会更大。

为了抓住这些机遇，作为美国国家空间委员会主席，我特别关注三个优先事项：扩大理工科人才培养规模；培养人才，以创新的方式全面应对气候危机；与私营部门密切合作，制定太空活动的管理规则和准则。

在这些事务中，我们非常荣幸能与法国合作。美国和法国的太空探索合作已逾六十年。我们共同取得了长足的进步。但在很多方面，我们才刚刚共同开始一段崭新的旅程。

总统先生，再次欢迎您，同时也非常期待我们现在正在进行的工作，包括我们今天要做的工作。再次欢迎马克龙总统。

法国总统马克龙：谢谢。非常感谢您，哈里斯副总统。谢谢您的接待，让我们能共同召开这次会议。

尊敬的各位部长、大使和政府官员，我要向你们表示由衷的感谢。我还要感谢美国国家航空航天局局长，谢谢您邀请我们来到美国国家航空航天局总部。感谢您对两国合作的大力支持。

您的讲话非常全面，我就不再赘述了，我希望我们能够继续交流。太空合作意味着科学研究与合作。在这一点上，美国和法国就是优良的典范。在过去几十年里，我们一起完成了很多工作，我认为我们可以一起完成更多的工作。我们还开展了很多合作项目。

我还想以个人的名义再次感谢您。一年前您曾经来访巴黎，当时我们进行了一次长谈，共同决定加强两国合作，就太空问题展开战略对话。如今，我们做到了，两国的科研负责人、学者和宇航员一起努力，已经取得了具体的成果。他们中的一些人今天就坐在这张桌子旁，我要感谢他们的投入和取得的成果。

我们希望在气候和环境方面加强合作，这也是我们有着共同目标的关键领域之一。先进研究和全球观测卫星(Argos)项目迄今已有 40 余年，准确来说是 45 年，是我们研究机构之间密切合作的最好证明。

除此之外，我们还将于 12 月 12 日发射地表水和海洋地形观测卫星，您刚才也提到这一点，它象征着两国新型合作的开启。

我们还就空间气候观测平台开展了合作。感谢美国加入我们的行列。这一倡议是在三年前发起的，而我们的合作将会加速这一倡议的进展。

还有大气观测系统。

我们可以共同发起这一倡议。

首先，我要感谢我们的研究机构和研究人员，感谢你们的奉献。

其次，我们的确进行了深入的探索。詹姆斯·韦伯太空望远镜显然是这种合作的最佳范例之一，彰显了我们如何通过合作、利用科学来实现我们共同的目标。

您还提到了阿尔忒弥斯一号、二号和三号飞行器。我们非常渴望参与这项工作，我们的法国团队也来到了现场。托马斯非常渴望也特别兴奋能成为团队的一员。

这对我们来说非常重要。您甚至可以提议让法国的领导人飞往月球。我们对此真的感到非常高兴。

索菲也被选中参与这项国际任务，继续跟进这一合作项目。

通过您的参与以及我们与美国国家航空航天局团队的讨论等等——我看到了这种伙伴关系有多牢固，以及我们如何在未来几年做更多的事情。

我认为，从科学视角和人类视角来看，所有月球、火星等探索发现对我们来说都是极其重要的。

我要感谢您的领导，并向您承诺我们将密切合作。

最后，关于太空活动准则，您刚才已经说得很清楚，我就不重复了。还要感谢您提出这些重要的议题。我也认为未来的太空将会是一个充满争议和冲突的场所。

5. 要点讲解

本文信息量较大，发言人发言过程中常用插入语和从句等对先前提到的信息进行补充说明，因此长难句较多且句式复杂。译员在翻译时，除了注重对信息的准确表达以外，可以运用诸多翻译技巧，如增译法、减译法、转换法、分译法、合译法及词序调整法等，使得译文更加通顺，更符合汉语的表达习惯。

善用增译法，通过增译连接词和副词等让译文衔接更加通顺，让译文更能体现发言人的意图。如在翻译“… we agree is an existential threat, satellite technology and the data… ”这句话时，原文的插入语和主句之间并无明显衔接词，直接翻译会使表达略显生硬，可以在主句前增加连词“而”与插入语进行衔接，使译文更通顺。

此外，发言人在提到参与美国《阿尔忒弥斯协定》的看法时，连用了两次“we are quite happy with that”。在翻译这个句子时，增译了“真的”“非常”等副词，既避免译文句子间的重复，又更好地还原了发言人对该协定十分感兴趣的语气。

对于原文中重复累赘的地方，可以运用减译法对译文进行处理，如在翻译“And this visit demonstrates the strength of our partnership, of our friendship, and our cooperation, and truly all as… ”这句话时，将与 our partnership 表达相同意思的 cooperation 省去，直接译成了“我们之间合作和友谊的力量”，使译文更简洁明了。

英文中多用名词、形容词，而中文里多用动词，可以巧用转换法，对原文词性进行转换，让表达更加地道。如在翻译“… I believe space remains a place of undiscovered and unrealized opportunity”这句话时，将 undiscovered 和 unrealized 转换为动词，并反义正译，译成“等待着人类去发现，去把握”，运用转换法处理后的译文比起直译更为通顺。

英语习惯先说重要信息，后说次要信息，而汉语的表达习惯与此相反，因此，可以使用词序调整的方式使译文的重点更为突出。如在翻译“And we did a lot during the past decades together, and… ”这句话时，译员将时间状语 the past decades together 这个次要信息提前，译为“在过去几十年里，我们一起完成了很多工作”，使译文更符合汉语的表达习惯。

面对原文复杂多变的句式和众多插入语，还需要注重对句式的调整，运用合译法对译文进行处理。如在翻译“… a set of principles that guide civil use in space. They contribute to… ”这两句话时，由于第二句话和第一句话紧密联系且主语相同，可以将两句话合起来翻译，译为“这项协定为民用太空活动制定了一系列原则，有助于……”，避免重复提到同一主语，使得译文表达更加简洁明了，逻辑清晰。

（二）王赤：中国空间科学的发展任重道远，希望2045年能与NASA比肩

1. 词汇表

中国科学院国家空间科学中心	National Space Science Center (NSSC), Chinese Academy of Sciences (CAS)
辐射带	radiation belt
日冕物质抛射	coronal mass ejections
太阳耀斑	solar flares
太阳和太阳圈探测器	Solar and Heliospheric Observatory (SOHO)
帕克太阳探针	Parker solar probe
电磁波段	electromagnetic spectrum

续 表

哈勃望远镜	Hubble Space Telescope
暗能量	dark energy
詹姆斯·韦伯空间望远镜	James Webb Space Telescope
系外行星	exoplanet
毅力号	Perseverance rove
慧星	comets

2. 背景知识

(1) 中国科学院国家空间科学中心(National Space Science Center, Chinese Academy of Sciences)

中国科学院国家空间科学中心是我国空间科学及其卫星项目和中国科学院月球与深空探测任务的总体性研究机构,我国空间科学领域的研究中心和创新高地,负责组织开展国家空间科学发展战略规划研究,具体负责中国科学院空间科学先导专项的组织与实施,开展空间科学及相关应用领域的创新性科学研究与技术发展和试验工作[①]。

(2) 太阳风-磁层相互作用全景成像卫星(Solar wind Magnetosphere Ionosphere Link Explorer,简称 SMILE)

太阳风-磁层相互作用全景成像卫星是拟于 2021 年发射的中欧联合科学卫星,2015 年 6 月 4 日,中国科学院与欧洲空间局联合公布了新遴选出的中欧联合空间科学卫星计划——"太阳风-磁层相互作用全景成像卫星计划",是继 2003 年双星计划之后,中欧又一大型联合空间科学探测项目,也是双方科学家开展深度国际合作新的里程碑。此次任务中,欧方拟支持 5300 万欧元,中方给予了大致同等强度的支持[②]。

(3) 爱因斯坦探针(Einstein Probe,简称 EP)

爱因斯坦探针计划是一台面向未来时域天文学和高能天体物理的小型科学探测卫星,是 X 射线天文卫星。

① 王赤. 加速空间科学发展 建设航天强国[EB/OL]. (2022-11-09)[2023-07-17]. http://cssar.cas.cn/ztzl2015/xxyd2015/202212/t20221230_6592174.html.

② 科学网. 中欧发布空间科学卫星计划"微笑"看地球[EB/OL]. (2015-06-04)[2023-07-17]. https://news.sciencenet.cn/htmlnews/2015/6/320265.shtm.

3. 原 文

中国空间科学的发展任重道远，希望2045年能与NASA比肩[①]

中国科学院院士
中国科学院国家空间科学中心主任
中国科学院大学地球与行星科学学院教授 王赤
2022年11月9日

各位来宾，女士们，先生们，非常高兴有机会来讲一讲我们国家的空间科学卫星计划的进展和展望。首先，我想讲一讲什么是空间科学：自1957年第一颗人造卫星上天以来，人类进入到了一个空间的时代。我们空间科学就是随着人类航天器进入到太空之后开展起来的一门新兴的交叉学科。所以我们空间科学的一个关键词，就是以航天器为主要的平台。我们这个航天器可以指我们的卫星，也可以指我们的空间站，还有我们的深空探测器。空间科学研究的范畴包括我们地球的空间，我们整个太阳系的空间，还有宇宙的空间，从近至远。

首先从地球空间起步。我们知道，地球的空间并不是一个太空，也并不是一个虚无缥缈的真空，它含有很多的物质，有很多的磁场、电场。所以我们的地球空间富含了我们的辐射带，也有在日地之间的各种结构，包括从太阳爆发的日冕物质抛射，还有整个地球的各种大尺度结构。我们观测的最近的恒星就是太阳，我们的SOHO卫星第一次从太阳的表面一直观测到我们太阳的内部的结构。帕克太阳探针第一次近距离观测到了我们的太阳，也发现了很多传统理论无法解释的物理现象。遥看遥远的天空和宇宙，人类从不同的电磁波段来进行观测，最有名也是大家最熟悉的，就是哈勃望远镜。它第一次精确地测量了我们宇宙的能量，发现了暗能量是怎么驱动宇宙不断膨胀。

最近发射的韦伯空间望远镜是哈勃望远镜的一个继承者，我们知道哈勃望远镜的口径才2米多，而韦伯空间望远镜的口径已经达到了6.5米。最近的一次科学发现也表明，在系外行星当中发现了二氧化碳。到目前为止，人类非常关心的一个问题就是地球是不是宇宙中唯一一个存在着智慧生命的星球？要了解这个根本的问题，就需要寻找地球2.0，寻找地球2.0，就要从寻找系外行星开始。我们现在已经找到了上千颗的系外行星，但是在这些系外行星当中，是不是具有跟地球一样的宜居环境，现在还是一个值得探索的科学问题。

回到我们的太阳系，火星是跟地球最为相近的一个行星。在火星上是不是存在着生物，或者在历史上存在生物，现在还不得而知。但现在添了一些火星上存在生命

① 王赤. 中国空间科学的发展任重道远，希望2045年能与NASA比肩[EB/OL].(2022-11-09)[2023-07-17]. https://i.ifeng.com/c/8KnQfc0vH3H.

的新证据：最近毅力号在火星上发现了大量的有机分子。我们还探测了彗星，探测了小行星，这都是太阳系起源时残留下来的物质。我们能够探测彗星、小行星，就窥视了太阳系小时候的样子。我们从地球出发，走向了月球，走向了火星，下一个目标就是日球层边界，或者说太阳系的边界。现在旅行者一号、二号已经跨过了日球层边界，进入到了星际空间。人类探索了太阳系的八大行星，现在目光已经瞄向了深邃的宇宙。到目前为止，我们一共发射了几百颗科学卫星，但实事求是地讲，中国的科学卫星还是屈指可数，我们的空间科学发展任重而道远。

下面我介绍一下我们空间科学先导专项。在2011年的时候，中国科学院率先行动，启动了一批先导专项。其中空间科学卫星是先导专项中，强度最大的一个空间科学专项。大家熟知的悟空号暗物质粒子探测卫星，是目前为止世界上探测能量值最高的卫星，达到了10个TEV。我们卫星探测的能量精度也是最强的，目前已经探测到了很多以往没有探测到的，在高能段的奇特物理现象。墨子号卫星是我们第一个量子科学实验卫星。我们的慧眼号硬X射线调制望远镜探测到了宇宙当中最强的磁场——10亿特斯拉，要知道在地球上，实验室中，最强的磁场也就到100个特斯拉的量级。

先导专项的一期也得到了国内外广泛地关注。目前我们正在实施先导专项的二期：太极一号微重力技术实验卫星，是我们国家进行空间引力波探测的第一步。我们已经获得了迄今为止在空间中最精确的一个激光测距的测量结果。我们的怀柔一号引力波暴高能电磁对应体全天监测器卫星已经成功发射，探测到了很多伽玛暴。前不久的10月9号，我们发射了先进天基太阳天文台，这也是我国第一颗综合性太阳观测卫星。我们第一次以一磁两暴作为目标，探测太阳的磁场、耀斑和日冕物质抛射事件，这就像太阳的闪电和暴雨一样，我们要了解太阳的磁场与太阳爆发活动之间的关系。目前工作都非常正常，我们预计在两三个月之内要发布第一批观测结果。明年先导专项会迎来爱因斯坦探针，这也是我们国家用X射线的波段来探测空间的一些包括沉寂黑洞爆发的现象，还有探测引力波暴的电磁高能对应体。

在2024年年底或者2025年，我们还将发射太阳风跟磁层相互作用的全景成像卫星SMILE，这是由中国与欧空局联合发起的空间科学卫星计划，也是第一次对地球空间大尺度的结构、磁层进行成像的观测。我们第一次对地球空间成像，地球空间就应该SMILE微笑。先导专项一直到2025年，每年都要有一颗卫星发射升空。可以看到，中国的空间科学卫星还处在一个起步阶段，我们也制定了一个中长期的发展目标：希望到2035年的时候，我们中国的空间科学能够跟欧洲空间局比肩；到2045年的时候，我们中国的空间科学能够与NASA比肩。

我们要瞄准一些前沿的科学问题，比如说宇宙的起源和演化，太阳系的起源与演化，以及太阳活动和我们人类活动的关系，瞄准这些重大科学问题开展前沿探索。我们也制定了四大科学主题：包括极端宇宙，就是要研究在极端宇宙条件下的物理规律；要研究时空涟漪，也就是空间引力波；要探测日地全景，太阳对太阳系的影响；还有宜居行星,也就是系外行星的探索。这也是2021年中国航天白皮书制定的四大科学主题。在航天强国建设当中,我们也凝练了五个重点方向、十个重点任务。

最后我也想说,我们空间科学的发展,是一项伟大的事业,而我们伟大的事业是始于梦想,基于创新,成于实干的。谢谢大家。

4. 参考译文

Address at China Aviation & Aerospace Forum①

Professor Wang Chi

Member of the Chinese Academy of Sciences

Director of the National Space Science Center of Chinese Academy of Sciences

Professor of the School of Earth and Planetary Sciences of

The University of Chinese Academy of Sciences

9 November 2022

Distinguished guests,

Ladies and gentlemen,

It's my great pleasure to share the progress and prospects of China's space science satellite program. First of all, I would like to talk about what space science is all about. Since the first satellite was launched into space in 1957, human beings have entered an era of space. Space science is a new interdisciplinary subject that has emerged with the launch of satellites. Therefore, a keyword for space science is spacecraft, which serves as a main platform. A spacecraft can be a satellite, a space station, or a deep space probe. The scope of space science research covers, in order of distance, the Earth, the solar system, and the universe.

Let's start with the Earth. As we know, unlike the outer space, the Earth is not empty, nor is it vacuum. It contains a lot of matter and many magnetic fields and electric fields. Therefore, the Earth's space is rich in radiation belts, as well as various structures between the Sun and the Earth, including coronal mass ejections from solar flares and various large-scale structures of the entire Earth. The nearest star we can observe is the Sun. Our SOHO (Solar and Heliospheric Observatory) satellite had a glimpse of the internal structure of the Sun for the first time from the surface of the Sun. The Parker Solar Probe watched the Sun up close for the first time and discovered many physical phenomena that traditional theories cannot explain. As for the sky and universe far away, humans observe them in different electromagnetic spectra. One of

---

① 编者译。

the most well-known ways is the Hubble Space Telescope. It accurately measured the energy of the universe for the first time and uncovered how dark energy drives the universe to expand continuously.

The recently launched James Webb Space Telescope is a successor to the Hubble Space Telescope. We know that the diameter of the Hubble Space Telescope is only more than 2 meters, while the diameter of the James Webb Space Telescope reaches 6.5 meters. The latest scientific discovery also shows that carbon dioxide has been found in exoplanets. So far, a fundamental question that humans are very interested in is whether the Earth is the only planet with intelligent life in the whole universe. To answer it, we have been looking for Earth 2.0. To find Earth 2.0, we need to start with searching for exoplanets. We have now found thousands of exoplanets, but whether they are hospitable to life like Earth is still a myth.

Next, let's come back to the solar system. Mars is the planet closest to the Earth. It is not yet known whether there is or was life on Mars. But now there is new evidence of life on Mars: Recently, the Perseverance rover discovered a large number of organic molecules on Mars. We have also detected comets and asteroids, which are remnants from the time when the solar system originated. By studying comets and asteroids, we have a glimpse of what the solar system looked like when it was young. We set out from Earth and visited the Moon and Mars. Our next target is the heliopause or the edge of the solar system. Voyager 1 and Voyager 2 have crossed the heliopause and entered interstellar space. Humans have explored the eight planets of the solar system, and now their eyes are on the deep universe. So far, we have launched hundreds of satellites for scientific purpose, but to be honest, China's satellites are still limited in number, and our space science still has a long way to go.

Now, let me introduce our space science pilot projects. In 2011, the Chinese Academy of Sciences took the lead in launching a batch of pilot projects. Among them, space science satellite project is the most intensive one. As you all know, the Dark Matter Particle Explorer (DAMPE) "Wukong" possesses the biggest detection power in the world, which is as high as 10 TEV. It also boasts the highest sensitivity and has captured many unusual physical events in the high-energy range that have not been discovered before. The Micius quantum satellite is our first quantum science experimental satellite. Our "Insight" Hard X-ray Modulation Telescope (Insight-HXMT) has detected the strongest magnetic field in the universe—1 billion Tesla. To put it into perspective, the strongest magnetic field on Earth, even in the laboratory, is only about 100 Tesla.

The first phase of the pilot projects has received extensive attention at home and abroad. We are currently at the second phase of the pilot projects: Taiji-1 microgravity technology demonstration satellite, which is the first step for our country to detect space gravitational waves. We have obtained the most accurate laser ranging result in space so far. Huairou-1 Gravitational wave high energy Electromagnetic Counterpart All-sky Monitor (GECAM) has been successfully launched and has detected many gamma-ray bursts. Not long ago, on October 9, we launched the Advanced Space-based Solar Observatory, which is China's first comprehensive solar observation satellite. For the first time, we set a target of exploring the Sun's magnetic field, flares, and coronal mass ejection events, which are like lightning and downpour on the Sun. We want to understand the relationship between the Sun's magnetic field and solar eruption activities. Everything has been going well, and we expect to publish the first batch of observation results within the next two or three months. Next year, the pilot projects will employ Einstein Probe (EP), an X-ray telescope, to capture some events in space, including dormant black hole eruptions and high-energy electromagnetic counterparts of gravitational wave bursts.

In late 2024 or 2025, we will launch thesolar wind Magnetosphere Ionosphere Link Explorer (SMILE). This is a space science satellite program jointly initiated by China and the European Space Agency. It is the first time to image large-scale structures and magnetospheres of Earth's space with SMILE. The pilot projects will continue until 2025, and a satellite will be launched every year. It can be seen that China's space science satellites are still in their infancy. We have formulated a medium to long-term development goal: we hope that by 2035, China's space science can be comparable to that of European Space Agency; by 2045, China's space science can catch up with NASA.

We need to focus on some state-of-art areas, such as the origin and evolution of the universe and the solar system, and the relationship between solar activity and human activity. We have identified four major scientific subjects: first, study physical laws under the conditions of extreme universe; second, learn about space-time ripples, or space gravitational waves; third, have an panoramic view of the Sun and Earth, and the Sun's impact on the solar system; fourth, search for habitable planets through the exploration of exoplanets. These are also the four major scientific subjects formulated in China's White Paper on Space Program in 2021. To build China into a major power in terms of space science, we have prioritized five goals and ten tasks.

Finally, I'd like to conclude by saying that the development of space science is a great cause, which begins with dreams, hinges on innovation, and can only be achieved through hard work. Thank you all.

5. 要点讲解

本篇演讲稿航空航天科技术语较多，大部分在词汇表里已经列出。

其中，需要注意“人造卫星”的中英文表达差异。“人造卫星”不要翻译成 artificial satellite，因为英文 satellite 本意就是指人造的卫星。中文经常说到“月亮是地球的卫星”，其中的“卫星”准确的英文表达是 natural satellite。

无主语的句子在中文的表达中十分常见，这种结构在这篇演讲稿中也有出现，如“要了解这个根本的问题，就需要寻找地球 2.0”，然而主语在英文的表达中是不可或缺的。译员在翻译这个无主语句子时，运用了增译法，补充了主语 We，使得译文更符合英文的表达习惯。无主语的句子还可以用被动语态或者形式主语来翻译，具体采用哪种方法可以根据上下文而定。

面对演讲稿中个别句子前置定语过长的问题，译员运用了转换法进行处理。如“我们空间科学就是随着人类航天器进入太空之后开展起来的一门新兴的交叉学科”这句话的定语就十分冗长。译员在翻译这句话时，先将前置定语转换为以 that 引导的后置定语，并将该前置定语中的动词“进入”转换为名词 launch，将形容词“新兴的”转换为动词 emerge。这样的处理方式使得译文的可读性更高。

除了转换词性，为了使译文更加通顺，译员还采用了词序调整法，对句子中的信息进行重组。如在翻译“在火星上是不是存在着生物，或者在历史上存在生物，现在还不得而知”这句话，译员将“现在还不得而知”提前，并将其用形式主语 It 引导，这样的词序调整能让读者更快地得出重点，符合英文中将句子重心提前的习惯，使得译文的表达更加地道、简洁。

对于“我们卫星探测的能量精度也是最强的”这句话中的“精度”，乍一想可能会想到 accuracy 这个词。但是接下来的一句话“目前已经探测到了很多以往没有探测到的，在高能段的奇特物理现象”解释了“精度”是指发现了以前没有探测到的现象，而 accuracy 这个词强调准确性，显然不符合此处语境。通过上下文线索，sensitivity 更加切合中文原意。

文中多次出现“探测”一词，但是英文习惯是同一意思可以尽量用不同的词去表达，避免呆板、重复。所以在翻译“探测”这一意思时，除了 detect，还可以根据上下文用 search，find，discover，capture 等词来替换。

遇到原文中伪并列结构的句子，可以先将各分句断句再翻译。如“我们观测的最近的恒星就是太阳，我们的 SOHO 卫星第一次从太阳的表面一直观测到我们太阳的内部的结构”这句话是一个伪并列结构的句子，句中有两个主语，一个是“恒星”，另一个是“SOHO 卫星”。对于这种多主语的句子，将这两个分句分开翻译，以避免译文表述不明或是过于冗长。

二、实战训练篇

(一) 2021 年波音公司首席执行官在虚拟年会上向股东们的致辞

1. 词汇表

Confident Travel Initiative	放心出行倡议
Federal Aviation Administration (FAA)	美国联邦航空管理局
T-7A Red Hawk Advanced Trainer	T-7A 红鹰高级教练机
Collier Trophy for aerospace excellence	航空航天领域科利尔奖
hot fire testing	热火测试
Space Launch System	太空发射系统
CST-100 Starliner	波音星际客机
Frontier Airlines	边疆航空
Royal Australian Air Force	澳大利亚皇家空军

2. 原 文

2021 Address to Shareholders Annual Meeting①

David L. Calhoun

President and CEO

The Boeing Company

20 April 2021

Thank you, Larry, both for the introduction and your board leadership. And let me add my welcome to all of you that are joining us today.

I hope you are all staying safe, healthy as we continue navigating the global pandemic. At this time last year, we were just beginning to truly understand its impacts, both personally and professionally. It's been a year like no other, and I could not be prouder of our team's dedication to supporting our customers, the communities in which we operate, their teammates and their families during these challenging times.

I've been touched by the incredible stories of compassion and care, as our people rushed to rally around those in need. In the very early days of the pandemic, even as we temporarily suspended operations at many of our facilities, our teams mobilized to

---

① David L Calhoun. 2021 Address to Shareholders Annual Meeting, Virtual[EB/OL]. (2021-04-20)[2023-07-17]. https://www.boeing.com/media/speeches/2021-shareholder-address.page.

manufacture 3D-printed face shields and used our own fleet to deliver massive quantities of personal protective equipment to the front lines.

The reality is that nothing could have prepared us for the sheer magnitude of the global crisis. The loss of life around the globe is beyond comprehension and deeply devastating. My heartfelt sympathies go to everyone who has lost a family member, a friend or a colleague to COVID-19.

Alongside the human pain caused by the virus, entire industry sectors suffered collapsing demand—with the global travel ecosystem at the top of the list. 2020 was a devastating year for our airline customers, as well as for airports, hotels, taxis, restaurants, conferences, events and all other experiences that revolve around travel and tourism.

And while it was tough, we saw the industry come together to support one another like never before. And thankfully, we view 2021 as a critical inflection point. While a full recovery is still likely several years away, we are seeing very encouraging signs of a turnaround.

And we are very thankful for the brilliant scientists who have worked tirelessly to create vaccines with remarkable speed. In our first impact assessment going back to May of last year, we never imagined that 50% of the U.S. adult population would be vaccinated a year later. And as governments around the world accelerate vaccine distribution, we begin to take small steps toward returning to daily life as we knew it pre-pandemic.

This past year, many of us have adopted new ways of working together. In fact, we're implementing virtual and hybrid work arrangements for many of our teams that will extend even beyond the safe return from telecommuting for the majority of teammates at our facilities.

But while this new dynamic creates opportunities to change what the future of work looks like, I know nothing can replace the value of the in-person human connection enabled by air travel. When we're in person, we forge stronger, more enduring connections, we spark creativity and innovation, and we strengthen trust. And for this reason, air travel has proven resilient time and time again—and we fully expect demand to return to historic long-term growth trends once this pandemic is behind us.

And until then, our Confident Travel Initiative, in partnership with experts across government, the health care industry and academia, has demonstrated that when flying, passengers are much less likely to contract COVID-19 than in everyday life. Boeing airplanes use high-efficiency filters to remove over 99.9% of viruses and particulates from the air. Similar to those used in hospitals, these filters exchange the volume of cabin air every two to three minutes.

The cabin design and the airflow system, coupled with other safeguards incorporated by our airline customers, like wearing face coverings and implementing more rigorous cleaning procedures, are part of the multilayered safety approach to help protect the health of passengers and crew throughout the air travel journey.

We're also working with governments and industry associations to help ensure that when people decide to travel, they know what to expect. We encourage that any new protocols use a data-driven and risk-based approach to minimize disease transmission risks between countries.

Standardized and secure methods to verify traveler information—whether a negative COVID-19 test result or proof of vaccination—should be a part of any solution to safely expand international travel. We encourage governments and industry stakeholders to agree on verification protocols that will enable safer air travel while maintaining global connectivity and propelling economic recovery.

As we work to navigate through this recovery in domestic and international travel with public health and safety at the forefront, we remain equally focused on our vital defense, space and security mission, helping protect our nation and allies around the world.

In support of meeting industry and customer needs as we move forward, we launched an internal transformation effort to adapt to the market impacts of COVID-19 and position our business to be leaner and sharper and more sustainable for the long term.

As part of this transformation:

- We're assessing all aspects of our infrastructure to make the most efficient use of our factories, our warehouses, our laboratories and our offices while reducing our environmental footprint.
- We're looking critically at our organizational structure—how we operate to reduce creeping bureaucracy and unnecessary layers.
- We're reevaluating our portfolio and investments to ensure that we're prioritizing our spending on core market opportunities, sustainability efforts and key next-generation technologies.
- We're supporting our suppliers who are managing their production stability in response to our own lower demand.
- And we're driving operational excellence into every corner of our company, so that we can create stability, enhance quality, ensure workplace safety, and reduce rework. Our recent 787 delivery pause is a demonstration of this, where we conducted comprehensive production inspections of our 787 airplanes while

focusing on driving stability in our production system so that we're positioned for the market recovery.

We also made difficult decisions to reduce the size of our workforce in line with our current market realities. As we go through that challenging process, we're not losing sight of our future. We've taken great care to ensure we have the team, the resources and the investments necessary to meet our customer commitments, to drive our improvement initiatives and innovate for the future.

As we transform, we're staying grounded, and grounded in and guided by our core values.

We will not forget the lives lost and where we fell short in the tragic Lion Air Flight 610 and Ethiopian Airlines Flight 302 accidents. Their memories drive us every day to be a better Boeing.

Our highest priority is ensuring the safety, the security and the protection of everyone we serve. You have our leadership team's promise that nothing is more important today—and nothing will be more important tomorrow. Delivering on that promise starts with a steadfast commitment to ensuring all of our teammates operate with complete transparency and the highest integrity.

Culture begins at the top, and our global leadership team is shifting more time, more attention to getting as close as possible to our day-to-day work. Together, we're fostering a culture of trust, one that encourages and recognizes transparency, accountability and integrity.

Boeing is now and will forever be an engineering company. We design, we build and sustain the world's most advanced technology, operating from the depths of the ocean to the far reaches of space. And to strengthen this foundational capability, we have brought together our 50,000 engineering teammates into a single, integrated global community to increase innovation, transparency, collaboration and accountability across all engineering designs and decisions.

We also made key leadership appointments. We hired a new vice president of Software Engineering, we named a new chief technology officer, and we appointed a chief aerospace safety officer.

In parallel, we are advancing our enterprise-wide Safety Management System that is grounded in timely data, analysis and insights to enhance how we manage and improve safety every day. It will be fully embedded in every aspect of how we design, how we build and how we support all of our products and our services, and it will help transparently inform our regulators and our customers every step of the way.

And beginning in 2021, product safety, employee safety and quality metrics are incorporated into our primary executive, management and employee annual incentive structures. This update further drives our focus on safety and quality across the enterprise at every level of the organization.

A key milestone on our journey was the safe return to service of the 737 MAX, beginning with the U. S. Federal Aviation Administration, the FAA, approval in November of 2020.

We continue to work with global regulators and customers to return that airplane to service worldwide. And since returning to operations, Boeing has delivered over 85 737 MAX aircraft to customers around the globe. Twenty-one airlines have returned their airplanes to service and have safely flown more than 23,000 revenue flights totaling more than 52,000 flight hours.

As we have discussed and have previously reported, Boeing identified a potential electrical issue for specific components in the 737 MAX flight deck. Boeing is working with the FAA to finalize the required action and to address the issue, and is working closely with both the FAA and our customers on implementation.

We have also resumed delivering 787s, following our comprehensive reviews to ensure each airplane meets Boeing's highest standards. While this work impacted our near-term financials, it was the right thing to do and is another demonstration of our unrelenting focus on quality and the long-term health of our business.

In our defense business, we've made important progress across a number of programs. For example, we delivered the first F-15EX Eagle Ⅱ for the U. S. Air Force ahead of schedule, and we've started production on the Air Force's new T-7A Red Hawk Advanced Trainer. And of course, we were honored to be awarded the 2020 Collier Trophy for aerospace excellence for the X-37B autonomous spaceplane.

We also took the time to ensure the safety and integrity of our key space systems as we advance toward important milestones in our nation's space program. We successfully completed the hot fire testing for NASA's Space Launch System rocket earlier this year, and we are excited for the next CST-100 Starliner launch as we prepare for its first mission to the International Space Station. We are proud of the decision—that was grounded in our safety-first focus—that NASA and our team made to delay our first crewed mission relative to the original schedule.

And in services, we adapted quickly to the impacts of COVID-19, as our balanced portfolio of digital offerings and government services provided key stability through the pandemic. For example, Frontier Airlines signed a 10-year digital services agreement for their fleet, and we secured a six-year support contract with the Royal Australian Air Force for their P-8As. Additionally, as e-commerce drove increasing cargo demand, we worked rapidly with customers to convert freighters to meet their growing cargo fleet needs. These achievements, and many others like them, reinforce the long-term strategy that underpins our services business and our commitment to meeting our customers' evolving life cycle needs.

As we safely deliver for customers today, we're also advancing technology that will define our next chapter. We anticipate that our R&D investments will lead to next-generation aircraft that offer higher performance while being more efficient, easier to maintain, easier to reconfigure.

Airplanes that we are delivering to customers today are already 15% to 25% more fuel efficient than their previous generation of airplanes, which helps our customers' economics while supporting important efforts to combat climate change. In addition, we've committed to deliver commercial airplanes that can fly using 100% sustainable fuels by 2030.

We expect the demand for carbon reduction to escalate considerably over the next decade, and we plan to help accelerate and lead our industry's focus on supporting bold climate goals.

Underscoring the criticality of this work, we appointed a chief sustainability officer in 2020 to help us further advance our approach to sustainability while staying focused on environmental, social and governance priorities.

As part of our commitment to sustainable aviation, Boeing has been exploring, developing and partnering in concepts for advanced aircraft that meet specific energy efficiency, environmental and operational goals for 2030 and beyond.

One such partnership is with Wisk, a joint venture between Boeing and Kitty Hawk. Wisk is the first company in the U. S. to develop and successfully fly an autonomous, all-electric vertical takeoff and landing aircraft and has completed over 1,500 full-scale flights since 2017 with zero incidents. Last year, Wisk and the New Zealand Government announced their partnership to trial Cora, the world's first autonomous air taxi, which will involve Boeing subsidiary Insitu Pacific.

In addition to commercial travel, sustainability and these advanced aircraft, we're making key investments to ensure we're providing our women and men in uniform with the technology advantage to stay one step ahead of evolving threats, achieve their critical security missions and return home safe every time.

We want future generations of Boeing's stakeholders to conclude that at this moment in our history, we listened, we adapted, we stayed focused, we remained tireless in our pursuit of continuous improvement.

We have also taken significant steps toward confronting racism and advancing our racial equity and inclusion efforts across Boeing.

We have expanded and reformulated the senior leadership team — our Executive Council — to ensure we have a far more diverse executive team shaping every aspect of our strategy, our operations and our culture. We've also developed a racial equity and inclusion action plan that raises the bar for progress on key measures of equity, key measures of inclusion for all our people and holds us accountable for clearing that bar.

As we prepare to release our company's first ever Global Equity, Diversity & Inclusion report this year, we recognize we still have lots of work ahead to live up to our aspirations. Our work to confront racism and discrimination across our ecosystem is core to our values and will be an enduring effort.

More broadly, we take pride in enabling travel experiences that create meaningful connections between people within and across countries and cultures. These in-person connections and conversations among colleagues, friends and family remind us that we are less divided than we might imagine.

From air and space travel to global security and global trade, Boeing plays an essential role in our world. Nearly every time a person or a package arrives on your household or company's doorstep, there's a high likelihood Boeing was involved in making that possible. Likewise, the active-duty members of our military in the U. S. and allied nations depend every day on our technology, our products and our services to help protect their nation, as well as their individual security, all while enabling them to achieve their mission on the ground, at sea or in the air, and now in space.

Through Boeing's purpose to protect, connect and explore our world and beyond, we're committed to understanding, meeting and exceeding the expectations of our stakeholders.

We've faced a reckoning over these last few years, but tough times have brought out the very best in our company and our people. And I'm heartened by the tremendous support we've received from our customers, from our suppliers and shareholders during this challenging period.

I want to take this opportunity to recognize Greg Smith. He's been an integral part of Boeing's ability to navigate the unprecedented challenges over the last year or two. Greg has decided to retire in July after an enormous contribution to Boeing over his 30-year career with the company. His stewardship of the company's financial position for nearly a decade and his leadership during the severe challenges our industry faced as a result of our global pandemic have been essential to positioning Boeing for a bright future.

As part of these efforts, he led the largest bond offering in the company's history, and he launched a comprehensive transformation program that will leave our business stronger and more resilient as we move forward. Greg has also driven Boeing to be a better and more competitive company through his oversight of the enterprise operations, sustainability and strategy efforts. He leaves a legacy of leadership and lasting impacts. I'm also grateful for Greg's commitment to support the upcoming transition and for his counsel as we select his successor. Please join me in thanking Greg for his enormous contributions and wishing him all the best in his future endeavors.

I want to conclude by saying thank you to our Boeing team. Your passion, your dedication, your resilience is inspiring—every day, all day. I am proud to be on your team, and because of you, I'm confident in our shared future.

Thank you, and that concludes my report, Larry.

(二) 习近平主席在第二届联合国全球可持续交通大会开幕式上的主旨讲话

1. 词汇表

港珠澳大桥	Hong Kong-Zhuhai-Macao Bridge
北京大兴国际机场	Beijing Daxing International Airport
中欧班列	China-Europe Railway Express
第二届联合国全球可持续交通大会	The Second United Nations Global Sustainable Transport Conference
南北合作	North-South cooperation
南南合作	South-South cooperation
联合国 2030 年可持续发展议程	2030 Agenda for Sustainable Development
中国国际可持续交通创新和知识中心	Global Innovation and Knowledge Center for Sustainable Transport

2. 原　文

与世界相交　与时代相通　在可持续发展道路上阔步前行

——在第二届联合国全球可持续交通大会开幕式上的主旨讲话①

中华人民共和国主席　习近平

2021 年 10 月 14 日

尊敬的古特雷斯秘书长，

各位同事，

女士们，先生们，朋友们：

很高兴出席第二届联合国全球可持续交通大会，同大家共商全球交通和发展大

① 习近平. 与世界相交 与时代相通 在可持续发展道路上阔步前行[EB/OL]. (2021-10-14)[2023-07-17]. https://www.gov.cn/gongbao/content/2021/content_5647344.htm.

计。首先,我谨代表中国政府和中国人民,并以我个人的名义,对会议的召开表示热烈的祝贺,对与会嘉宾表示热烈的欢迎!

交通是经济的脉络和文明的纽带。纵观世界历史,从古丝绸之路的驼铃帆影,到航海时代的劈波斩浪,再到现代交通网络的四通八达,交通推动经济融通、人文交流,使世界成了紧密相连的"地球村"。

当前,百年变局和世纪疫情叠加,给世界经济发展和民生改善带来严重挑战。我们要顺应世界发展大势,推进全球交通合作,书写基础设施联通、贸易投资畅通、文明交融沟通的新篇章。

第一,坚持开放联动,推进互联互通。小河有水大河满,大河无水小河干。各国只有开放包容、互联互通,才能相互助力、互利共赢。我们要推动建设开放型世界经济,不搞歧视性、排他性规则和体系,推动经济全球化朝着更加开放、包容、普惠、平衡、共赢的方向发展。要加强基础设施"硬联通"、制度规则"软联通",促进陆、海、天、网"四位一体"互联互通。

第二,坚持共同发展,促进公平普惠。各国一起发展才是真发展,大家共同富裕才是真富裕。在新冠肺炎疫情冲击下,贫富差距恶化,南北鸿沟扩大。只有解决好发展不平衡问题,才能够为人类共同发展开辟更加广阔的前景。要发挥交通先行作用,加大对贫困地区交通投入,让贫困地区经济民生因路而兴。要加强南北合作、南南合作,为最不发达国家、内陆发展中国家交通基础设施建设提供更多支持,促进共同繁荣。

第三,坚持创新驱动,增强发展动能。当今世界正在经历新一轮科技革命和产业变革,数字经济、人工智能等新技术、新业态已成为实现经济社会发展的强大技术支撑。要大力发展智慧交通和智慧物流,推动大数据、互联网、人工智能、区块链等新技术与交通行业深度融合,使人享其行、物畅其流。

第四,坚持生态优先,实现绿色低碳。建立绿色低碳发展的经济体系,促进经济社会发展全面绿色转型,才是实现可持续发展的长久之策。要加快形成绿色低碳交通运输方式,加强绿色基础设施建设,推广新能源、智能化、数字化、轻量化交通装备,鼓励引导绿色出行,让交通更加环保、出行更加低碳。

第五,坚持多边主义,完善全球治理。当今世界,各国前途命运紧密相连,利益交融前所未有。要践行共商共建共享的全球治理观,集众智、汇众力,动员全球资源,应对全球挑战,促进全球发展。要维护联合国权威和地位,围绕落实联合国2030年可持续发展议程,全面推进减贫、卫生、交通物流、基础设施建设等合作。

不久前,我提出了全球发展倡议,旨在加快落实联合国2030年可持续发展议程,推动实现更加强劲、绿色、健康的全球发展,构建全球发展命运共同体,希望各方积极参与。

女士们、先生们、朋友们!

新中国成立以来,几代人逢山开路、遇水架桥,建成了交通大国,正在加快建设交通强国。我们坚持交通先行,建成了全球最大的高速铁路网、高速公路网、世界级港口群,航空航海通达全球,综合交通网突破600万公里。我们坚持创新引领,高铁、大飞机等装备制造实现重大突破,新能源汽车占全球总量一半以上,港珠澳大桥、北京大兴国际机场等超大型交通工程建成投运,交通成为中国现代化的开路先锋。我们坚持交通天下,已经成为全球海运连接度最高、货物贸易额最大的经济体。新冠肺炎疫情期间,中欧班列、远洋货轮昼夜穿梭,全力保障全球产业链供应链稳定,体现了中国担当。

女士们、先生们、朋友们!

中国将继续高举真正的多边主义旗帜,坚持与世界相交,与时代相通,在实现自身发展的同时,为全球发展作出更大贡献。

我愿重申,中国构建更高水平开放型经济新体制的方向不会变,促进贸易和投资自由化便利化的决心不会变。中国开放的大门只会越开越大,永远不会关上!

中国将继续推进高质量共建"一带一路",加强同各国基础设施互联互通,加快建设绿色丝绸之路和数字丝绸之路。我宣布,中方将建立中国国际可持续交通创新和知识中心,为全球交通发展贡献力量。

女士们、先生们、朋友们!

让我们携手走互联互通、互利共赢的人间正道,共同建设一个持久和平、普遍安全、共同繁荣、开放包容、清洁美丽的世界,推动构建人类命运共同体!

预祝大会圆满成功!

(三)习近平在香港特别行政区政府欢迎晚宴上的致辞

1. 词汇表

高性能计算机	high-performance computer
载人航天	manned spaceflight
探月工程	lunar exploration program
量子通信	quantum communication
北斗导航	BeiDou Navigation Satellite System
载人深潜	manned deep-sea submersible
"沪港通"	Shanghai-Hong Kong Stock Connect
"深港通"	Shenzhen-Hong Kong Stock Connect
"债券通"	Bond Connect
亚投行	Asian Infrastructure Investment Bank

2. 原　文

在香港特别行政区政府欢迎晚宴上的致辞①

中华人民共和国主席　习近平

2017 年 6 月 30 日，香港

行政长官梁振英先生，

候任行政长官林郑月娥女士，

同胞们，朋友们：

大家晚上好！时隔 9 年，重临美丽的香江，同大家共同庆祝香港回归祖国 20 周年，我感到十分高兴。在此，我代表中央政府和全国各族人民，向在座各位并通过你们，向全体香港居民致以诚挚的问候和良好的祝愿！

——岁月不居，时节如流。转眼间，香港特别行政区迎来 20 周年的生日。回想当年，游子回归母亲怀抱的一幕幕感人场景仍历历在目。我们还记得，香港政权交接时，中华人民共和国国歌雄壮奏响、中华人民共和国国旗和香港特别行政区区旗冉冉升起的神圣庄严。我们还记得，中国人民解放军驻港部队进驻时，香港同胞冒着瓢泼大雨热烈欢迎的兴奋激动。我们还记得，在喜迎回归的日子里，神州大地张灯结彩、男女老少载歌载舞、举国同庆的幸福欢乐。这些历史画面已成为全体中国人的集体记忆。

——春耕夏耘，万物生长。20 年间，"一国两制"在香港的实践，就像一棵幼苗，在风雨中茁壮成长，结出了累累硕果。《中华人民共和国宪法》和《中华人民共和国香港特别行政区基本法》确立的特别行政区制度有效运作，民主政治依法推进，政府效能、法治水平等多项指标均比回归前大幅提升；经济平稳增长，竞争力和自由度在全球名列前茅；社会大局保持稳定，各项事业长足发展，人均预期寿命位居世界前列。这些成绩的取得，离不开中央和祖国内地的大力支持，但主要还是香港特别行政区政府和广大香港同胞团结奋斗的结果。在这里，我要为香港同胞点赞！为所有作出贡献的香港社会各界人士点赞！

——承前启后，继往开来。"一国两制"是中国的一个伟大创举。在统一的国家之内，国家主体实行社会主义制度，个别地区依法实行资本主义制度，这在过往的人类政治实践中还从未有过。前人用超凡的勇气探索和突破，后人要以坚定的信念实践和发展。前进道路并不平坦，但我们实行"一国两制"的初心不会改变，决心不会动摇。我们要以"长风破浪会有时，直挂云帆济沧海"的信心，以"千淘万漉虽辛苦，吹尽狂沙始到金"的恒心，推动"一国两制"在香港的实践取得更大成就。

---

① 新华社香港. 习近平在香港特别行政区政府欢迎晚宴上的致辞[EB/OL]. (2017-07-01)[2023-07-17]. http://www.xinhuanet.com/politics/2017-07/01/c_1121244552.htm.

在这里，我想对香港同胞讲“三个相信”。

第一，相信自己。中国人是了不起的。我们有5000多年源远流长的文明历史，是世界古代文明中唯一没有中断而延续至今的。在有史籍记载的多数时间里，中华民族在经济、科学、文化、艺术等诸多领域都走在世界前列，为人类文明进步作出过巨大贡献。尽管中国在近代以后落后了，但新中国成立以来，在中国共产党领导下，经过几代人艰苦卓绝的奋斗，中华民族已经巍然屹立在世界民族之林。香港从一个默默无闻的小渔村发展成为享誉世界的现代化大都市，是一代又一代香港同胞打拼出来的。香港同胞所拥有的爱国爱港、自强不息、拼搏向上、灵活应变的精神，是香港成功的关键所在。我要特别指出的是，香港同胞一直积极参与国家改革开放和现代化建设，作出了重大贡献。对此，中央政府和全国人民从未忘记。香港同胞不仅完全有能力、有智慧把香港管理好、建设好、发展好，而且能够继续在国家发展乃至世界舞台上大显身手。

第二，相信香港。香港发展具有很多有利条件和独特优势。香港经济高度自由开放，人员、货物、资金等要素自由流动，这是吸引国际资本、留住本地资本的重要因素。香港法律、会计、监管等制度同国际接轨，服务业完备，政府廉洁高效，营商环境便利，深得外来投资者信任。香港是重要的国际金融、航运、贸易中心，是连接内地和国际市场的重要中介，是国家“引进来”“走出去”的双向服务平台。迄今，香港仍是内地最大的外来直接投资来源地和境外融资平台，同时也已成为内地最大的境外投资目的地和全球最大的离岸人民币业务中心。更为重要的是，香港享有“一国两制”的制度优势，不仅能够分享内地的广阔市场和发展机遇，而且经常作为国家对外开放“先行先试”的试验场，占得发展先机。“沪港通”“深港通”以及即将开通的“债券通”都在香港试点。香港只要巩固和提升这些优势，就一定能够留住并吸引各方投资和人才，在经济全球化和区域合作中把握机遇，促进本地创新创业，开发新的增长点，续写狮子山下发展新故事、繁荣新传奇！

第三，相信国家。不论是过去、现在还是将来，祖国始终是香港的坚强后盾。经过近40年改革开放，中国实现了从站起来到富起来再到强起来的伟大飞跃。目前，我国是全球第二大经济体，世界第一制造大国和货物贸易大国、第一外汇储备大国，是全球经济增长的最大贡献者。我国科技实力日益强大，高性能计算机、载人航天、探月工程、量子通信、北斗导航、载人深潜等尖端成就相继问世，高铁走向世界，自行研制的大飞机首飞成功。我们的朋友圈越来越大，倡导的亚投行已批准成员达80个，发起的“一带一路”倡议有100多个国家、地区和国际组织积极参与。一个多月前在北京举行的“一带一路”国际合作高峰论坛取得圆满成功，中国倡议引领世界发展，为推动构建人类命运共同体注入强劲动力。祖国日益繁荣昌盛，不仅是香港抵御风浪、战胜挑战的底气所在，也是香港探索发展新路向、寻找发展新动力、开拓发展新空间的机遇所在。国家好，香港会更好！

同胞们、朋友们，

正如香港一首流行歌曲中唱到，“自信好要紧，应该放开胸襟，愿望定会一切都变真”。只要我们相信自己、相信香港、相信国家，坚持全面准确贯彻落实“一国两制”、“港人治港”、高度自治的方针和《中华人民共和国香港特别行政区基本法》，聚精会神搞建设，一心一意谋发展，齐心协力、团结奋斗，就一定能够开创香港更加美好的明天。

我提议：

为国家繁荣富强和中华民族伟大复兴，

为香港长期繁荣稳定，

为在座各位朋友及家人的健康，

干杯！

第八章　现场应急情况处理

口译活动现场是一个动态互动的过程，不可能每个环节都按照事前计划分毫不差，突发情况时有发生。突发情况可能是设备故障、发言者偏离议程、表达混乱，或是临时增加发言，等等。面对意外，译员只有沉着冷静，灵活处理，才能迅速恢复秩序，化解危机，确保口译工作顺利进行。本章将通过实例展示译员在面对突发情况时如何灵活应对、快速判断，以及与各方协作，确保任务圆满完成。

第一节　现场应急情况处理办法

虽然译员在每次口译任务之前都会做充分的准备工作，但是活动现场千变万化，总是会遇到各种各样意料之外的情况。这就需要译员应急处理的经验和技能。下面分享几个语言类和非语言类的应对策略。

1. 语言类的应对策略

在口译现场最常遇到的一种情况是由于发言人声音小或者现场突然出现噪声等其他原因，译员没有听清楚某个词或者某句话。一般情况下(除了很正式严肃的场合)，译员可以向发言人询问，但是一定注意要有针对性地向发言人提问，询问具体某个词或者某句话是什么，而不是只说“I beg your pardon.”或者“Sorry. Would you please say it again?”如果译员没有针对性地提问，发言人只好原原本本再说一遍，导致拖延活动时间，影响沟通效果。

翻译内容涉及各行各业，译员在口译现场难免会遇到一些自己不熟悉的词语或者行业术语，这个时候可以通过不同的方式解围。例如，在技术交流时，如遇双方沟通不畅的情况，有时可以通过在纸上写写画画的方式促进双方的理解。在技术展会上，可能遇到一下子没有反应过来的新技术和新设备的名字，可以先用手指一指设备，让观众有个大概感知，然后通过之后更详细的介绍来说明具体的技术或设备。

有些时候，译员如果不知道某些词语，就可以用解释的方法来表达同样的意思。例如，发言人提到 Trinidad and Tobago，译员如果一时想不起来中文名字，就可以先解释说这是加勒比海的一个岛国。如果译员确实不知道讲话中的关键词语，也不好解释，这时可以求助于现场的专业人员或者电子词典。例如，在一次电力领域的会议上，发言人突然提到了某个核电站使用放射性金属元素 thorium 作为核燃料，如果译员不知道 thorium 的中文，可以小声问下旁边的专业技术人员或者在电脑上、手机的电子词典上查询，但是要注意的是，查询术语的时候，译员还要同时兼顾聆听发言人讲话、短时记忆和笔记，否则很有可能为了一个词，丢掉了下面一整句话。

有时候，因为不了解背景情况，对一些专有名词不太熟悉，这时可以采用音译法或者让发言人解释一下。例如在翻译过程中，发言人提到 Challenge Bibendum，如果译员不知道背景情况，很有可能也不知道 Bibendum 是什么。其实 Bibendum 是米其林轮胎人的名字，这里音译为“必比登”就可以了。

2. 非语言类的应对策略

除了语言类的应对策略，还有一些非语言类的应对策略。因为情况各异，这里仅举几个例子。

发言人讲错话。例如，发言人说：“第一，×××。第二，×××。第四，×××。”这里明显是发言人把第三误说成了第四。如果译员明确知道发言人说错了，可以在传译时及时更正，但是不必当众指出发言人犯的错误。需要注意的是，如果译员无法确定发言人是否真的说错，则不要轻易更正，因为有可能是个人理解偏差。

发言人由于紧张或者兴奋等原因，讲话时滔滔不绝，似乎忘记了翻译这回事儿。这时译员可以用目光、手势甚至语言来提醒发言人暂停一下，让译员把一部分信息先传递过去。

发言人讲完，没等翻译，台下就响起了掌声。在这种情况下，译员可以等掌声停止，现场稍微安静后再开始传译。

口译搭档突然感到身体不适，译员需要为全局考虑，及时顶替搭档位置。如果剩下的工作时间较长，译员可以在茶歇或者不需要翻译的时间寻找其他合适的口译搭档来会场顶替。

同声传译设备出现故障，译员需要协助客户解决问题，很有可能需要临时将翻译形式切换成交替传译或者耳语式传译。

口译现场涉及演讲嘉宾、工作人员、译员、观众、场地、灯光、音响、技术等很多人员和设施，每个环节都有可能出现意外。所以，即使译员在会前做了充足的准备，活动现场可能还是会出现一些意想不到的情况。本章给出了一些具体的案例，但是无法穷尽所有的意外情况。因此，译员需要临危不乱、保持镇定，灵活地根据现场情况去处理问题、化解危机。

第二节 篇章讲解与练习

一、重点解析篇

（一）欧盟驻华使团团长在"2022与世界对话·中国航空航天论坛"上的致辞

1. 词汇表

China Aviation and Aerospace Forum	中国航空航天论坛
AC352 helicopter	AC352 直升机
C919 narrow-body aircraft	C919 窄体大飞机
bilateral aviation safety agreement	双边航空安全协议
EU-China Aviation Partnership Project(APP)	欧盟-中国航空伙伴关系项目
airworthiness	适航性
European Green Deal	欧洲绿色协议
European Space Programme	欧洲空间计划
SMILE solar physics mission	SMILE 太阳物理任务
Einstein Probe X-ray astrophysics mission	爱因斯坦探针 X 射线天体物理任务
Prevention of an Arms Race in Outer Space	《防止外空军备竞赛》
Transparency and Confidence-building Measures	《透明度和建立信任措施》

2. 背景知识

(1) 欧盟-中国航空伙伴关系项目(APP)

2016 年 2 月，中国民用航空局局长冯正霖与欧洲航空安全局(EASA)局长帕特里克·奇在北京共同签署中欧民航合作伙伴项目，标志着新一期中欧民航合作项目正式启动。该项目实施期为 5 年(2015—2020 年)，重点涉及适航、空中交通管理、航空安全、航空安保、通用航空、节能减排、经济管理、立法与执法等领域合作，旨在促进中欧民航技术合作，开展民航政策对话[①]。

① 新浪财经. 中欧启动 1000 万欧元民航合作项目[EB/OL]. (2016-02-24)[2023-08-14]. https://finance.sina.com.cn/stock/usstock/c/2016-02-24/doc-ifxpvysx1623461.shtml.

(2) 欧洲绿色协议(European Green Deal)

欧洲绿色协议是欧盟的长期发展战略，旨在到2050年实现欧洲气候中立，其中核心主要有两个：应对气候变化和可持续发展转型。在具体实施路线的规划中，基本上涵盖了所有的经济层面，包括工业、交通、农业、能源、建筑等，重点聚焦在清洁能源、循环经济、数字科技、生物多样性等方面，推动欧盟加快从传统模式转向可持续发展模式。此外，欧盟也出台了Fit for 55一揽子计划来支持《欧洲绿色协议》落到实处①。

(3) SMILE太阳物理任务(SMILE solar physics mission)

SMILE (Solar wind Magnetosphere Ionosphere Link Explorer)，即太阳风-磁层相互作用全景成像卫星计划，是由中科院空间中心空间天气学国家重点实验室和英国伦敦大学学院共同提出的，SMILE计划将利用创新的X射线和紫外成像仪器，首次对太阳风和地球磁层之间的相互作用进行全球成像，这将对人类进一步了解太阳活动对地球等离子体环境和空间天气的影响具有重要的科学意义和应用价值②。

3. 原　文

Speech at China Aviation & Aerospace Forum③

Jorge Toledo

EU Ambassador to China

Zhuhai, 8 November 2022

Dear Mr. Mayor of Zhuhai,

Dear attendees present with us here today in Beijing,

Dear viewers online and on Phoenix TV,

Good morning.

It is a pleasure to address the China Aviation and Aerospace Forum today, which is taking place in parallel to the Airshow in Zhuhai. It is doubly befitting to speak to you

① 搜狐网.双碳研究|欧洲绿色协议：加快可再生能源部署[EB/OL].(2023-04-02)[2023-08-14]. https://www.sohu.com/a/662245705_488177.

② ESA. ESA gives go-ahead for Smile mission with China[EB/OL].(2019-05-03)[2023-08-14]. https://www.esa.int/Science_Exploration/Space_Science/ESA_gives_go-ahead_for_Smile_mission_with_China.

③ Jorge T. Ambassador Toledo speaks at China Aviation & Aerospace Forum[EB/OL].(2022-11-08)[2023-08-14]. https://www.eeas.europa.eu/delegations/china/ambassador-toledo-speaks-china-aviation-aerospace-forum_en?s=166#:~:text=On%208%20November%2C%20Ambassador%20Toledo%20addressed%20China%20Aviation,safety%2C%20climate%2C%20and%20peaceful%20use%20of%20outer%20space.

as ambassador of the European Union, as aviation and aerospace are by their very nature cross-border endeavours, while the European Union itself has always been a project to transcend borders.

Sino-European cooperation in the sky and in space has long been of practical and symbolic relevance. The first cooperation between China and the European Space Agency, ESA, dates back 43 years to 1979, while cooperation between the EU and China on civil aviation goes back 23 years to 1999.

Chinese aviation has soared to great heights in recent decades, with passenger numbers growing from just 13 million in 1989, to 660 million in 2019, just before the pandemic, making it the largest aviation market in the world at that time.

Throughout the years, the European aircraft manufacturing industry has been, and continues to be, a reliable partner to China. The European Union is China's largest supplier of aviation products, with a well-deserved reputation for the quality and reliability of both its goods and its services.

The Chinese aviation industry too has made great achievements in recent years and I congratulate the recent certification of the AC352 helicopter and the C919 narrow-body aircraft, both of them with the proud participation of European firms in their development.

But whether the final product is Chinese, or European, or from anywhere else in the world, aircraft are a quintessentially global product for a globalized world. Much like the aviation industry's final products, its supply chains cross borders easily. This is why the European Union remains committed to an open world economy, also in the field of civil aviation.

In government-to-government cooperation, there have also been significant milestones between the EU and China in recent years, most notably the signing of a bilateral aviation safety agreement between the two countries in 2019, aiming to boost the two economies' cooperation on aviation safety and environmental compatibility.

For some years now, experts and decision makers from both sides collaborate in the EU-China Aviation Partnership Project or "APP", to cooperate on topics as important as airworthiness, operations and air traffic management. Let me highlight in particular the cooperation on sustainable aviation, perhaps the largest challenge facing the industry in the coming decades. While Europe aims to be a forerunner in the field, in line with its ambitions as part of the European Green Deal, we highly welcome collaboration with China on the topic.

Aviation is key to realizing Europe's ambitions on connectivity, bringing together people, countries and livelihoods worldwide. Having flown to China from Europe twice since becoming ambassador less than three months ago, I have experienced for myself that Chinese aviation is going through a difficult time. While air traffic in Europe is back up to around 87% of its pre-pandemic level, Chinese flights remain very restricted. We look forward to the recovery of predictable air links between Europe and China. Back in 2019, there were over 80 airline routes between China and the European Union.

I move up finally, and literally, from aviation into space. Chinese and European astronomers and poets alike have, for millennia, looked up at the sky with wonder and admiration. It is thereforefitting that European astronaut Samantha Cristoforetti quoted the ancient Chinese calligrapher Wang Xizhi from the International Space Station saying that "Looking up, I see the immensity of the cosmos; bowing my head, I look at the multitude of the world."

Looking from space at the multitude of the world, both China and Europe will see many common challenges that the world faces today. But space also provides ample opportunities to help the world address many of these challenges. This is why the European Space Programme, focuses in particular on the use of space technology to help fight climate change, stimulate technological innovation, and provide socio-economic benefits to citizens. The EU knows that China too shares these goals as it pushes on into space.

China and the EU are already cooperating on space topics, such as frequency interference prevention and relaying of distress calls from emergency beacons. The European Space Agency cooperates with the China National Space Administration on a number of scientific missions, notably the joint SMILE solar physics mission and the Einstein Probe X-ray astrophysics mission.

At the same time, the EU is strongly committed to prevent an arms race in outer space. We therefore welcome that, at the latest General Assembly in September, all Member States of the UN, including of course the EU and China, voted in favour of two Resolutions: one on the *Prevention of an Arms Race in Outer Space, and another on Transparency and Confidence-building Measures*.

An increasingly important topic in the use of outer space is space traffic management, as key orbits become increasingly polluted with space debris. As the representative of one important space player to another, I invite China to share its ideas on space traffic management with the EU, so that we may tackle it together.

We need not look any further than the European and Chinese flags to see that both our people aim for the stars. And while there will be competition between the two sides on the way up, I certainly hope that there will also continue to be fruitful cooperation along the way.

Thank you.

4. 参考译文

在"2022与世界对话·中国航空航天论坛"上的致辞[①]

欧盟驻华使团团长庹尧诲

2022年11月8日,珠海

尊敬的珠海市市长先生,

各位来宾,

凤凰卫视的观众朋友们:

早上好!

很高兴今天在"中国航空航天论坛"上发言,本论坛与珠海航展并行举办。而我认为,我以欧盟驻华大使的身份发言也十分契合主题,因为航空航天就其本质而言是跨国事业,而欧盟本身就是一个跨国组织。

中欧在航空航天领域的合作长期以来都兼具现实意义和象征意义。中国和欧洲航天局(ESA)的首次合作是在43年前,即1979年;而中欧在民用航空领域的首次合作是在23年前,即1999年。

近几十年,中国航空业飞速发展,乘客人数从1989年的1300万,增长到2019年的6.6亿,使得中国成为当时全球最大的航空市场。

多年来,欧洲一直是中国在飞机制造业可靠的合作伙伴。欧盟是中国最大的航空产品供应商,商品和服务质量高、可靠性强,享有盛誉、名不虚传。

中国航空业近年来也取得了巨大的成就。AC352直升机和C919窄体大飞机最近获得认证,我对此表示祝贺。这两种飞机的研发都有欧洲公司的参与,我感到很自豪。

但是,无论最终产品是中国的、是欧洲的,还是来自世界其他地方,飞机都是典型的全球化产品。与航空业的其他成品一样,其供应链很容易跨越国界。这就是为什么欧盟始终致力于建设开放型世界经济,在民用航空领域也是如此。

近年来中欧在政府间合作方面也取得了里程碑式的进展。最值得一提的是,双方在2019年签署了双边航空安全协议,这一协议旨在推动两大经济体在航空安全和环境适应性方面的合作。

近几年,双方的专家和决策者在欧盟-中国航空伙伴关系项目(APP)下,就适航性、运营和空中交通管理等重要议题进行合作。我要特别强调我们在可持续航空领域的合作,可持续航空也许是航空业在未来几十年面临的最大挑战。欧洲的目标是

① 编者译。

成为该领域的先行者,这符合其作为"欧洲绿色协议"(European Green Deal)成员的雄心,我们非常期待和中国在该领域进行合作。

实现欧洲互联互通,航空是关键,它将世界各国、各国人民和人民生计联系在一起。我就任驻华大使还不到三个月,到任以来,我曾两次从欧洲飞往中国,亲身体会到中国航空业,尤其是国际航空,正在经历困境。欧洲的航班已恢复到疫情前水平的87%左右,而中国的航班仍然非常受限。我们期待着中欧之间的航线恢复稳定。2019年,中欧之间的航线数曾达80多条。

说完航空,我们再向上说说航天。几千年来,中国和欧洲的天文学家和诗人都怀着惊叹与敬慕之情仰望天空。正如欧洲宇航员萨曼莎·克里斯托福雷蒂(Samantha Cristoforetti)在国际空间站时引用中国古代书法家王羲之所说"仰观宇宙之大,俯察品类之盛",是何等恰切。

从太空俯察地球,中国和欧洲都看到了当今世界面临的许多共同挑战。同时,太空也提供了大量的机会来帮助我们应对许多挑战。这就是为什么欧洲空间计划(European Space Programme)尤其关注利用空间技术应对气候变化、促进技术创新、为人民带来社会经济福祉。欧盟明白,随着中国向太空推进,中国也拥有同样的目标。

中国和欧盟已经在空间议题上进行了合作,例如预防频率干扰和转发紧急救援无线电信标的求救信号。欧洲航天局与中国国家航天局在一些科学任务上开展合作,特别是SMILE太阳物理任务和爱因斯坦探针X射线天体物理任务。

欧盟坚决致力于防止外空军备竞赛。在9月举行的联合国大会上,所有会员国,包括欧盟成员国和中国,投票赞成两项决议:一项是关于《防止外空军备竞赛》(*Prevention of an Arms Race in Outer Space*),另一项是关于《外空透明与建立信任措施》(*Transparency and Confidence-building Measures*)。我们非常欢迎这一举动。

在利用外太空方面,一个越来越重要的话题是太空交通管理,因为越来越多的太空碎片在污染关键轨道。作为欧盟代表,我邀请中国与欧盟分享其关于太空交通管理的想法,以便共同解决这个问题。

从欧盟和中国的旗帜就能睹见,我们的人民同样仰望星空。虽然双方在前进的道路上会有竞争,但我真诚希望我们继续进行富有成效的合作。

谢谢。

5. 要点讲解

在本篇致辞稿英译汉的过程中,可以采取增译和省译的翻译方法,使得译文更加符合汉语表达习惯。在增译方面,例如:将" It is doubly befitting to… "译为"**我**认为我……"(增加主语);将"… has long been of practical and symbolic relevance"译为"……**兼**具现实意义和象征意义"(增加副词);将"… bringing together people… "译为"……**它**将……联系在一起"(增加主语);将"I certainly hope that… "译为"**但**我真诚希望……"(增加连词)等。而从省译方面看,例如:将" **with** passenger numbers growing… "译为"乘客人数从……增长到……"(省略介词的翻译);将"… in 2019, **just before the pandemic**… "译为"2019 的……"(省略重复信息);将"**At the same time**, the EU is strongly committed to prevent… "译为"欧盟坚决致力于防止……"(省略连接词)等。

从语言形式转换的角度来看,可以通过词性转换让译文更加符合汉语表达思维,例如:将 the recovery of predictable air 译为"航线恢复稳定"(名词译为动词)等。

此外,还可以通过调整语序来使译文更加流畅。例如:将"… in recent decades"译为"近几十年……"(将时间状语提前);将" I congratulate…"译为"……我对此表示祝贺"(将动词短语调整在后);将"In government-to-government cooperation… in recent years. "译为"近年来中欧在政府间合作……"(时间状语提前,并转变主语);将"provide… to citizens"译为"为人民带来……"(间接宾语提前)等。

(二) 中国商用飞机发展三部曲

1. 词汇表

干线飞机	trunk liners
运十飞机项目	Y-10 Aircraft Program
麦道飞机	McDonnell aircraft
AE100	"Asia Express" aircraft
中国商飞公司	Commercial Aircraft Corporation of China, Ltd. (COMAC)
支线飞机	regional aircraft
窄体干线飞机	narrow-body trunk aircraft
宽体飞机	wide-body trunk aircraft
涡扇支线飞机	turbofan regional aircraft
ARJ21 支线飞机	ARJ21 regional aircraft
发动机转子	engine rotor
高平尾飞机失速	high horizontal tail stall
最小离地速度	minimum unstick speed
起落架摆阵	landing gear shimmy
C919 大型客机	C919 large passenger aircraft
低阻流线型	low-resistance streamlined
承载式风挡	load-support windshield
超临界机翼	supercritical wings

续 表

全电传飞控	full fly-by-wire
模块化航电	integrated modular avionics
中国东方航空	China Eastern Airlines
C929 宽体客机	C929 wide-body passenger aircraft
工程定义	model-based definition (MBD)
高模量碳纤维复合材料	high-modulus carbon fiber composite materials
流体力学	fluid mechanics
固体力学	solid mechanics
麦道公司	McDonnell Douglas Corporation
空客公司	Airbus

2. 背景知识

(1) AE100(“Asia Express” aircraft)

AE100 飞机是一款新型的民用飞机,其采用了先进的材料和工艺,具有很好的空气动力性能和飞行稳定性。其最大起飞重量为 45 吨,可搭载 120 名乘客,最大航程可达 5500 千米。AE100 还配备了先进的航空电子系统,包括自动导航、机载通讯和气象雷达等,能够提高飞行安全性和舒适度。同时,该飞机也采用了先进的发动机技术和燃油管理系统,使其在航空燃油效率方面处于领先地位①。

(2) 中国商飞公司(Commercial Aircraft Corporation of China, Ltd.)

该公司于 2008 年 5 月 11 日在上海成立,也是统筹干线飞机和支线飞机发展、实现我国民用飞机产业化的主要载体。

(3) C919 大型客机(C919 large passenger aircraft)

C919 是中国首款完全按照国际通行适航标准自行研制的单通道大型干线客机,具有中国完全的自主知识产权。最大航程超过 5500 千米,性能与国际新一代的主流单通道客机相当。C 是 China 的首字母,也是中国商用飞机有限责任公司英文缩写的首字母,而第一个“9”的寓意是天长地久,“19”代表的则是我国首型大型客机最大载客量为 190 座②。

---

① 沈晓杰.空中快车“AE-100”世人注目你的腾飞——中国跻身世界航空工业的身影[J].江苏航空,1996(4):4-6.

② 中国政府网.从一个机型到一个产业:C919 成为我国航空产业发展“新引擎”[EB/OL].(2022-10-01)[2023-08-14]. https://www.gov.cn/xinwen/2022-10/01/content_5715329.htm.

(4) C929 宽体客机(C929 wide-body passenger aircraft)

C929 是中国商用飞机有限责任公司研制的远程宽体客机,采用双通道客舱布局,航程为 12000 千米,座级 280 座。可以广泛满足全球国家间、区域间航空客运市场需求。①

3. 原 文

中国商用飞机发展三部曲②

飞机设计专家,中国工程院院士　吴光辉

1. 发展大飞机具有重大战略意义

大飞机,是指 150 座级以上、起飞重量 100 t 以上的运输类飞机(俗称干线飞机)。在干线飞机中,150～200 座级的商用飞机具有特殊的重要意义,满足了航空市场对短程航线、低载客量的需求,因此最受航空公司的青睐。这个座级的机型也是干线飞机制造的起点机型和基本机型。

飞机制造业是国家战略性高技术产业,是国民经济发展的重要引擎,对科学技术的发展具有极其重要的推动作用。发展大飞机,具有重大战略意义。

未来 20 年,全球经济将保持年均 2.0%的增长率,航空运输周转量是现在的 2.3 倍,预计全球客机机队规模将达到 5 万架左右,是现有机队的 2.1 倍,中国机队规模将超过 1 万架,国内外商用飞机市场需求强劲。这样一个巨大的市场需求,是大飞机发展的强劲动力,也给大飞机发展带来了巨大机遇。发展大飞机,能更好地满足经济社会发展和人民出行的需要,也必将成为一个新的潜力巨大的经济增长点。

飞机研制和发展具有"高风险、高投入、长周期"的特征,行业门槛极高,目前世界上只有少数国家的几个大型企业可以进入。大飞机是高端装备制造业的代表,发展大飞机对于提高自主创新能力、增强国家核心竞争力,对于转变经济发展方式、推动供给侧结构性改革、建设制造强国具有巨大作用。民用航空产业具有产值高、产业链条长、辐射面宽、联带效应强的特点,发展大飞机带动作用巨大。一方面,可以带动新材料、现代制造、先进动力、电子信息、自动控制、计算机等领域的集群性突破,另一方面可以带动基础科学,如流体力学、固体力学、计算数学、热物理、化学、信息科学、环境科学等诸多基础学科的重大进展;同时,还可以创新航空工业的体制机制,整合利用全球资源,开展国际合作,提高航空工业的制造能力和管理水平。

---

① 中国商飞. C929 飞机[EB/OL]. (2023-05-30)[2023-08-14]. http://www.comac.cc/cpyzr/kuanti/.

② 吴光辉. 中国商用飞机发展三部曲[J]. Engineering, 2021, 7(4): 28-33.

2. 中国人的大飞机梦从未停歇

中国航空工业有60多年的历史，大飞机的制造很早就起步了，但20世纪70年代以后它就停了下来，大型客机项目几上几下，始终没有走完一个完整的型号研制过程，始终没有形成真正的商用飞机产业。

1970年8月，运十飞机项目开始启动研制。经过十年的艰辛探索，运十飞机于1980年首飞。遗憾的是，由于种种原因，运十飞机项目于1985年终止。运十飞机一共飞行了164个飞行小时，120个起落，先后飞抵多个国内主要城市，七次进藏运输物资，在当时中国的经济条件和技术水平下可谓创造出了一个奇迹。1985年开始，我国又与麦道公司合作生产麦道飞机，1996年开始，与空客公司合作研制AE100，但都无果而终。中国人的大飞机梦一次次燃起希望，又一次次陷入失落，但我们始终没有放弃，始终坚守一个信念：一定要有自己的大飞机。2008年，中国商飞公司成立，标志着中国大型客机研制项目正式启动，也表明中国商用飞机产业站在了一个新的历史起点，步入一个新的发展阶段。

3. 中国商用飞机发展三部曲

中国商飞公司通过支线飞机型号研制、窄体干线飞机产业发展、宽体飞机拓展形成全系列产品三部曲，构建完整的研发体系和产品谱系，探索独具特色的商用飞机发展路径。

3.1　ARJ21支线飞机

ARJ21是我国自行研制的具有自主知识产权的新型涡扇支线飞机，载客78～97座，航程2225～3700千米，主要用于满足从中心城市向周边中小城市辐射型航线的使用要求。

通过ARJ21飞机的研制，国内首次系统完整地建立了民机适航设计和验证技术体系，解决了系统间互联安全性评估技术难题，攻克了双发动机失效、轮胎爆破、鸟撞、发动机转子爆破等特殊风险验证的技术难关，突破了结冰、污染跑道、大侧风、高温高寒等极端复杂气象条件下的分析和试验验证技术，掌握了包括高平尾飞机失速、最小离地速度、起落架摆振、飞控故障模拟和功能可靠性等多项验证试飞关键技术。我国首次走完了喷气支线客机设计、制造、试验、试飞、交付、批产、运营等阶段全过程，积累了重大创新工程的项目管理经验，初步探索了一条"自主研制、国际合作、国际标准"的国产商用飞机技术路线，初步建立了新时期我国商用飞机产业体系、技术创新体系和项目管理体系。

ARJ21飞机2002年4月国家批准立项，2008年11月28日在上海成功首飞，2014年12月30日获得中国民航局颁发的型号合格证，2015年11月29日首架机交付首家用户成都航空公司，2016年6月28日首架机首航，2017年7月9日获得生产许可证进入批产阶段。2022年12月18日交付首家海外客户印尼翎亚航空，2023年4月18日完成印尼首航，迎来了批量化生产、规模化运营、系列化发展新阶段。

截至 2023 年 12 月 6 日，中国商飞公司已向已累计向中国国航、中国东航、中国南航、成都航空、天骄航空、江西航空、华夏航空、印尼翎亚航空等客户交付 119 架 ARJ21 飞机(包括 2 架 ARJ21 客改货飞机)。先后开通航线 400 余条，通航城市 140 余座，安全运送旅客超过 1000 万人次。

3.2　C919 大型客机

C919 大型客机是我国自行研制的具有自主知识产权的中短程窄体客机。围绕“更安全、更经济、更舒适、更环保”和“减重、减阻、减排”的设计理念设计。座级为 158～192 座，航程为 4075～5555 千米，可满足航空公司对不同航线的运营需求。

C919 大型客机在工程技术上走出了一条拥有完全自主知识产权的商用飞机研制的正向设计之路。采用异地协同机制，基于模型的工程定义(MBD)，实现了产品设计与制造高度并行和广域协同，实现了无纸数字化制造技术的应用；低阻流线型机头设计、承载式风挡设计、超临界机翼和先进的气动布局；第三代铝锂合金、高模量炭纤维复合材料和钛合金等新材料的大规模应用；全电传飞控和综合模块化航电等系统集成；先进前沿技术推进系统的应用，促进窄体飞机新一轮发展。上述新技术的采用，进一步提升了 C919 大型客机的四性(安全性、经济性、舒适性和环保性)，极大地促进了中国和全球航空工程技术的发展。

C919 大型客机项目 2007 年立项，2008 年开始研制。2015 年 11 月 2 日首架机总装下线。2017 年 5 月 5 日，首架机在上海浦东国际机场成功首飞。2022 年 9 月 29 日，获颁中国民航局型号合格证，11 月 29 日取得生产许可证，12 月 9 日交付首家用户中国东方航空。2023 年 5 月 28 日投入商业运营，正式进入民航市场。C919 大型客机国内外用户达到 30 家，订单总数达 1061 架，东方航空为全球首家用户。目前已交付首家用户东方航空公司 2 架，正在执飞“上海虹桥—成都天府”空中快线。截至 2023 年 11 月 5 日，累计安全运行超 1500 飞行小时，载客超过 5 万人次，运营状况良好。

3.3　C929 宽体客机

C929 宽体客机基本型标准商载航程为 12000 千米，标准三舱 280 座。可以广泛满足全球国际间、区域间航空客运市场需求。目前已确定总体技术方案，正在开展初步设计和机载系统供应商全球招标工作。宽体客机作为现代商用飞机发展的重要组成部分，在商用飞机产业未来发展的技术创新、体系创新及人才队伍建设中有着重要的推动作用。

经过十余年的探索实践，中国商飞公司基本走过了喷气式客机产品研制的全过程，初步掌握了大飞机研制规律、研制方法和研制技术，初步形成了从支线飞机到中短程宽体客机的产品谱系，初步奠定了公司长远发展所需的人才、技术、管理等能力基础，初步带动了相关基础科学、航空工业和相关产业发展，实现了我国商用飞机从无到有的历史性跨越，开启了我国民用飞机产业从弱到强的新征程。

4. 参考译文

A Trio of Commercial Aircraft Developments in China[①]

Wu Guanghui

Academician of the Chinese Academy of Engineering

Chief Designer of the C919 Large Passenger Aircraft

1. The development of large passenger aircraft holds great strategic significance

The term "large passenger aircraft" refers to transport airplanes with more than 150 seats or over 100 t takeoff weight (also known as trunk liners). Among these, commercial airliners with 150—200 seats meet the aviation market's requirements for short-haul routes and relatively low passenger capacity. As a result, large passenger aircraft are prevalent within airlines and are competitive in the aviation market. This kind of aircraft is also the entry level and basic type for the development of trunk liners.

As a national strategic high-tech industry, the aircraft manufacturing industry serves as a powerful engine for national economic growth and plays a vital role in promoting the development of science and technology. The development of large passenger aircraft is of great strategic importance.

In the next 20 years, the global economy will maintain a stable growth of 2.0% on average. The turnover of air transport will be 2.3 times greater than it is now, and the number of passenger aircraft worldwide is expected to double, rising to about 50,000. In China, the number of passenger aircraft will increase to over 10,000. Commercial aircraft are in great demand in both the domestic and international markets, and this powerful drive will bring about extensive opportunities for the development of large passenger aircraft. Moreover, the development of large passenger aircraft can satisfy people's travel needs and contribute to economic and social development. Given this great potential, large passenger aircraft development will surely become a new turning point for economic growth.

Aircraft development and manufacturing is a business characterized by high risk, high investment, and a long cycle, making the entry barrier high. Only a limited number of large enterprises from very few countries around the world can enter into this industry. Large passenger aircraft are a representative of the high-end equipment manufacturing industry. The development of large passenger aircraft in China plays a

---

① 编者译。

significant role in improving independent innovation, enhancing core competitiveness, optimizing the economic growth model, promoting supply-side structural reform, and positioning China as one of the world leaders in the high-end manufacturing industry. The civil aviation industry features high output, long industrial chains, an extensive range of radiation, and a robust leading effect. The development of large passenger aircraft will significantly promote the development of the civil aviation industry. For one thing, aircraft development encourages breakthroughs in sectors such as new materials, modern manufacturing, advanced power systems, electronic information, automation, and computer technology; for another, it leads to significant progress in basic sciences, such as fluid mechanics, solid mechanics, numerical mathematics, thermal physics, chemistry, information science, and environmental science. Furthermore, the development of large passenger aircraft can promote innovation in the system and mechanisms of the aviation industry, integrate and make use of global resources, promote international cooperation, improve the manufacturing capacity and management level of the aviation industry, and also cultivate a team of aviation talent with international first-class level and promote the development of education.

2. Chinese engineers have never stopped pursuing their dream of large passenger aircraft

The manufacturing of large passenger aircraft in China began over 60 years ago. However, this endeavor was suspended in the 1980s. With several large passenger jet programs launched and then canceled, no program survived the entire development cycle, and all ended with no definite progress made.

In August 1970, the Y-10 Aircraft Program began. The Y-10 aircraft took its maiden flight in 1980, after 10 years' endeavor. Unfortunately, the Y-10 Aircraft Program was canceled in 1985 for various reasons. The Y-10 aircraft underwent a total of 164 flight hours and 120 takeoffs and landings, flew to major cities in China, and transported cargo to Tibet, China seven times, thus achieving a miracle under the economic conditions and technological level at that time in China. In 1985, China started a cooperation with the McDonnell Douglas Corporation to manufacture McDonnell aircraft; in 1996, China began to work with Airbus on the development and manufacturing of "Asia Express" (AE100) aircraft. However, all these efforts ended in vain. Again and again, Chinese engineers have ignited their ambition to produce large passenger aircraft, and have failed time and time again. However, Chinese engineers have never given up and have continued to hold fast to the goal of developing our own large passenger aircraft. In 2008, the Commercial Aircraft Corporation of China, Ltd. (COMAC) was established, officially marking the launch of a large passenger aircraft program in China. This initiative indicated that the Chinese commercial aircraft industry stood at a new historical starting point and was entering into a new development phase.

3. A trio of commercial aircraft developments in China

COMAC's development trio includes regional aircraft, narrow-body trunk aircraft, and wide-body trunk aircraft. By building a complete research and development (R&D) system and product portfolio, COMAC is exploring the development of a unique roadmap for commercial aircraft.

3.1 The ARJ21: regional jet for 21st century

The ARJ21 aircraft is a new turbofan regional aircraft independently developed by China, owning independent intellectual property right. It has a layout of 78 to 97 seats and a range of 2,225 to 3,700 km. This type of aircraft is designed to meet the operational needs of spoke routes from central cities to surrounding smaller cities.

With the development and manufacturing of ARJ21 regional aircraft, China can systematically establish its airworthiness design and verification system of civil aircraft for the first time. Furthermore, this advance enables China to solve the technical problems of assessing safety in inter-system connection and to overcome the technical difficulties in verifying unique risks such as dual engine failure, tire blowout, bird strike, and engine rotor bursting. Moreover, China is now able to make breakthroughs in analysis and testing technology under very complicated weather conditions such as freezing, contaminated runways, strong crosswinds, and extreme temperatures, and to master several key flight verification technologies such as high horizontal tail stall, minimumunstick speed, landing gear shimmy, flight control fault simulation, and performance reliability. For the first time, China has journeyed through the entire development cycle of a regional jet, including design, manufacturing, development testing, flight testing, delivery, batch production, and operation. This accomplishment allowed us to accumulate extensive experience in the management of a significant innovation project; to explore for the first time a technology route for domestically made commercial aircraft, in line with the principle of "independent development, international cooperation, and international standard"; and to establish a preliminary commercial aircraft industry system, technological innovation system, and project management system in the new era.

The ARJ21 program obtained official approval in April 2002. It completed its maiden flight in Shanghai on November 28th, 2008. On December 30th, 2014, it received the type certificate issued by the Civil Aviation Administration of China (CAAC). The first ARJ21 was delivered to Chengdu Airlines on November 29th, 2015, and made its first commercial flight on June 28th, 2016. After obtaining the production certificate on July 9th, 2017, the aircraft was mass produced. On December 18th, 2022, the ARJ21 aircraft was delivered to its first overseas customer TransNusa. Its maiden flight in Indonesia took place on April 18th, 2023, marking a new stage of mass production, large-scale operation, and serialized development.

As of December 6, 2023, COMAC has delivered a total of 119 ARJ21 aircraft (including 2 ARJ21 passenger-to-freighter conversions) to various customers, such as Air China, China Eastern Airlines, China Southern Airlines, Chengdu Airlines, Genghis Khan Airlines, Jiangxi Air, China Express, and TransNusa. With over 400 air routes connecting more than 140 cities, ARJ21 has safely transported 10 million passengers to their destinations.

3.2 The C919: large passenger aircraft

The C919 large passenger aircraft is China's self-developed, medium-range, narrow-body jetliner, owning independent intellectual property right. Adhering to the philosophies of "safer, more economic, more comfortable, and more environmentally friendly" and "reduction of weight, resistance, and emission", the C919 large passenger aircraft is designed to have a layout of 158 to 192 seats and a range of 4075 to 5555 km, meeting the operational needs of airlines on different routes.

With the C919 large passenger aircraft, COMAC has embarked on the path of top-down design for commercial aircraft development and manufacturing with independent intellectual property rights in engineering technology. The use of a distributed collaboration mechanism and model-based definition (MBD) makes it possible to conduct product design and manufacturing in a highly synchronized and collaborative way, while employing paperless digital manufacturing technology. The C919 aircraft features a low-resistance streamlined nose, a load-support windshield, supercritical wings, and an advanced aerodynamic configuration. New materials such as third-generation aluminum-lithium alloys, high-modulus carbon fiber composite materials, and titanium alloys are widely used on the C919 aircraft. Full fly-by-wire, integrated modular avionics, and other systems are integrated on the C919 aircraft. The use of a propulsion system developed through state-of-the-art technology has prompted a new round of development of narrow-body aircraft. All the new technologies mentioned above improve the performance of the C919 large passenger aircraft in four aspects: safety, economy, comfort, and environmental friendliness. This achievement dramatically propels the development of aeronautical engineering technology forward, both in China and across the globe.

The C919 large passenger aircraft program was officially approved in 2007, and its development program was launched in 2008. On November 2, 2015, the assembly of the first aircraft was completed. On May 5, 2017, its maiden flight took place at Shanghai Pudong International Airport. On September 29, 2022, C919 received its CAAC type certificate, followed by the issuance of the production certificate on November 29. The first C919 aircraft was delivered to China Eastern Airlines on December 9, 2022. It was then put into commercial passenger service on May 28, 2023, marking its official entry into the civil aviation market. A total of 1061 orders from 30 customers, both at home and abroad, have been received for the C919 large passenger aircraft, with China Eastern Airlines as the launch customer. Two C919 aircraft have been delivered to China Eastern Airlines and are currently deployed on the "Shanghai Hongqiao—Chengdu Tianfu" air express line. As of November 5, 2023, the accumulated safe flight time of C919 has exceeded 1,500 hours and over 50,000 passengers have been safely transported, reflecting a sound operational status.

3.3 The C929: wide-body aircraft

The C929 wide-body passenger aircraft has a standard payload range of 12,000 kilometers and is designed with a standard three-cabin configuration accommodating 280 seats. Its introduction can meet the demands of both international and regional air passenger transport markets worldwide. The overall technical scheme has been finalized, and preliminary design and global tendering for on-board system suppliers are currently underway. Wide-body passenger aircraft are an integral part of modern commercial aircraft development, and play a vital role in promoting technological innovation, systems innovation, and talent training for the future development of the commercial aircraft industry.

After more than 10 years' exploration and practice, COMAC has undergone the entire development and manufacturing process of passenger jetliners. In doing so, it has effectively mastered the law, methods, and technologies of this process; preliminarily established a product family from regional aircraft to short- and medium-range wide-body passenger aircraft; set up an initial management system for talent and technologies for long-term development; and driven the development of relevant basic sciences, the aviation industry, and other related sectors. COMAC has realized a historical leap in commercial aircraft development in China: Starting from almost no basis, it has embarked on a new journey to lead China's civil aircraft industry from a position of weakness to one of strength.

5. 要点讲解

本篇文章是吴光辉院士在期刊上公开发表的文章,有中文和英文两个版本。

从翻译方法来看,本篇译文主要采用了增译和省译两种翻译方法。增译主要表现在主语和关联词的增加上，英文可以适当增加用“of”连接的主语以让表达意义更加明确，通过增加介词“to”以表示目的或原因，也通过增加 Moreover 和 In doing so 等词以增加逻辑关系，让读者更好地理解原文意思。而省译则主要表现在省略重复信息上，比如，在翻译“中国航空工业有 60 多年的历史，大飞机的制造很早就起步了”这句话时，第二个短句的重复信息就被省译掉了。

从语言形式转换的角度来看，本篇译文多为：①转换为被动语态，例如将“大飞机的制造很早就起步了，但 20 世纪 70 年代以后它就停了下来……”译为“... this endeavor was suspended in the 1980s.”；②将形容词译为动词，例如将“重要的推动作用”译为“promoting... ”等。

从长短句的转换角度来看，本篇译文通常采用拆句法和合并法。一方面，当原文一个句子中主语不同而表达信息过多时，英文通常会将其拆分为两个或多个短句并加以不同的主语，以便于读者理解。例如：“飞机研制和发展……目前世界上只有少数几个国家……”这句话中，二者主语不同，英文便用两个主语将句子拆分开来，分别进行翻译。另一方面，在中文里，用逗号相连的几个短句中有表示转折的意思时，英文通常会在转折处将其拆分为独立的两句，并用 However 来相连，以便让目标语受众更好地理解文章内容。例如将“与空客公司合作研制 AE100，但都无果而终”中表转折意义的“但”译为了“... aircraft. However... ”。同时，在长句中的两个短句也可以翻译整合为一组名词短语，例如将“采用……，基于……”翻译为“The use of... and... makes... ”，这样就将原文的内容进行了充分的整合。

从语序调整的角度来看，本篇译文通篇采用顺句驱动法，但在微观层面上，译文也进行了适当调整，例如将汉语名词＋形容词的词组，翻译为英语形容词＋名词词组；再如将位于汉语句子中间的状语译在英语句子开头，这样的表达方法也更符合英语的表达习惯。

二、实战训练篇

（一）“我们一起飞得更高”旋翼机安全会议

1. 词汇表

rotorcraft	旋翼机
FAA (Federal Aviation Administration)	美国联邦航空管理局
Compliance and Airworthiness Division	合规和适航部门
Certification and Flight Standards	认证和飞行标准
SMS (safety management system)	安全管理系统
Part 121 airline operation	航空规章 121 部

续表

Flight Data Monitoring and Safety Reporting	飞行数据监控和安全报告
Just Culture	公正文化
Air Taxis	空中的士
Air Tour Operator	航空旅行社
PMA parts	零部件制造人批准书
FAA's Safety Team (FAAST Team)	美国联邦航空管理局安全小组
Helicopter Safety Advisory Council	直升机安全咨询委员会
U. S. Helicopter Safety Team	美国直升机安全小组
crash resistant fuel system	防撞燃料系统

2. 原　文

"Together, We Go Higher"①

Speech at the Rotorcraft Safety Conference

Stephen M. Dickson

Administrator of the Federal Aviation Administration

27 October 2020

Hello everyone, and thank you for joining us today for the FAA's Rotorcraft Safety Conference. I'd like to thank Lance Gant, Director of the FAA's Compliance and Airworthiness division, and his entire team for continuing this safety conversation in the virtual environment until we can meet again in person.

I'd also like to thank Steve, Lance, and Wayne, for opening the conference with an excellent discussion about the FAA's certification and flight standards priorities for rotorcraft.

Certification and Flight Standards are two of the many components that support the FAA's broad safety mandate, which is to provide the safest, most efficient aerospace system in the world.

Rotorcraft are essential to the efficiency and productivity of that aerospace system.

We saw this in the spring without a doubt when COVID-19 caused a large percentage of our air transportation network to go dormant—but not rotorcraft. Helicopter operations were back to normal levels by mid-May.

Whether for, police, EMS, utilities, corporate shuttles, or literally hundreds of

---

① Stephen M. Dickson. "Together, We Go Higher" [EB/OL]. (2020-10-27) [2023-08-14]. https://www.faa.gov/speeches/together-we-go-higher.

other purposes—rotorcraft are essential. No other flying machine can do the same thing. Everyone here knows that, and you want what's best for the industry—that's why you're here.

My reason for being here, and the reason the FAA is having this conference, is to both recognize this unique industry and to put our heads together—government, industry and academia—to figure out how to move the ball forward on safety.

To realize the full potential of any sector of aviation, safety has to be its top priority. It's no secret that the airline industry is the gold standard when it comes to unprecedented safety levels.

One of the key elements to the success story is the collaboration, partnering, and sharing of information and data between everyone who has a role in the system—the FAA, manufacturers, pilots, mechanics, controllers, flight attendants, and many others. I also include survivors in this list, as they have experienced what can happen when we don't get it right.

We are increasingly using safety management systems, or SMS, to formalize and streamline this flow of information and data within an organization. As you know, SMS is a required element of Part 121 airline operations, and we're progressively deploying the practices throughout the aerospace industry.

Part and parcel of SMS are the practices of Flight Data Monitoring and Safety Reporting. These are proactive, data-driven approaches to oversight that prioritize safety above all else. To be successful, these programs rely on a Just Culture that places great value on front-line employees raising and reporting safety concerns.

With a Just Culture, pilots and aviation workers feel empowered to report honest mistakes and issues without fear of retribution. That atmosphere gives workers the freedom to report and provide their management with data they can use to get a heads-up on what might be an accident in the making.

We are encouraging operators to adopt and use Flight Data Monitoring as feedback into their training programs, and ideally, make it part of an SMS process.

When we integrate safety management principles into the design and manufacturing processes, we will ensure a systems approach to safety by coordinating risk management processes and feedback loops between design, manufacturing, operation, and maintenance.

You can see that we're firm believers in the power of SMS. In fact, right now, the FAA is targeting spring 2022 to publish a proposed SMS rule that will apply to Air Taxis, Air Tour Operators, Repair Stations, and PMA parts providers. We're also working on an SMS for airports.

Of course, you don't have to wait for the rules. By voluntarily implementing an SMS, an operator can identify hazards and head off incidents or accidents by putting safety risk management processes in place. The key is being able to identify and understand the risks in your operation, and that's what an SMS provides.

For the rotorcraft sector specifically, there are a variety of outlets for sharing information to make all of us safer.

We have "Go Local" Workshops, where we take the FAA's Safety Team, or FAAST Team, and industry safety experts directly to local pilots to discuss certain accident scenarios as a starting point to educate pilots on decision-making.

We had to suspend these in-person meetings temporarily due to COVID-19, but the good news is that we're close to launching virtual workshops where participants will vote in real time on how pilots should react to challenges during a precarious helicopter flight.

A great way to share your experiences and learn about the best practices of others is to participate in our newly instituted helicopter InfoShare program. InfoShare, if you're not aware, is a program we started in partnership with the airline industry, but its success is leading other sectors, including business aviation, and now rotorcraft, to adopt the same model.

Another avenue for sharing best practices is through the Helicopter Safety Advisory Council, which has developed recommended practices for oil and gas industry rotary-wing operators that are easily adaptable to other helicopter sectors.

I know it's cliché to say "we need to think outside of the box," but for the rotorcraft sector, that's what I really need all of us to do right now. For 15 years now, the helicopter fatal accident rate has remained roughly the same. As I said earlier, we've got to move the ball toward the safety goal line—zero fatal accidents.

No accident, and more so, no fatal accident, is acceptable. That's why we support the U. S. Helicopter Safety team, which has made zero fatal accidents its primary mission. This government and industry group, that includes some of our FAA safety professionals, is taking a scientific approach, urging adoption of safety proposals based on data.

We're also strongly advocating for operators to make voluntary, safety upgrades where beneficial, including helicopter occupant protection features.

Why is that so important? Because blunt force trauma injuries are linked to more than 90 percent of helicopter fatalities.

For new helicopter designs, certification rules require potentially lifesaving protection through crash resistant seats and surrounding structures. But the thousands of helicopters in our legacy fleet aren't required to have these features. Why not consider retrofitting these upgrades?

Other retrofit safety options we'd like to see, include crash resistant fuel systems. As required in our 2018 Reauthorization, the FAA is requiring new production helicopters built after April 5, 2020, to have these systems out of the box. But we would really like to see these same systems available and operators voluntarily installing them on our legacy helicopter fleet.

I think you can see that we already have many options available to help improve the safety record of the rotorcraft industry, and that we're always looking for new ideas.

That's where you come in. Please use this conference to recalibrate and recommit to helicopter safety, and tell your friends who couldn't join us. Now is the time. We have the critical mass to make real change.

Safety has to be the top priority, our North Star.

But you don't have to take my word for it. Listen in on the next session where you'll hear from three people who will remind us why safety is so important. Like many others who have lost family and loved ones in aviation accidents, Dave & Amanda Repsher and Karen Mahany have become catalysts for change in the industry.

When you hear their stories, they will drive home why all of our efforts to increase safety are truly necessary.

When we make the aviation industry as safe as it possibly can be, we save lives in the process, and we make progress, perhaps slowly but steadily and surely, in our quest to achieve zero fatal accidents.

Thank you, and have an excellent conference.

(二) 在首届北斗规模应用国际峰会上的致辞

1. 词汇表

北斗规模应用国际峰会	International Summit on BDS Applications
北斗卫星导航系统	BeiDou Navigation Satellite System (BDS)
北斗一号、北斗二号、北斗三号“三步走”	a three-step development process of BDS-1, BDS-2 and BDS-3

续表

万物互联	Internet of Everything
国家航天局	National Space Administration
国防科技大学	National University of Defense Technology

2. 原　文

在首届北斗规模应用国际峰会上的致辞(节选)[①]

2021 年 9 月 16 日，长沙

各位领导、各位来宾，女士们、先生们、朋友们：

大家上午好！在习近平总书记考察湖南一周年、北斗三号全球卫星导航系统建成并开通全球服务一周年之际，首届北斗规模应用国际峰会在星城长沙隆重举行。首先，我谨代表湖南省委、省政府，对峰会召开表示热烈祝贺！代表组委会对参加峰会的线上线下各位嘉宾表示诚挚欢迎！对长期以来关心支持湖南发展的各国各界朋友表示衷心感谢！

北斗卫星导航系统是党中央决策实施的国家重大科技工程和重要时空基础设施。以习近平同志为核心的党中央高度重视北斗卫星导航系统的建设和应用，去年 7 月 31 日习近平总书记亲自宣布北斗三号全球卫星导航系统正式开通，标志着我国建成了独立自主、开放兼容的全球卫星导航系统，这是我国攀登科技高峰、迈向航天强国的重要里程碑。这次，习近平总书记又专门为本次峰会发来贺信，高度评价中国卫星导航事业取得的丰硕成果，深刻阐明北斗规模应用对于中国和世界发展的积极影响，明确提出北斗服务全球、造福人类的美好愿景和中国方案，为我们办好本次峰会和推进北斗规模应用提供了根本遵循。我们一定深入学习领会，抓好贯彻落实。

仰望星空，北斗璀璨。从 1994 年正式立项开始，我国北斗卫星导航系统经历了北斗一号、北斗二号、北斗三号“三步走”发展历程，实现了从无到有、从有到优、从区域到全球的历史跨越。北斗三号全球卫星导航系统正式开通以来，运行稳定、服务优质、覆盖广泛，呈现出产业“蒸蒸日上”、地域“百花齐放”、行业“千帆竞发”的蓬勃态势。随着新一轮科技革命和产业变革不断深化，北斗正以前所未有的速度、广度和深度引领我们加快迈向万物互联的智慧时代。作为一名曾在航天系统工作 32 年的老兵，2000 年我有幸参与组织了北斗一号首星发射，先后将北斗一号 A 星、北斗一号 B

① 湖南省人民政府. 在首届北斗规模应用国际峰会上的致辞[EB/OL]. (2021-09-17)[2023-08-14]. https://www.hunan.gov.cn/hnszf/hnyw/sy/hnyw1/202109/t20210917_20595906.html.

星成功送入预定轨道;2013 年我在国家航天局工作时,着力推动空间技术、空间科学、空间应用一体发展,其中空间应用就包括北斗应用推广。这些年不管在哪里,始终难以忘怀的是追逐航天梦想的激情燃烧岁月,始终期盼并致力于推动卫星应用产业做强做大。

"复移小凳扶窗立,教识中天北斗星"。湖南与北斗有着深厚的历史不解之缘,2000 多年前屈原在汨罗江畔发出了"日月安属,列星安陈"的"天问";长沙马王堆出土的西汉《五星占》,记载了世界上最早描绘彗星形态的图集。毛泽东同志在诗词中更是抒发了"可上九天揽月、可下五洋捉鳖"的豪迈情怀和"欲与天公试比高"的雄心壮志。秉承源远流长的"问天"情结,一代代中华儿女投身祖国航天事业,为我国北斗卫星导航系统建设发展作出了积极贡献,其中涌现出一大批湘籍航天领军人物。国防科技大学作为中国卫星导航重大专项导航技术专家组单位,参与了北斗一号、二号、三号系统建设全过程,成为北斗系统建设的重要力量和技术创新引领者。

(三)外交部部长王毅在《全球发展报告》发布会上的致辞

1. 词汇表

中国国际发展知识中心	Center for International Knowledge on Development (CIKD)
《全球发展报告》	*Global Development Report*
2015 年联合国发展峰会	United Nations Sustainable Development Summit in 2015
2030 年可持续发展议程	2030 Agenda for Sustainable Development
可持续发展目标	SDGs
2030 年议程国别方案	a national plan on the implementation of the 2030 Agenda

2. 原 文

在《全球发展报告》发布会上的致辞[①]

外交部部长　王毅

2022 年 6 月 20 日

各位同事,女士们,先生们:

非常高兴出席今天的发布会,祝贺中国国际发展知识中心发布首期《全球发展报告》。

发展是人类社会的永恒主题,是时代进步的重要标尺。习近平主席在 2015 年联

① 外交部.王毅国务委员兼外长在《全球发展报告》发布会上的致辞[EB/OL].(2022-06-20)[2023-08-20].https://www.mfa.gov.cn/web/wjbz_673089/zyjh_673099/202206/t20220620_10706285.shtml.

合国发展峰会上指出，对各国人民而言，发展寄托着生存和希望，象征着尊严和权利。也正是在发展峰会上，各国领导人一致通过2030年可持续发展议程，为全球发展事业制定了行动蓝图，开启了国际发展合作新航程。

七年弹指一挥间，2030年议程已近半。在各方共同努力下，全球发展事业取得积极进展，但也面临百年变局带来的空前挑战。全球疫情、地区冲突、大国博弈、逆全球化等复杂因素相互交织，全球减贫进程严重受挫，超过8亿人仍然生活在极端贫困之中。更令人担忧的是，发展问题在国际议程中日益被边缘化，发达国家援助义务远未落实，全球发展资源缺口巨大，如期实现可持续发展目标不容乐观。

在此历史关头，习近平主席站在全人类福祉的高度，在联合国提出全球发展倡议，以构建全球发展共同体为目标，秉持发展优先、以人民为中心等理念，推动加快落实2030年议程，致力实现更加强劲、绿色、健康的全球发展。倡议一经提出，便获得100多个国家的热烈响应和坚定支持。

全球发展倡议吹响了聚焦发展的"集结号"，推动发展问题回归国际核心议程；倡议铺设了促进发展的"快车道"，为各方对接发展政策和深化务实合作搭建有效平台；倡议提供了落实2030年议程的"加速器"，为汇聚各方资源、破解发展难题、促进协同增效注入强劲动力。本周，习近平主席将主持全球发展高层对话会，和新兴市场和发展中国家领导人一起，共商全球发展大计，推动国际发展合作"再出发"。

女士们，先生们！

中国高度重视实现共同发展。我们率先制定落实2030年议程国别方案，发布多期进展报告，向世界无私分享中国理念、中国方案、中国智慧。今天中国国际发展知识中心发布的《全球发展报告》，梳理了2030年议程落实进展与挑战，旗帜鲜明地指出，面对百年变局和世纪疫情，国际社会应当更加聚焦发展合作和全球伙伴关系建设，迎接数字变革和绿色转型，实现人类共同可持续发展。报告还立足于中国和世界各国积累的有益经验，从八个方面对落实2030年议程提出政策建议。这是中方落实全球发展倡议的一项重要举措，将为各国发展提供有益借鉴，为全球发展事业提供智力支持。

不忘来时路，方可致远途。作为世界上最大的发展中国家，中国始终致力于促进全球发展合作。我们将坚持真正的多边主义，秉持开放包容的伙伴精神，积极分享发展知识和经验，同各方一道，携手落实全球发展倡议，加快推进2030年议程，共建全球发展共同体。

谢谢大家！

实战训练篇参考译文

第一章实战训练篇参考译文

美国国家航空航天局[①]

美国国家航空航天局(NASA)是美国的民用航天部门，也是太空探索的全球领军者。该机构拥有一支多元化的人员队伍，由近18 000名公务员组成。此外，它与更多的美国承包商、学术机构以及国际商业伙伴开展合作，为人类福祉探索、发现并拓展知识。2021财年，NASA的年度预算为232亿美元，占美国联邦总预算不到0.5%，却在全美提供了超过31.2万个工作岗位，创造了超643亿美元的经济总产出。

NASA在全美有20个中心和设施，以及唯一在太空设立的国家实验室。NASA致力于研究地球(包括气候)、太阳、太阳系及更远区域。我们进行研究、测试和开发，以推动航空学发展，包括电力推进和超音速飞行等领域。我们开发并资助航天技术，支撑未来探索，造福地球生命。

NASA还引领一项"月球到火星"探测计划，包括与美国工业界、国际伙伴和学术界合作开发新技术，开展月上科学研究，并计划于不久后实现载人登月。该计划又称阿尔忒弥斯计划，将有助于为人类探索火星做准备。除这些主要计划外，本机构还分享所学，充分利用掌握的信息改善人类生活。例如，公司利用NASA的发现和技术为大众开发新产品。为了确保本机构和美国未来的成功，NASA还支持科学、技术、工程和数学(STEM)方面的教育工作，重点是增加未来劳动力的多样性。

欧洲航天局[②]

欧洲航天局(ESA)是一个政府间组织，创建于1975年，致力于推动欧洲航天能力的发展，并确保太空投资为欧洲和世界公民带来益处。

ESA有22个成员国：奥地利、比利时、捷克、丹麦、爱沙尼亚、芬兰、法国、德国、希腊、匈牙利、爱尔兰、意大利、卢森堡、荷兰、挪威、波兰、葡萄牙、罗马尼亚、西班牙、瑞典、瑞士和英国。斯洛文尼亚是准成员国。

ESA已与六个欧盟成员国建立了正式合作关系。加拿大基于一项合作协议参

---

① 编者译。

② 同①。

与了ESA的某些计划。

ESA协调其成员国的财政和智力资源，可以开展远超任一单一欧洲国家能力范围的计划和活动。

ESA开发所需的发射器、航天器和地面设施，使欧洲处于全球航天活动的最前沿。

现如今，它发射用于对地观测、导航、电信和天文等领域的卫星，向太阳系深处发送探测器，并在人类太空探索方面开展合作。

2023 New Year Address by President Xi Jinping①

Greetings to you all. The year 2023 is approaching. From Beijing, I extend my best New Year wishes to all of you.

In 2022, we successfully convened the 20th National Congress of the Communist Party of China (CPC). An ambitious blueprint has been drawn for building a modern socialist country in all respects and advancing the great rejuvenation of the Chinese nation on all fronts through a Chinese path to modernization, sounding a clarion call of the times for us forging ahead on a new journey.

The Chinese economy has remained the second largest in the world and enjoyed sound development. GDP for the whole year is expected to exceed 120 trillion yuan. Despite a global food crisis, we have secured a bumper harvest for the 19th year in a row, putting us in a stronger position to ensure the food supply of the Chinese people. We have consolidated our gains in poverty elimination and advanced rural revitalization across the board. We have introduced tax and fee cuts and other measures to ease the burden on businesses, and made active efforts to solve the most pressing difficulties of high concern to the people.

Since COVID-19 struck, we have put the people first and put life first all along. Following a science-based and targeted approach, we have adapted our COVID response in light of the evolving situation to protect the life and health of the people to the greatest extent possible. Officials and the general public, particularly medical professionals and community workers, have bravely stuck to their posts through it all. With extraordinary efforts, we have prevailed over unprecedented difficulties and challenges, and it has not been an easy journey for anyone. We have now entered a new phase of COVID response where tough challenges remain. Everyone is holding on with

① Ministry of Foreign Affairs of the People's Republic of China. Full Text: 2023 New Year Address by President Xi Jinping [EB/OL]. (2022-12-31) [2023-08-20]. https://www.fmprc.gov.cn/mfa_eng/zxxx_662805/202212/t20221231_10999475.html

great fortitude, and the light of hope is right in front of us. Let's make an extra effort to pull through, as perseverance and solidarity mean victory.

Comrade Jiang Zemin passed away in 2022. We pay high tribute to his towering achievements and noble demeanor, and cherish the great legacy he left behind. We will honor his last wishes and advance the cause of socialism with Chinese characteristics in the new era.

Wave upon wave, the mighty river of history surges forward. With the persistent efforts of one generation after another, we have taken China to where it is today.

Today's China is a country where dreams become reality. The Beijing Olympic and Paralympic Winter Games concluded with a resounding success. Chinese winter sports athletes gave their all and achieved extraordinary results. Shenzhou-13, Shenzhou-14 and Shenzhou-15 soared into the heavens. China's space station was fully completed and our "home in space" is roving in the deep-blue sky. The people's armed forces marked the 95th birthday and all service members are marching confidently on the great journey of building a strong military. China's third aircraft carrier Fujian was launched. C919, China's first large passenger aircraft, was delivered. And the Baihetan hydropower station went into full operation… None of these achievements would have been possible without the sweat and toil of the numerous Chinese people. Sparks of talent are coming together, and they are the strength of China!

Today's China is a country brimming with vigor and vitality. Various pilot free trade zones and the Hainan Free Trade Port are booming, innovations are gushing out in the coastal areas, development is picking up pace in the central and western regions, the momentum for revitalization is building in the northeast, and there is greater development and affluence in the border regions. The Chinese economy enjoys strong resilience, tremendous potential and great vitality. The fundamentals sustaining its long-term growth have remained strong. As long as we stay confident and strive for progress while maintaining stability, we will realize the goals we have set. On my visit to Hong Kong earlier this year, I was deeply glad to see that Hong Kong has restored order and is set to thrive again. With determined implementation of One Country, Two Systems, Hong Kong and Macao will surely enjoy long-term prosperity and stability.

Today's China is a country that keeps to its national character. In the course of 2022, we encountered various natural disasters including earthquakes, floods, droughts and wildfires, and experienced some workplace accidents. Amid those disconcerting and heartbreaking scenes, there have emerged numerous touching stories of people sticking together in face of adversity or even sacrificing their lives to help others in distress. Those heroic deeds will be forever etched in our memories. At every turn of the year, we always think of the great character of resilience that the Chinese nation has carried forward through millennia. It gives us still greater confidence as we continue our way forward.

Today's China is a country closely linked with the world. Over the past year, I have hosted quite a few friends, both old and new, in Beijing; I have also traveled abroad to communicate China's propositions to the world. Changes unseen in a century are unfolding at a faster pace, and the world is not yet a tranquil place. We cherish peace and development and value friends and partners as we have always done. We stand firm on the right side of history and on the side of human civilization and progress. We work hard to contribute China's wisdom and solutions to the cause of peace and development for all humanity.

After the 20th CPC National Congress, my colleagues and I visited Yan'an. We were there to relive the inspiring episode in which the Party's central leadership overcame extraordinary difficulties in the 1930s and 1940s, and to draw on the spiritual strength of the older generation of CPC members. I often say, "Just as polishing makes jade finer, adversity makes one stronger." Over the past 100 years, the CPC has braved wind and rain, and forged ahead against all odds. That is a most difficult yet great journey. Today, we must press on courageously to make tomorrow's China a better place.

Going forward, China will be a country that performs miracles through hard work. Here I want to quote Su Shi, a renowned Chinese poet, "Charge at the toughest and aim at the farthest." It means to take on the biggest challenges and go after the most ambitious goals. Long as the journey is, we will reach our destination if we stay the course; difficult as the task is, we will get the job done if we keep working at it. As long as we have the resolve to move mountains and the perseverance to plod on, as long as we keep our feet on the ground and forge ahead with our journey by making steady progress, we will turn our grand goals into reality.

Going forward, China will be a country that draws its strength from unity. Ours is a big country. It is only natural for different people to have different concerns or hold different views on the same issue. What matters is that we build consensus through communication and consultation. When the 1.4 billion Chinese work with one heart and one mind, and stand in unity with a strong will, no task will be impossible and no difficulty insurmountable. The people on both sides of the Taiwan Strait are members of one and the same family. I sincerely hope that our compatriots on both sides of the Strait will work together with a unity of purpose to jointly foster lasting prosperity of the Chinese nation.

Going forward, China will be a country that has great expectations of its younger generation. A nation will prosper only when its young people thrive. For China to develop further, our young people must step forward and take on their responsibilities. Youth is full of vigor and is a source of hope. Youngsters should keep their country in mind, cultivate keen enterprise, and live youth to the fullest with great drive, to prove worthy of the times and the splendor of youth.

To the many people who are still busy working at this very moment, I salute you all! We are about to ring in the New Year. Let us welcome the first ray of sunshine of 2023 with the best wishes for a brighter future.

May our country enjoy prosperity and our people live in harmony. May the world enjoy peace and people of all countries live in happiness. I wish you all a happy New Year and may all your wishes come true.

Thank you.

第二章实战训练篇参考译文

中国民航发展论坛①

美国联邦航空局前代理副局长

卡尔·E·伯利森

2019 年 5 月 1 日

感谢中国民用航空局的盛情邀请,我很荣幸能有机会参加此次盛会,代表美国联邦航空管理局,即 FAA,来到北京。

中美两国乃至所有国家都处在创新、自信而崭新的时代,这需要我们之间进行更密切的合作,需要我们用革命性的技术打破过时的概念。我们同中国民航局在这条路上并肩作战,我们拥有相似的安全理念,我们的双边合作也将充实我们的安全信息。

首先,让我们回顾一下 FAA 和中国民航局自 1979 年中美建交以来,在过去 40 多年取得的合作成果。在双方共同努力下,中美商业航空服务于 1981 年 1 月重启。首次航班是从北京飞往旧金山的中国民航 981 次航班,该航班的飞机型号为波音 747SP。

大家想一想,1981 年 981 航班那时,平均每天只有 2 个来往于中国大陆和美国的航班。如今,根据国际航空运输协会的最新数据,中美日均航班为 100 个。

除了航班活动飞速发展以外,我们两国在几乎所有的航空领域也取得了巨大成就,其中包括运营、飞机认证、空中交通管理,以及最重要一点——航空安全。我们感

---

① 编者译。

谢中国民航过去40年来与我们的合作,我们期待下一个40年的到来。有一点是肯定的,那就是40年后的航空业和如今相比将会有天壤之别,但我们关于安全的基本原则不会改变。

那么我们以及全世界的航空监管部门应该如何将这群千差万别的新用户引进这个世界上最安全的交通模式中?我们应该如何明智、安全、高效,且在合理的时间内完成这一任务?

思考这些挑战时,我不禁想到了很久以前我和儿子们在一个暴风雨夜里的对话。

那是八月一个湿热的夜晚,华盛顿暴风雨肆虐,家里断电了。我便领着我儿子们出来,走进装着纱窗的门廊里,让他们看倾盆大雨、狂风和闪电汇成的表演。我的大儿子十二岁,胆子很大,他恳求我:"爸爸,我能出去到雨里玩吗?"我的小儿子则安静地在我大腿上坐了很久。最后我问他对于这场暴风雨有什么看法,他说:"爸爸,一个七岁的小孩会不会因为暴雨死掉呢?"这就是现实——两个有着相同遗传基因的男孩子,在观察一个相同的事件时却有着截然相反的想法。

他们对世界的不同看法像极了我们航空航天事业中面临的挑战。新技术的快速发展让更多的自动操作成为可能。一些人看到了发展、新市场和新服务带来的非凡机会。其他人则认为这有可能损害安全性能,提高系统的风险等级。作为一个监管部门,我们的工作是找到前进的道路。

如今,其中一个能让我们快速找到前进道路的领域便是无人机系统,以及城市空中交通工具,又名空中的士。

截至上个月,FAA已经注册了近140万个无人机系统,其中近40万个用于商业运营。无人机注册数量是载人飞行器的四倍,而我们注册无人机的时间还不到四年。如今,我们已经发放了约13万份遥控驾驶员执照。其中许多是小型、半自动无人机系统,使用智能手机在指尖便可操作。这类广受青睐的无人机来自中国的大疆公司,该公司目前是全世界最大的民用无人机制造商。

空中的士如今已不再是天方夜谭,这个概念很快就将成为现实。到那时,在座的各位都将不禁思考,是否该搭乘一辆空中的士,远离车水马龙的束缚。

亿航公司的总部设在中国,它是新兴城市空中交通市场的领导者,也是第一家使用全自动、双座、电动空中的士在试飞中载客的企业。虽然还没有得到中国民航局的认证,但亿航正在为中国、美国、加拿大和其他国家的各类飞行表演活动提供飞行器。其他新生空中交通公司,如Uber Elevate和Kitty Hawk等,也正在研发类似产品,参与市场竞争。

亿航公司的运营理念如同从科幻小说中横空出世:一名乘客搭乘全自动飞行器,然后在触摸显示屏上的列表中选择目的地,点击一个按钮起飞,之后再点击另一个按钮降落。空中的士在航线上全自动飞行,但地面指挥控制中心会对其进行全程监控。而以上所有环节都在试飞中演示了。

不难看出监管部门会有哪些担心的事项:一是认证无人驾驶载客飞机的安全性;

二是找到将空中的士运营整合进传统空中交通中的方法；三是监管众多的空中的士。

让我们共同面对吧：同我们几十年来对载人飞行器的成功管理相比，这些挑战不可谓不艰难。我们的挑战不仅在于无人机的使用，更在于产品周期一般不是用年，而是用月来衡量的。

因此，我想谈一谈我们应该如何制定发展路线，整合这些新产业。我们应当搜集数据、创建框架、建立基础设施，以确保各环节的安全。

我们的风险管理策略、法规的制定、政策的推行——以及最重要的——我们在传统航空领域达到安全水平的方式，都需要先获得数据，然后正确理解和管理风险。这就是为什么甘达庆（Derek Kan）先生今天上午谈到的无人机系统融合试点计划，即IPP，是如此重要。

IPP使我们能够与各种用户和政府组织合作，安全地测试和验证无人机的先进操作。这将帮助我们获得数据，通过实际应用来制定无人机系统的法规、政策和指导方针。而且，正如甘先生所指出的，在IPP支持下，我们向Alphabet旗下的Wing Aviation公司颁发了第一个航空承运人证书，允许其在弗吉尼亚州布莱克斯堡（Blacksburg，Virginia）农村地区运送包裹。我们预计其他几家公司在未来几个月内也将获得这一批准证书。

FAA专注于不断扩大无人机系统操作范围。为了使这些操作能够安全地进行，我们已经推进了一些关于小型无人机系统的规则制定，包括外部标记，在人群上空和夜间的飞行，以及安全和可靠的操作概念，并将很快发布一个关于远程ID的规则。这些规则将共同构成我们推进这个新行业发展的框架。

我们不仅仅需要监管框架，还需要一个能够管理载人和无人驾驶飞机的空中交通管理系统，其中多数无人机将是高度自主的。正如甘达庆先生今天上午所指出的，我们正在与NASA和工业界密切合作，开发我们需要的概念和工具来完成这项任务。从这些在高密度城区日益复杂的试飞中所学到的东西将有助于我们向这些服务提供者的发展提出要求。

我们能以多快的速度将新的创新和技术引入服务？在商业航空领域历史上，没有任何一项新技术在我们有信心将其安全地整合到我们的空域之前就已经推出了。虽然我一直在谈论未来，但我们不能忘记需要应对当前的挑战。

去年，在波音737 MAX飞机上发生了两起悲剧，我想代表FAA向这两起事故遇难者的朋友和家属表达衷心的慰问。航空打破了边界，在世界各地带来了机会，但这也意味着我们都要分担彼此损失的悲伤。

我们会继续审查正在进行的事故调查中涉及的一切证据。在成为国际航空业领导者的路上，我们的行动将以数据为基础，并不断解决航空风险。我们永远都会勇敢承认自身的缺陷，这也是我们坚持不懈地倡导保持行业透明、追求与日俱进的原因。事实上，这也是我们为什么要在本月邀请世界各地的领导来到达拉斯，并向他们全面介绍事故后所采取的措施。我想在此说明，FAA复飞波音737 MAX的时间线十分明确，当我们确信导致事故的问题已经得到解决，飞机可以安全运行时，我们便会批准复飞。

展望未来，展望即将进入航空业的新用户和新技术，我们可以肯定的是，任何一项技术或创新会带来帮助还是伤害，人们必然会持有不同的意见，就像我的两个儿子对那晚暴风雨怀有的感受一样。

创新是不会慢下来的。这对FAA和世界各地的其他民航监管机构来说是一个挑战。FAA致力于找到一条既能维护民众安全又能让继续开拓创新、谋求机遇的道路，我们将与中国民航局等国际伙伴合作，以共同应对这些变幻莫测的挑战，并从中互相学习。

谢谢大家！

International Science and Technology Cooperation Initiative①

Science, Technology and Innovation (STI) is an important engine for the development of human society and an important means to address global challenges. At present, the world is undergoing profound changes unseen in a century, and mankind is facing more and more major challenges. More than ever, human society needs international cooperation, openness and sharing, and it needs STI cooperation to explore solutions to global issues, address the challenges of the times together, and to jointly promote peaceful development. To promote open, fair, equitable and non-discriminatory international science and technology cooperation, develop science for the benefit of all regardless of borders, and jointly build a global science and technology community, China puts forward the following International Science and Technology Cooperation Initiative:

—Staying committed to the pursuit of science. We should approach science and pursue truth in the most appropriate way as it should be. We should uphold scientific integrity, respect research ethics, foster a vision of tech for good, and improve globalscience and technology governance. We should strengthen protection of intellectual property rights, and develop and govern emerging technologies in a more inclusive and prudent manner.

—Pursuing innovation-driven development. We should enhance global STI collaboration, build a global innovation network, promote the adoption and diffusion of emerging technologies, and bolster technological innovation cooperation among enterprises to inject new impetus into world economic recovery and growth. Countries should work together to promote connectivity in the digital age, accelerate global

① 编者译。

transition toward green and low-carbon development, and achieve sustainable development for all.

—Adhering to open cooperation. We should be committed to open science regardless of borders and with no barriers imposed, ensure the free flow of STI personnel and resources around the world, increase personnel exchanges and cooperation, and foster anopen and free international ecosystem for science and technology cooperation. We will resolutely oppose restrictions or obstacles to science and technology cooperation which undermine the common interests of the international community.

—Adhering to equality and inclusiveness. We should uphold the principles of mutual respect, fairness, equality and non-discrimination, and encourage all countries and research entities to participate in international science and technology cooperationon an equal footing. We firmly oppose politicizing, instrumentalizing or weaponizing science and technology cooperation, and oppose hegemonic and bullying acts through science and technology under the pretext of national security.

—Upholding solidarity and coordination. In the face of urgent global challenges in such fields as climate change, health, environmental protection, energy security and food security, countries should work together to boost collaboration and mutual learning among STI entities, advance the implementation of major international science programs and projects, and achieve breakthroughs in major scientific a technological challenges concerning the future of mankind.

—Pursuing win-win outcomes for all. We should uphold genuine multilateralism, explore a new model of win-win global cooperation on STI, and ensure that the STI achievements are shared by all. China will set up a global scientific research fund and increase scientific and technological assistance to other developing countries, so that scientific and technological progress will benefit mankind as a whole.

Speech on the Achievements of Applications of the Manned Spaceflight Project①

Chen Dong

Astronaut of the Chinese People's Liberation Army Astronaut Corps

26 September 2018

Tiangong-2 was launched about two years ago. The 33-day space stay has been the longest one among all the Shenzhou spacecraft missions. Our motherland has invested a

① 编者译。

lot of human, material and financial resources to send us into space. In order to make good use of the precious opportunity, as astronauts, we raced against time and made every effort to complete 38 experiments with high standards. This also fully reflected the status, role and value of humans in space activities.

Each of these experiments was instructive and practical. In the early stages, we made lots of efforts on conception, scheme design and modification, and equipment processing. We completed these experiments in weightless environment. After the flight, experts spent a lot of time analyzing and processing these experimental data. That was a long process before valuable results were finally obtained. Here, I wish to express my gratitude to the teams and experts of each experiment for their hard work.

Now I will share with you my thoughts from the following three aspects:

1) **Support astronauts.** We have only completed six manned flights so far, not many in number and not long in time duration. More data and experience await to be accumulated to enable astronauts to adapt to the weightless environment and complete their work efficiently. Among the 38 experiments, eight are related to human health, which, I believe, are very important and necessary. For example, astronauts' vision will be affected as time goes by. Besides, the noise in the experimental cabin will impact astronauts' hearing, sleeping and mood, which will ultimately affect work. These experiments focus on changes of body functions in weightless environment, so as to come up with preventive and protective measures. Humans are the mainstay of manned spaceflight. In the future, it is also humans that will leave Earth to find new home planets. Therefore, we need to continue relevant experiments on human health.

2) **Make the most of astronauts.** Ground scientists have designed sound experiments, and we astronauts are like their eyes and arms in space. We need to make good use of the time in space. For example, this time we demonstrated to President Xi Jinping human-machine coordination in-orbit operation and maintenance technology of the robotic arm, which reflected the coordination of humans and machines. We also conducted plant cultivation verification experiments, such as sowing, watering, fertilizing, and thinning seedlings, recording the growth of plants every day until finally bringing back nine heads of lettuce. We can proudly say that we helped realize the space dreams of ground technology personnel through our own hands. Aside from conducting scientific experiments and technical verification, we also attached great importance to science education. Many science popularization activities were also completed during this mission. For example, the experiment of silkworm breeding in space designed by Hong Kong children aroused strong interest among primary and secondary school students. Through the astronauts' demonstrations, more people can

learn about space and even participate in space activities in the future.

3) **Trust astronauts.** Someone believes that robots can be our eyes and hands and wonders whether robots can replace us in the future. I want to say that no one can underestimate human potential. For subsequent space station missions, the system will become larger and it is inevitable that there will be malfunctions that require repairs. During the Shenzhou-11 mission, we conducted in-orbit repair experiments. I believe that humans have more advantages than robots in adapting to changing situations and turning danger into safety. I also hope scientists and engineers have faith in astronauts. We have done our best in the experiments to obtain perfect data.

Some experiments are challenging for us, for they require strong professional knowledge, but we can make suggestions on improving procedures and tools based on our previous weightless experience. In fact, by doing so, we also engage in the experiment design. Meanwhile, we also strengthen learning and constantly improve our ability, striving to become research astronauts. Later, we will have the third batch of astronauts. They all have strong professional backgrounds. The future space station mission will last longer with more experiments. We also hope that we can independently design experiments and complete them in space by ourselves in the future. That must be marvelous.

I believe that with the joint efforts of astronauts in space and ground scientists, we will be able to complete experiments and make good use of the results to serve the public, so that space technology can play a greater role in creating a better life.

第三章实战训练篇参考译文

在国际民航组织创新博览会上的闭幕词[①]

国际民航组织秘书长

胡安·卡洛斯·萨拉萨尔先生

2022年9月25日,蒙特利尔

75年来,在国际民航组织践行使命的过程中,创新发挥了重要作用。监测并把握技术和工艺进步所带来的影响是建立国际标准的重要前提。

我们行业采用并整合了最新的技术和解决方案,增强了我们全球网络的能力,使其能更安全、更可靠、更高效和更可持续地运行,并为世界各地带来更多的社会经济利益。

这些益处反过来又增强了各国投资和实现可持续发展目标的能力。目前,在联

① 编者译。

合国 2030 年可持续发展议程下,全球都在努力实现这一目标。

我们生活在一个非凡的时代,创新渗透至研发的每一环节。

生活在这样的时代,要记住我们行业可以从清洁能源推进、电池科学或数字革命(如区块链和机器学习)等新发现中受益,就像可以从新机身结构、航空学新研究或旅客服务技术等航空创新中直接受益一样。

事实证明,21 世纪为科学发现提供了有利的多样化环境,为非传统的飞机或发动机制造国提供了极好的机会。

但是,为了充分挖掘新兴产业和初创企业的潜力,政府和业界需要密切合作,创造可靠和优越的投资环境,并配合实施有效的国家孵化器计划。

这一创新民主化的过程是航空业乃至所有行业和部门所经历的发展方式转变,它强调了为什么需要努力发展动态的全球分析和认证过程。

今年的创新博览会强调了在将航空创新从概念变为现实的过程中,创新者和监管者所面对的一系列挑战。

许多制约因素在不同地区都十分常见,专家讨论中也提到这一点。

这突出表明,监管机构需要采取新的、创造性的方式,与创新者有效互动,并以新的方式与更广泛的利益相关群体沟通。

例如,在城市空中交通领域,我们听到许多见解,探讨更具包容性的治理体系如何提高社会接受度,推动部署。

我们还了解到目前正在进行的重大转型,许多民航局已经在调整工作方式以及与创新者的沟通方式。

一些政府正在建立公私委员会,促进相互理解,其中包括帮助初创公司熟悉政府对其的监管范围。

许多孵化器和加速器计划也越发重视提高监管意识,将其视为提升创新生态体系活力的一个重要方面。

作为秘书长,我也在努力改革 ICAO,使其更加迅速灵活地应对新情况,并推动创新,快速谨慎地造福我们的行业及其所服务的国家。

在确保国际航空安全、安保和可持续性方面没有捷径可走,但这并不意味着我们不能坚持更高效、负责任地评价和规范新技术和规程。

同样重要的是要认识到,这些创新只有在应用它们的监管者面前才是有效的,因此,ICAO 将在这一领域更加积极主动地支持各国。

我们在创新战略里明确了目标,并将在此基础上给予创新者更多支持,并寻求新的方法,为各国创造最大的附加值。

我们还将高度关注并确保所有国家都能平等、迅速地获得最新的解决方案,这既是为了全球协调的目的,也是为了确保当我们的航空网络变得更环保、更高效和更有韧性时,没有一个国家被落下。

此次创新博览会使我们认识到,ICAO 作为航空运输领域公私合作的纽带所发挥的作用。这也使创新博览会成为开展全球对话、发展伙伴关系和采取创新行动的天然平台。

作为联合国系统的一个机构，我们有责任展望航空创新的愿景，充分包容相关的社会优先事项，包括性别平等、社会经济发展机会，可持续繁荣和复原力等方面。

近年来，鉴于航空运输在国际人员和货物流动中发挥的基础作用，我们还与联合国打击恐怖主义和犯罪流动的机构进行了非常密切的协调，并作为重要合作伙伴协助其开展工作。

这些优先工作影响到我们的规划和决策事项，涉及乘客数据的收集和传输，包括纸质或电子护照、签证和健康证明在内的公民身份管理系统，以及存储和保护相关数据所需的安全骨干网。

在创新博览会的第二天，我们讨论了航空安全和便利化的最新发展，也探讨了上述一些观点。

美国运输安全管理局局长佩科斯克先生在开幕式上指出航空安检采用更开放的架构是有潜力的，可能有助于各国在航空安全方面实现更安全、有韧性和可互操作的体系。

然而，从开发和实施的角度来看，这些开放式架构自身也面临一系列挑战。

航空安全领域的进一步合作侧重于机器学习能力的最新发展，以及人类和人工智能在该领域的相互联系。

各方取得了广泛共识，即航空安全的未来将需要在两者之间实现有效平衡和互补。

最后，我们讨论了实施无缝衔接和非接触式旅行的挑战，重点关注可互操作的、安全的健康和身份数字凭证，智能数据共享，批准使用生物识别数据以及重新调整操作流程。

来自政府和行业的政策制定者和开发者需要更好地合作，以完成这些优先事项，最终在机场实现更高效的旅客服务。

最后，我想呼吁所有创新者和监管者重视提高女性在创新领域的贡献和显示度。

采用包容和多样的方法是因为我们服务的人群是多样化的，将有助于发现并减少误导性偏见的风险，这些偏见会对人工智能和数据分析的应用产生不良影响，而许多未来的解决方案将依赖于这些应用。

2024 年，联合国未来峰会将在纽约举行，届时将邀请整个联合国系统、各国和公民社会制定并商定一项新的“未来公约”。

这个新公约的目标是加快实现可持续发展目标的进程，并确保在考虑和采取下一步措施来保护子孙后代免受当前生存威胁时更加协调一致。

ICAO 将确保国际航空业在这些讨论中有强大的声音，并确保各方都能认识到创新正在使民间社会发挥更大的作用。

在第 41 届大会召开前夕，能够接待这么多杰出的演讲者和与会者，我们感到由衷的高兴。相信过去三天的热烈讨论会写入大会成果。我要感谢为本次博览会做出贡献的创新展商以及聚焦未来的专家和与会者。我们还要感谢赞助商，为我们充分交流提供场地和茶点。

谢谢大家!

Building on Past Achievements and Jointly Creating a Brighter Future of China-GCC Relations

—Keynote Speech at the China-GCC Summit①

H. E. Xi Jinping

President of the People's Republic of China

Riyadh, 9 December 2022

Distinguished Colleagues,

Secretary General Dr. Nayef Falah Al Hajraf,

Good afternoon!

At the outset, I wish to express heartfelt appreciation to the Kingdom of Saudi Arabia for hosting the China-Gulf Cooperation Council (GCC) Summit. It gives me great pleasure to join you in planning for the future of China-GCC relations.

The friendly exchange between China and GCC countries goes back nearly two millennia in history. Throughout those years, the two peoples interacted with each other continuously along the ancient Silk Road inspired by the "Eastern wisdom" of peace, harmony and truth. In 1981, China established contact with the GCC upon its inception. Forty plus years on, the two sides have written together a splendid chapter of solidarity, mutual assistance and win-win cooperation.

The leapfrog growth of China-GCC relations is attributed to the profound mutual trust. China and GCC countries have all along supported each other's sovereignty and independence, respected each other's development paths, upheld equality between countries regardless of their size, and stood firm in defending multilateralism.

The leapfrog growth is attributed to the high degree of complementarity. China has a vast consumer market and a complete industrial system, while the GCC, with rich energy and resources, is embracing diversified economic development. This makes the two sides natural partners of cooperation.

The leapfrog growth is attributed to the empathy of the two peoples. Both belonging to the family of Eastern civilizations, China and GCC countries have similar

---

① Ministry of Foreign Affairs of the People's Republic of China. Building on Past Achievements and Jointly Creating a Brighter Future of China-GCC Relations[EB/OL]. (2022-12-14)[2023-08-20]. http://ws.china—embassy.gov.cn/eng/xwdt/202212/t20221214_10990137.htm.

culture and values. And the peoples enjoy close bonds of friendship.

The leapfrog growth is also attributed to the solidarity between the two sides in times of adversity. Against regional and international uncertainties as well as challenges such as the financial crisis, the COVID-19 pandemic and major natural disasters, the two sides have come to each other's aid and navigated through the difficulties together.

Colleagues,

In response to the profound changes unseen in a century, GCC countries have strengthened themselves through unity, achieved economic growth despite COVID-19, and worked vigorously for political settlement of regional hotspots and thorny issues. They have made the GCC the most dynamic regional organization in the Middle East and the Gulf. China highly commends you for all this. Standing at a historical crossroads, we should carry forward the tradition of China-GCC friendship, and take the establishment of the China-GCC strategic partnership as an opportunity to enrich the strategic substance of this relationship.

—We should be partners for greater solidarity. We need to further consolidate political mutual trust and firmly support each other's core interests. We need to jointly uphold the principle of non-interference in internal affairs, practice true multilateralism, and defend the common interests of all developing countries.

—We should be partners for common development. We need to better synergize development strategies and leverage our respective strengths to cultivate driving forces for development. China looks forward to working with all parties to advance and implement the Global Development Initiative (GDI) and follow through on the 2030 Agenda for Sustainable Development, with a view to promoting regional development and prosperity.

—We should be partners for common security. China will continue to firmly support GCC countries in safeguarding their security, and support the efforts by regional countries to resolve differences through dialogue and consultation and to build a Gulf collective security architecture. China welcomes the participation of GCC countries in the Global Security Initiative (GSI) in a joint effort to uphold regional peace and stability.

—We should be partners for cultural prosperity. We need to enhance interactions between our peoples, increase cultural exchanges, draw on each other's fine cultural achievements, and promote the rich values of Eastern civilizations, so as to contribute our share to the development and progress of human civilizations.

Colleagues,

In the next three to five years, China is ready to work with GCC countries in the following priority areas：

First，setting up a new paradigm of all-dimensional energy cooperation. China will continue to import large quantities of crude oil on a long-term basis from GCC countries，and purchase more LNG. We will strengthen our cooperation in the upstream sector，engineering services，as well as storage，transportation and refinery of oil and gas. The Shanghai Petroleum and Natural Gas Exchange platform will be fully utilized for RMB settlement in oil and gas trade. The two sides will work more closely on clean and low-carbon technologies involving hydrogen，energy storage，wind and photovoltaic power and smart power grids，as well as localized production of new energy equipment. We will jointly establish a China-GCC forum on peaceful use of nuclear technology and a China-GCC nuclear security demonstration center. China will provide 300 training opportunities to GCC countries on peaceful use of nuclear energy and technology.

Second，making new progress in finance and investment cooperation. China and GCC countries could collaborate on financial regulation and facilitate the entry into China's capital market for GCC companies. China will work with the GCC to set up a joint investment commission and support cooperation between sovereign wealth funds from both sides in various forms. The two sides could explore setting up a China-GCC forum on industrial and investment cooperation，strengthen investment cooperation on digital economy and green development，and build a working mechanism on investment and economic cooperation. The two sides could start currency swap cooperation，deepen digital currency cooperation and advance the m-CBDC Bridge project.

Third，expanding new areas of cooperation on innovation，science and technology. China is ready to build big data and cloud computing centers with GCC countries，strengthen 5G and 6G technology cooperation，build together innovation and entrepreneurship incubators，and implement ten digital economy projects in such areas as cross-border e-commerce and communications network. A China-GCC cooperation mechanism in meteorological science and technology will be set up，and the two sides could convene a seminar on climate response.

Fourth，seeking new breakthroughs in aerospace cooperation. China will carry out a string of cooperation projects with GCC countries in remote sensing and communications satellite，space utilization，and aerospace infrastructure. The two sides could select and train astronauts together，and China welcomes GCC astronauts to its space station for joint missions and space science experiments with their Chinese colleagues. China welcomes GCC countries' participation in payloads cooperation in its aerospace missions including Chang'e and Tianwen，and will consider establishing a

China-GCC joint center for lunar and deep space exploration.

Fifth, nurturing new highlights in language and cultural cooperation. China will cooperate with 300 universities, middle and primary schools in GCC countries on Chinese language education, work with GCC countries to set up 300 Chinese language smart classrooms, provide 3,000 "Chinese Bridge" summer/winter camp opportunities, and set up Chinese language learning and testing centers and online Chinese classes. The two sides could hold a China-GCC language and culture forum, and compile a bilingual library for people-to-people and cultural exchanges and mutual learning.

Colleagues,

China and GCC countries share the glorious mission of developing and revitalizing our nations. Our relations are both time-honored and young. Let us build on our achievements and work together to deliver an even brighter future of China-GCC relations.

Thank you.

Adhering to Multilateralism and Jointly Creating a Better Future —Lead Remarks at the First Plenary Session of the 14th ASEM Foreign Ministers' Meeting①

H. E. Wang Yi

State Councilor and Minister of Foreign Affairs

Of the People's Republic of China

Madrid, 16 December 2019

Mr. Chair,

Colleagues,

I am delighted to join you at the 14th ASEM Foreign Ministers' Meeting. Let me first thank our host Spain for its thoughtful arrangements for our meeting.

We live in a world of profound changes unseen in a century. The rise of unilateralism is undercutting international rules and rule of law. The spread of protectionism is weighing down global growth. Rampant power politics is threatening world peace and stability. Accounting for nearly 60 percent of the world's population and GDP, we as Asian and European countries have, for years, been both beneficiaries and upholders of multilateralism. To adhere to multilateralism has long been a shared commitment and collective action of our leaders.

---

① Forum on China-Africca Cooperation. Adhering to Multilateralism and Jointly Creating a Better Future [EB/OL]. (2019-12-18)[2023-08-20]. http://www.focac.org/eng/zfzs_1/201912/t20191218_7086029.htm.

The theme of today's meeting, "Together for Effective Multilateralism", embodies our shared aspiration for greater unity and coordination. I trust that under the stewardship of our Chair, the 53 ASEM members gathered here today will join hands in sending a strong message and rallying powerful energy in support of multilateralism.

Colleagues,

China has always been a firm defender and active practitioner of multilateralism. China was the first country to put its signature on the UN Charter, and has since joined almost all inter-governmental international organizations and more than 500 international instruments, bringing our practices fully in line with international rules. China faithfully honors every treaty it has signed and earnestly delivers on every commitment it has made.

We have fulfilled our commitments under the *Paris Agreement* on climate change three years in advance and met the poverty alleviation targets of the United Nations 2030 Agenda for Sustainable Development ten years ahead of schedule. Next year, China will eliminate absolute poverty across the country and complete the building of a society with moderate prosperity in all aspects. Even as dark clouds of unilateralism and protectionism are hanging over the horizon, they will never block the sun as long as countries take concerted actions to practice multilateral in a firm and effective way.

In this connection, I would like to propose the following:

In pursuing multilateralism, we need to aim for promoting peace and development. Multilateralism is the "golden key" to addressing global challenges. No matter what approach it takes, all acts of multilateralism have as their ultimate objective the promotion of world peace and development. Asian and European countries should lead by example in defending multilateralism. We should firmly stand for the international system built around the United Nations and the WTO-centered multilateral trading regime.

We support necessary reforms to the WTO and efforts by parties to resume the functioning of its appellate body. The new round of scientific and technological revolution should provide a new driving force for common development and a new opportunity shared by all countries. We need to encourage innovation cooperation to foster new drivers of growth, and reject technological monopoly, digital hegemony, and the attempt to create technological divide or decoupling of development. As artificial intelligence, 5G, bio-tech and other advanced technologies will shape the development of the global economy, countries are duty-bound to provide a fair, equitable and non-discriminatory environment for foreign businesses to invest, operate and pursue cooperation.

The Climate Change Conference in Madrid is again shining the spotlight on climate change, highlighting the urgency for global cooperation. In 2020, China will host the 15th meeting of the Conference of the Parties to the Convention on Biological Diversity. We hope the conference will produce ambitious outcomes and make important contributions to the international response to climate change.

In pursuing multilateralism, we need to adhere to international law and universally recognized norms governing international relations. Multilateralism is an important cornerstone of the existing international order. The international order shall not be trampled upon, nor shall international agreements be willfully discarded. We must work together to uphold the purposes and principles of the UN Charter and honor the commitment of upholding the international rule of law.

We are opposed to acts of putting one's own interests above those of others, arbitrary distortion of and a selective and utilitarian approach to international law, which smack of hegemonism. Bullying tactics like long-arm jurisdiction and unilateral sanctions should be rejected. The phase one trade agreement that has been reached between China and the United States is designed to address the concerns of both sides. It is a result of equal consultations on the basis of mutual respect, serves the common interests of the two peoples, and meets the shared aspiration of the world. It will help boost confidence in the world economy and stability of global trade.

In pursuing multilateralism, we need to live up to the overarching principles of fairness, justice and win-win cooperation. Equality and mutual respect are important tenets to which ASEM holds dear. We must abandon ideological prejudices and the outdated cold-war mentality, respect other countries' choice of development paths that suit their national conditions, and oppose imposing one's own values on others. As countries in Asia and Europe aspire for win-win cooperation and shared development, we should enhance the synergy between initiatives such as the Belt and Road Initiative, the Eurasian Economic Union, the EU Strategy on Connecting Europe and Asia, and the Master Plan on ASEAN Connectivity. Next year, China will host the second UN Global Sustainable Transport Conference. We hope that it will further boost and contribute to the connectivity and common development in Asia, Europe and beyond.

All in all, multilateralism is a prevailing trend of world development. It serves the shared interests and well-being of all Asian and European countries and the rest of the international community. China will continue to work with countries in Asia and Europe to advocate multilateralism and build a community with a shared future for mankind. Let us join hands in creating an even brighter future for people across the world.

Thank you!

第四章实战训练篇参考译文

欧洲强力支持太空发展，以此对抗气候危机[①]

一项针对欧洲公民的民意调查显示，人们认为欧洲应该在太空技术发展中表现出责任感、领导力和自主性，且发展中最优先事项应当是应对气候变化。

接受调查的欧洲公民中，有九成表示太空技术最重要的应用是收集气候变化的相关信息，和掌握地球上正在发生的情况。

欧洲航天局局长约瑟夫·阿施巴赫说："这项调查显示，欧洲公民强烈支持投资太空技术发展以改善地球上的生活，同时，欧洲对太空发展有着更高远的志向。此前，各国领导人齐聚第27届联合国气候变化大会(COP27)，欧洲人希望能看到太空发展在监测和减缓气候变化方面更多的应用。我们现在必须采取行动，增加欧洲在太空发展方面的自主权、领导权和责任感"。

一项由欧空局22个成员国2.1万余公民参与的调查于今日发布。该项调查由民意调查机构Toluna和Harris Interactive在9月25日至10月6日期间进行。

卫星数据能够为联合国气候变化系统所确定的一半以上基本气候变量提供信息。太空发展不仅能够帮助科学家、决策者和政治领导人监测、了解、建模、预测气候变化等因素引起的危机，更关键的是能够使其采取行动应对以上危机。

绝大多数受访者表示，太空对于进一步拓展人类对宇宙的认知，和通过无时无刻无所不在的通信连接和导航信号改善地球上的生活至关重要。

然而，在未来，大多数欧洲市民希望，政府能够更加重视监测和减缓气候变化，而非仅限于更好地了解宇宙，以及受益于卫星支持下的通信连接和导航技术。

超过80%的受访者还表示，欧洲应该整合其空间活动资源，以同其他航天国家竞争，此外，欧洲的空间活动不得受其他航天大国决定的支配。

的确，自2019年最近一次调查以来，越来越多的声音希望欧洲太空发展能实现资源整合和独立自主。

法国、德国、意大利、西班牙和英国的受访者中，约84%的市民赞成欧洲应联合起来与美国、俄罗斯、中国、印度和巴西竞争，较之前上升了14个百分点，81%的人则表示欧洲的太空活动应该独立，比之前上升了17个百分点。

在进军外太空方面，大多受访者认为清理太空垃圾应成为最优先事项，约86%的市民认为该事项很重要，排在之后的为组织机器人火星测探任务，占77%；载人登月，占71%；载人登陆火星，占70%。

然而自2019年的调查以来，太空探索在人们心中的重要性大大增加。在法国、德国、意大利、西班牙和英国的受访者中，对机器人探索火星的支持率上升了18个百

① 编者译。

分点，对宇航员登月和登陆火星的支持率分别上升了18个百分点和16个百分点。

这次投票适逢欧空局理事会部长级会议召开前夕。该部长级会议将于11月22日和23日在巴黎举行，欧空局成员国、联系国和合作国将受邀参会，这次会议为欧洲太空发展注入活力，同时确保太空发展继续服务于民生。

Unleash the Potential and Boost the Momentum[①]
—Interview with Minister of Industry and Information Technology

Mr. Jin Zhuanglong (Excerpt)

Looking back at 2022, Mr. Jin Zhuanglong, Minister of Industry and Information Technology admitted that various unexpected factors have brought huge challenges. In order to fully maintain the stability and smoothness of the supply chain, the Ministry of Industry and Information Technology (MIIT), together with various regions and departments, has timely introduced a series of measures to invigorate the industrial economy in a timely manner.

Mr. Jin used a series of data to summarize the achievements made in the industrial sector last year. From January to November 2022, the added value of China's industrial enterprises above designated size increased by 3.8% year on year. The added value of high-tech manufacturing and equipment manufacturing increased by 8% and 6.2% respectively. The proportion of manufacturing industry in GDP remained stable and played a role of "ballast stone" in stabilizing the macro economy. In the meanwhile, the quality of development has steadily improved, and a number of landmark achievements have been made in new energy vehicles, large passenger aircraft, photovoltaic and other fields.

This year marks a critical year for the implementation of the "14th Five-Year Plan". In 2023, China's industrial economy still faces many difficulties and challenges. The impact of the external environment is deepening, and pressures posed by shrinking demand, supply shock, and weakening expectations are still relatively large. Nevertheless, Jin Zhuanglong is still confident that the overall economic operation will improve. His confidence comes from China's complete industrial system and the scale and technological advantages formed in some important areas.

Let's take a look at the country's economic lifeline—manufacturing. The key to building a modern industrial system is to revitalize manufacturing. After years of development, China's manufacturing industry has accounted for about 30% of the

① 编者译。

global total, ranking first in the world for 13 consecutive years. Manufacturing is the foundation of the real economy, and the digital economy is another important engine driving China's economic growth and an important advantage in building a modern industrial system. After several years of vigorous development, China's digital economy has stabilized as the world's second largest overall, and the role of the digital economy in empowering the real economy is becoming increasingly prominent.

China has already had the world's largest and technologically advanced network infrastructure. It has built and launched 2.28 million 5G base stations, with more than 50,000 5G application cases accumulated. The gigabit optical network has the ability to cover 500 million households, achieving "gigabit coverage in every city", "5G coverage in every county", and "broadband coverage in every village".

A solid information and communication infrastructure has laid a good foundation for accelerating the growth of the digital industry. From January to November 2022, China's software business revenue reached 9.46 trillion yuan, a year-on-year increase of 10.4%; the annual transaction scale of mobile payments reached 527 trillion yuan; and the global share of Chinese companies in areas such as smartphones, optical communication equipment, and optical fiber cables reached 50%. The digitization of industries has accelerated comprehensively. The industrial Internet has been widely integrated into 45 national economic categories. There are 248 influential industrial Internet platforms, and the number of industrial equipment connections on key platforms exceeds 80 million units. The number of industrial apps is nearly 300,000, and the popularity rate of digital R&D design tools for manufacturing enterprises has reached 76%. The CNC rate of key processes has reached 57.2%.

Opening Speech at the Meeting of the Advisory Council of The Belt and Road Forum for International Cooperation, 2020①

Foreign Minister Wang Yi

Beijing, 18 December 2020

Members of the Advisory Council,

Good evening.

It gives me great pleasure to meet with old friends. At the outset, I would like to

① Ministry of Foreign Affairs, the People's Republic of China. Opening Speech by State Councilor Wang Yi At the Meeting of the Advisory council of The Belt and Road Forum for International Cooperation, 2020[EB/OL]. (2020-12-20)[2023-08-20]. https://www.fmprc.gov.cn/eng./wjb_663304/zzjg_663340/ggjjs_665228/xwlb_665230/202012/t20201220_598898.html.

express warm congratulations on the opening of this year's meeting of the Advisory Council of the Belt and Road Forum for International Cooperation.

Despite the impact of the pandemic this year, international cooperation on the Belt and Road Initiative (BRI) has continued to demonstrate strong resilience and vitality. Our cooperation has played an important role in helping countries fight the virus, stabilize the economy, and protect people's livelihoods. It has also contributed to the implementation of the UN 2030 Agenda for Sustainable Development.

The world today is going through a pandemic and global changes, both unseen in a century. The life, safety and health of people around the world are under threat. With global industrial and supply chains disrupted and international trade and investment contracting, the world economy slipped into recession. Unilateralism and protectionism have been on the rise. Yet no matter how the situation may evolve, the international community's need for the Belt and Road remains unchanged; partners' support for the Belt and Road remains unchanged; and China's resolve to advance Belt and Road international cooperation remains unchanged.

Members of the Advisory Council,

President Xi Jinping has proposed making the Belt and Road a model of cooperation, health, recovery and growth. This has charted the course for Belt and Road international cooperation.

To make the Belt and Road a model of cooperation, we need to promote openness and win-win cooperation. Despite the onslaught of COVID-19 this year, BRI partners have convened 30-plus international video conferences under a dozen or so BRI multilateral platforms to coordinate policies and advance practical cooperation on finance, taxation, energy, green development, anti-corruption and other areas. The joint statement issued at the High-level Video Conference on Belt and Road International Cooperation sent a strong message of collectively fighting the virus, strengthening connectivity and promoting economic recovery, and reaffirmed to the world the strong confidence in Belt and Road international cooperation.

The BRI is an initiative that is completely open and above board. It has followed the principles of openness, inclusiveness and transparency, and of extensive consultation, joint contribution and shared benefits. It is a platform for economic cooperation, not a geopolitical tool. It is a chorus by BRI partners, not a solo show by China. We support Chinese and foreign companies in participating in cooperation projects in an open and transparent manner following market rules and the principle of fairness and sharing benefits and risks.

China is ready to continue advancing bilateral, trilateral and multilateral cooperation in various fields with all parties to build an open world economy, and maintain the smooth functioning of global industrial and supply chains. Not long ago, we signed the Regional Comprehensive Economic Partnership and announced our readiness to favorably consider joining the Comprehensive and Progressive Agreement for Trans-Pacific Partnership. All this is testimony to China's firm commitment to free trade.

To make the Belt and Road a model of health, we need to further advance international cooperation against COVID-19. As the virus is still rampaging globally, countries face a daunting task of COVID-19 containment. For BRI partners, our immediate priority is to defeat the virus through solidarity and cooperation. This year, China and BRI partners have worked closely on COVID-19 control and advanced the development of a Health Silk Road. We have coordinated pandemic response measures, shared information and diagnostic and therapeutic experience, and provided each other with urgently needed medical supplies. Together, we have kept land, sea and air links open to facilitate the transportation of life-saving supplies and built a solid line of defense against the virus. Over eight million pieces of emergency medical supplies have been delivered by the China-Europe Railway Express, and over 1,700 tons of medical supplies from China have been delivered around the world through the Silk Road in the Air.

Vaccines are the focus of international cooperation for the next stage. China has joined the COVAX initiative led by the WHO. Important progress has been made in phase Ⅲ trials carried out by China's vaccine companies and firms in BRI partner countries such as Russia, Egypt, Indonesia, Pakistan and the UAE. China will make vaccines a global public good when they become available for deployment. They will contribute to vaccine accessibility and affordability in developing countries and deliver benefits to BRI partners.

To make the Belt and Road a model of recovery, we need to promote parallel progress in COVID-19 response and economic and social development. China has established fast tracks with many countries, BRI partners in particular, and green lanes with interested parties to facilitate personnel flows and smooth delivery of goods. Progress has been made in projects such as the Jakarta-Bandung high speed railway, the China-Laos railway, and the Budapest-Belgrade railway. The metro train in Lahore, a project under the China-Pakistan Economic Corridor, is now open to traffic. With rigorous containment measures in place, many Belt and Road projects have kept going with no layoffs. A stream of new projects have been launched. Against the backdrop of the global economic recession, China's input into the Belt and Road has not decreased but kept a momentum of growth. In the first three quarters this year, China's non-financial direct investment in countries along the Belt and Road increased by nearly 30% year-on-year. China has also taken an active part in the G20's Debt Service Suspension Initiative (DSSI), ranking the first among G20 members in terms of deferral amount. We have provided support and help to the best of our ability for COVID-19 response and economic recovery in many of our BRI partners.

As we speak, world economic recovery is still facing many uncertainties. International logistics is facing an imbalance in supply and demand. We will continue to work with others to enhance the cooperation frameworks of fast tracks and green lanes, strengthen land, sea and air multi-modal transport, and further enhance the efficiency and quality of key international corridors.

To make the Belt and Road a model of growth, we need to unleash new drivers of the world economy. The BRI partners are actively developing Silk Road e-commerce, digital transport corridors, cross-border fiber optic cables for information transmission and the China-ASEAN Information Harbor. Multilateral platforms including the BRI International Coalition for Green Development, the BRI Green Investment Principles and the BRI Green Investment Fund are growing in depth and vitality.

Digital Silk Road is a priority area for BRI cooperation in the next stage. China has put forth the Global Initiative on Data Security designed to contribute to data security and development and cooperation of digital economy, and promote collective efforts toward a cyberspace featuring peace, security, openness and cooperation. Issues related to technology have been politicized by certain countries who are attempting to preserve their monopolistic position in high tech areas. Such moves are not constructive. What has happened shows that extensive consultation, joint contribution and shared benefits represents the right way toward better global digital governance.

China is committed to environment-friendly development and is actively involved in international cooperation on climate change. It has announced that it will strive to peak CO_2 emissions before 2030 and achieve carbon neutrality before 2060. By 2030, China will lower its CO_2 emissions per unit of GDP by over 65% from the 2005 level, increase the share of non-fossil fuels in primary energy consumption to around 25%, increase the forest stock volume by 6 billion cubic meters from the 2005 level, and bring its total installed capacity of wind and solar power to over 1.2 billion kilowatts. All this will catalyze the development of a green Silk Road.

Members of the Advisory Council,

Confronted by the most serious pandemic in a century, China has taken swift, resolute and extraordinary measures to contain the virus and ensure economic and social development. In the first three quarters, China's economy resumed positive growth. Production and demand continue to pick up; employment is on the whole stable; fiscal and financial performance is steady; and market expectations have kept improving. China is expected to become the only major economy to register positive growth this year. It is estimated that China's economy will contribute over one third to global growth.

The momentum of steady recovery of the Chinese economy is good news for the world. China will continue to deepen reform and expand opening-up. A growing China will help drive global growth more strongly and make greater contributions to global economic recovery.

The recently concluded Fifth Plenum of the 19th CPC Central Committee adopted the *Recommendations for Formulating the 14th Five-Year Plan for Economic and Social Development and the Long-Range Goals for* 2035, unveiling a blueprint for the next stage of China's development. Guided by a deep understanding of the new development stage, China will apply the new development philosophy, and foster a new development paradigm with domestic circulation as the mainstay and domestic and international circulations reinforcing each other.

The paradigm is aimed at promoting mutual openness and mutually reinforcing development between domestic and international circulations. It will fully unlock China's market potential and facilitate greater opening-up. China has a population of 1.4 billion and a middle-income group exceeding 400 million. This year, the value of its domestic retail market will surpass 6 trillion US dollars. In the coming decade, total import of goods into China is estimated to top 22 trillion US dollars. The high-quality development of the Chinese economy will provide stronger impetus and create broader space for high-quality Belt and Road cooperation.

Members of the Advisory Council,

The BRF Advisory Council is an important platform for multilateral Belt and Road cooperation. As seasoned political leaders, experts and scholars, you all have a lot to contribute in terms of policy recommendations, academic research and international cooperation. I look forward to your insights on high-quality Belt and Road cooperation, and encourage you to communicate the true story of Belt and Road cooperation to the world and bring more wisdom and resources to bear on Belt and Road cooperation.

第五章实战训练篇参考译文

在联合国世界旅游组织第24届全体大会上的主旨演讲[①]

国际民航组织秘书长　胡安·卡洛斯·萨拉萨尔先生

2021年11月30日，马德里

今天，能在此发表讲话，我深感荣幸，希望能够为各位的主旨演讲和讨论抛砖引玉。

我谨向联合国世界旅游组织秘书长波洛利卡什维利先生致以最诚挚的感谢，感谢他盛情邀请我在航空运输业和全球旅游业如此紧要的发展关头发表讲话。

首先，我想强调尽管2021年国际旅行经历了巨大的波动，但在夏季仍取得了显著进展。

国际短途航线尤其明显，比如大家非常熟悉的欧洲航线，以及亚太、美洲和加勒比地区的国际航空运输。

几天前我刚在坎昆参加国际机场理事会大会，谈到该地区座位容量在过去一个月激增14%以上。

世界各地都出现了类似的迹象，全球最新的客运量趋势也表明，在年末假日期间，国内外的订票量将进一步增加。

我想大家都很清楚，这些早期复苏迹象范围逐渐扩大，其意义深远，对旅行和旅游部门的创新性和灵活性是重大考验，将考验我们如何共同转型，重建美好未来。

我还想着重指出，复苏初期是合作和多边主义发挥作用的关键期，既可以增强、加快伙伴间合作，也可以就合作领域和方式达成共识。

我认为这也凸显出联合国机构的根本作用，对于我们取得符合《2030年议程》和《共同议程》的全球集体成果来说无比重要。

这在一定程度上解释了为何现如今UNWTO和ICAO比以往任何时候都更紧密地合作。

我们近日新签署了一份谅解备忘录，明确全面的优先合作领域，这将在具体的复苏领域给我们提供指导，如实现旅客体验的现代化、在全面的"端到端"基础上实现旅行和旅游的绿色化。

我们还将为12月7日的国际民航日发表一份联合声明，这将是ICAO历史上第一次为该节日发表联合声明。这一天也将开启以"重新连接世界"为主题的交流活动。

一个月前，在ICAO COVID-19高级别会议上通过的《部长宣言》清楚地表明了各国和所有行业利益相关者坚定的政治承诺和决心，大家都希望安全、高效地恢复航

---

① 编者译。

线网络，并增加其韧性和可持续性。

为实现该目标，ICAO 需要密切监督航空业安全恢复，管理复杂的路径，恢复全标准合规，并与行业运营商和国家监管部门合作，它们正面临着空前的融资和资源挑战。

正如刚才所强调的，我们还需要确保各阶段乘客体验的现代化，优先解决航空安全和旅客健康问题。

在后疫情时代，旅客到达机场时，会期待一次安全、无接触、健康的飞行体验。为此，我们需要与各国、各供应商合作，利用新的筛查方法和技术来实现这一目标。

幸运的是，数字化的流程会在入境点、安检口及海关关卡为乘客带来更便利的体验，同时也会增加接待人数。但这些也引起了网络安全方面的担忧，需要认真加以解决。

我们的目标是提升公众对航空旅行和旅游的信心，使乘客在受到更好保护的同时，享受更高效、更舒适的体验。

ICAO 最近制定了个人健康证明可视数字印章的新规范，提升了乘客对非接触和安全航空流程的信心。

可视数字印章提供了全球信任并接受的证明和认证，一方面保护了数据和隐私，另一方面大大提升了乘客对国际旅行的信心。

我们在全球推广可视数字印章的同时，也将考虑航空业在后疫情时代的新常态，综合协同旅客安全、航空流程和健康筛查的目标。

当然，我们为恢复而做出的所有努力，是减少并最终消除碳足迹的必要条件。

近来，无论是在最近召开的第 26 届联合国气候变化大会、七国集团、二十国集团活动中，还是在 ICAO 的 COVID-19 高级别会议上，许多国家都强烈呼吁航空业去碳化。

这反映了一个事实，即气候危机是我们面临最严峻的威胁之一。从自然灾害到经济衰退，从疾病暴发到冲突争端，全球变暖及其连锁反应影响着每个人，每个地方，包括航空业和旅游业。

因此，如果我们想减轻最严重的影响，现在必须大胆采取行动，显然，一切照旧是不够的。

我非常支持联合国秘书长古特雷斯最近的呼吁，他号召航空业迅速实现碳中和。目前航空运输业已制定详细的 2050 年净零承诺，因此，我对实现这一目标持乐观态度。

通过更省油的飞机技术、更优化的操作、可持续的航空燃料以及 ICAO 在《国际航空碳抵消和减排计划》下设立的透明、负责任的碳抵消方案，我们已经取得了相当大的进展。

在今年和去年开展的活动中，ICAO 一直在评估航空减排方面的创新举措。很显然，创新对于实现航空业的气候目标至关重要，就像我们努力重新设计飞机，或者使乘客体验现代化一样。

ICAO 的关键作用是通过有效的全球标准化来理解和促成创新，并确保所有新技术和操作都能安全、可靠、可持续地融入我们的全球网络中。

鉴于现在事物发展速度令人瞠目结舌，我们也面临着从零开始转型的挑战，使 ICAO 更加灵活、高效和结果导向，以更好地应对这个充满惊人想象力和发明创造的新时代。

今天在座的每一位几乎都可以期待，有一天，自动驾驶的空中出租车将你从机场送到酒店或其他旅行目的地。

但是，从在社交媒体上看到这些惊人新飞行器的视频，到真正坐上这样的飞机，需要详细、全面的研究，以及复杂的标准化和认证程序。

我想强调的是，疫苗接种正在使航空旅行和旅游重新成为可能。然而，为了让各国从重振的旅行和旅游市场中获益，我们需要迅速审查并在条件允许的情况下取消对已接种两针疫苗的乘客实施出入境限制。

在新冠疫情大流行之前，每年约有 14 亿游客跨境，其中远超半数的人乘坐飞机抵达目的地。

这些航空旅客是许多国家的重要经济生命线，尤其是那些最不发达国家、内陆发展中国家和小岛屿发展中国家。

因此，我们共同承担着庄严的职责，那就是让旅客顺利到达目的地，重新实现世界互联互通，从而使全球可持续发展的共同目标回到正轨，使数百万人的生计得以恢复。

最后，我代表 ICAO 由衷感谢 UNWTO 在 ICAO 理事会航空恢复工作组中和最近的高级别会议上，对航空业应对和恢复工作的大力支持。

我认为，在面对这些挑战中，我们的组织展现出了惊人的团结和高效。现在，我们已经具备指导未来合作的坚实基础和强大动力。

未来几个月很重要，我期待看到共同取得的进展。

谢谢大家！

Remarks by Chairman of COMAC He Dongfeng①

First of all, I want to express my sincere thanks to leaders at all levels, all sections of society and international friends for your long-time attention, support and assistance to COMAC.

COMAC was established in Shanghai, a promising land for reform and opening-up, on May 11th, 2008. This event symbolized the start of independent development of China's large commercial aircraft and brought hope for the innovative development of

① 编者译。

China's trunk liner. In the past ten years, especially sincethe 18th National Congress of the Communist Party of China, we have always been following the guidance of Xi Jinping Thought on Socialism with Chinese Characteristics for a New Era, thoroughly studying and implementing the spirits of major instructions of General Secretary Xi Jinping on trunk liner career, upholding the banner of scientific and technical innovation, making concerted efforts, and tackling difficult problems. We have made a historic leap in China's business jet industry from nothing to something, and started a new journey for China's trunk liner industry to develop from weak to strong.

Looking back, we have seized every minute, spared no effort to work hard, and taken the first step in the long "march" of developing trunk liner. Looking into the future, we are full of confidence that COMAC will develop into a new phase and start a new journey. We will set high aims and have lofty aspirations, firmly implement the overall arrangement of "Three Steps", strive to achieve the phase objectives of "Three Ones" and two ambitious objectives, persist in building COMAC into an aviation enterprise of "Four World Classes", and make contribution to building modern socialist country.

We will always drive development through innovation, promote innovation in management, technology, products and business mode, strive to achieve autonomous control of key core technology, and unceasingly enhance the core competence of main manufacturer to build brands with international influence.

We will always embrace open cooperation, play a leading role in the industry, promote the construction of the system of trunk liner industry, and work with global partners to build a career community with shared benefits, shared achievements and shared glory, a life community of mutual understanding, mutual support, win-win cooperation and risk sharing, and a dream community that cares about, supports and promotes the trunk liner career.

We will always insist on the principle of being customer centered and the quality guideline of "mastery design, fine manufacture, service in good faith and constant perfection seeking" to provide commercial aircraft with safety, economy, comfort and environment-friendly characteristics for customers.

We will carry forward the legacy of the trunk liner community in servicing the nation with aviation and "never give up", foster respect for model workers and promote quality craftsmanship, work consistently, perseveringly and steadfastly to "develop trunk liner based on frugality", and build a clean COMAC to enable the development of the trunk liners.

Let us remain true to the original aspiration, remember the mission, and make new and greater contributions to realizing the Chinese Dream of national rejuvenation.

Shoulder the Mission of Youth and Pursue the Dream of Serving the Country —In the Aerospace Industry Speech at the 2022 Opening Convocation①

Wang Yunpeng
President of Beihang University
Academician of the Chinese Academy of Engineering
Beijing, 10 September 2022

Dear students, parents and faculty,

Good morning!

Today, we are holding the inauguration ceremony of the new school year at both campuses to welcome 3,871 undergraduate students and 6,650 graduate students to join the Beihang family and start a new chapter in your lives. On behalf of Secretary Zhao Changlu and all the faculty and students of the University, I wish to extend a warm welcome to you. I also want to express my heartfelt gratitude to your parents and teachers who have guided you over the years.

I have noticed that the youngest freshman among you is Luo Weigao from Shenyuan College, who is only 15 years old; and the youngest graduate student is Zhu Chen from the School of Computer Science, who is only 18 years old. There are six students each with the most common name "Zhang Hanwen" and "Yang Fan". There are 108 students whose names contain the elements of "aviation" and "space". Today is the birthday of 37 students including Zhao Jianhui from the School of Aeronautics and Engineering and Xin Zhuoyan from Shi'e College. I wish you all a happy birthday. While you come from all over the country, from today on, you will share a common name—Beihanger.

The year 2022 is destined to be an extraordinary one. Together, we witnessed those "unparalleled" beautiful moments at the Beijing Winter Olympics, which demonstrated the confidence and style of a great country to the world. As the 20th National Congress of the Communist Party of China is around the corner, people of all ethnic groups across the country are striding towards a new journey of building a modern socialist country in all respects and advancing towards the second centenary goal. We are also navigating unprecedented global changes and once-in-a-century pandemic. The complex international situation has brought various risks and

① 编者译。

challenges. You have overcome the impact of the epidemic, studied at home, and attended online classes. You have grown up in a difficult environment and finally realized your dream in Beihang. I want to give each of you a big thumb-up.

2022 also marks the 70th anniversary of Beihang's founding. Beihang was born in the flames of the War to Resist US Aggression and Aid Korea and grew up in the wave of reform and opening up and socialist modernization drive. It has been advancing on the path of national rejuvenation and building of Socialism with Chinese Characteristics for a New Era. For 70 years, generations of Beihangers have inherited the red genes, adhered to the motto of "Integrating Virtue with Brilliance and Combining Knowledge with Practice", cultivated top-notch talent, and encouraged them to "serve the country in the aerospace industry" with such qualities as "patriotic dedication, daring to be the first, solidarity and hard work, taking responsibility and being practical" on the journey of building a world-class university with Chinese characteristics.

Patriotic dedication is the unwavering pursuit of Beihangers. In the early days of the University's founding, a large number of outstanding scientists represented by Shen Yuan, Tu Shou'e, Lin Shi'e, and Lu Shijia returned from overseas studies and founded Beihang. Over the past 70 years, the University has trained more than 230,000 outstanding professionals. A large number of academicians, chief designers and industry elites have set sail from here and pursued their careers.

Daring to be the first is within the genes of Beihangers. From China's first light passenger plane "Beijing No. 1", Asia's first sounding rocket "Beijing No. 2", China's first drone "Beijing No. 5", a large number of core technologies such as 3D printing, electromagnetic compatibility and high-temperature coating materials are directly applied to national key models. Thanks to unceasing pursuit of excellence for 70 years, 15 achievements have won the first prize in national-level science and technology awards since the beginning of the new century.

Beihangers are also well known for their solidarity and hard work. In the early days of the University, all faculty members and students worked hard for one hundred days before sending "Beijing No. 1" into air. More than 200 faculty members and students from the innovation team worked together throughout the year to create a world record of continuous controllable flight. After 70 years of vicissitudes, the team working culture has always been embraced by Beihangers in their efforts to strive for success through solidarity and hard work.

The value of "taking responsibility and being practical" has been passed down among Beihangers. Cen Zheng, the commander-in-chief of the model rocket who made a speech just now has been working hard for over 30 years for China's aerospace industry. He led the team to develop the Long March 3A series, which is known as China's "gold medal rocket" and has guaranteed major tasks such as the lunar exploration project and BeiDou network. As we have done in the past 70 years, Beihangers love what we do and always strive to excel at what we do.

General Secretary Xi Jinping emphasized that "realizing the Chinese dream is a historical relay race, and young people should strive to be the first on the path of realizing national rejuvenation." You are a generation that advances together with the new era. You are going to witness achieving of the second centenary goal of the country. "Heaven will bestow great responsibility upon those who are capable." Take on the mission of youth with courage, pursue your dreams of serving the country in aerospace, and be the one who stands out! As you are about to take over the "relay baton" of serving the country in aerospace, I have three expectations for you.

Firstly, set a lifelong goal of serving the country. As the ancients put "Without an ambition, one cannot achieve anything great". A person's ability to make achievements depends first on whether their aspirations are lofty and their ideals are firm. Throughout history, anyone who scored great achievements regarded serving their country and people as their lifelong pursuit.

Academician Qi Faren, a space expert and chief designer of the Shenzhou spacecraft in China, was studying in Dalian in 1950 when the War to Resist US Aggression and Aid Korea broke out. His heart ached, when the motherland's territory was repeatedly bombed and strafed by American planes. He made up his mind to study aviation in the future, build airplanes and protect his country. After taking the college entrance examination, he chose aviation as his major and finally entered Beihang University. He led the development of many types of Dongfanghong series satellites and the overall scheme design of the Shenzhou spacecraft, making significant contributions to China's nuclear bombs, the missile, and the space satellite as well as manned spaceflight.

With beliefs in the heart, one can go far. You might have different ideals and career choices. Facing the influence of various social trends, it is inevitable that there will be doubts and confusion. I hope that you will firmly hold on to your ideals and beliefs, integrate your ideals with the national development, and achieve a career while taking roots in the motherland and serving the people.

Secondly, cultivate practical skills and shoulder the historical mission of national rejuvenation. You are about to enter the golden stage of learning in your life. The information age has accelerated the generation and upgrading of knowledge. It is particularly important to lay a solid foundation of knowledge and develop practical skills to cultivate the ability to meet the requirements of the times.

Lay a solid foundation in the pursuit of knowledge. Only through diligent study can one become knowledgeable and accumulate small gains into great achievements. During his studies, Luo Yang, the "Aviation Industry Model" and the late chief commander of the J-15 fighter jet, worked hard and excelled in both theoretical mechanics and mathematics exams. His undergraduate thesis drawings basically reached the level of engineering practical application and were still commended by his supervisor Academician Wang Jun.

Enhance ability in the pursuit of innovation. Innovation has always been a beacon of excellence for Beihang University. During his doctoral studies, Zhu Daoqian, a student at the School of Integrated Circuits, pursued technological innovation and proposed a new solution. Heprovided an important technical path for supporting national key R&D programs and proved that innovation has no age limit.

Grow your skills through practice and hard work. "Although the Tao is close at hand, it is not until you practice it that you can reach it; although the matter is small, it is not until you do it that you can achieve it." Skills and abilities do not grow on their own. There is knowledge everywhere. Only by combining theories with practical action can we increase what we cannot do and unleash tremendous power.

Learning is like a bow, and talent is like an arrow. Achieving good results is never an easy task. I hope that you will cherish your time, study diligently, and strive to become outstanding talent who can take on great responsibilities and accomplish great things. "If you want to be new every day, do new things every day."

Thirdly, build a strong and courageous character, cultivate the spirit of dedication. The road of life has smooth paths as well as rough ones, and it has never been a plain sailing. You are now in a "the period of rapid growth", and only by experiencing more setbacks and difficulties can you become a strong person in life.

Those who are strong in life are always willing to selflessly dedicate themselves and maintain optimism. During the Beijing Winter Olympics, 428 student volunteers from Beihang University served the games at the "National Alpine Ski Centre" with full enthusiasm despite the harsh weather on the mountain. Among you is Liu Xin, a graduate student who has not succumbed to the fate of illness for 24 years. He was committed to invention and creation during his undergraduate period, won multiple national awards in four years, and obtained more than ten patents. He was admitted to Beihang University with excellent results. Let us give a round of applause to him.

Hardship and adversity make one stronger. In any case, the spirit of tenacity and perseverance cannot be lost, and the quality of being willing to dedicate oneself, unite and help each other cannot be absent. I hope that you can learn to correctly view success or failure, enlighten your minds and sharpen your wills from difficulties and setbacks, so that life can be sublimated and transcended.

Endeavour and actions are the most valuable assets for youth. You are about to kick start your college journey. I hope you will not waste your time and live up to your youth. Let your youth bloom in places where the motherland and the people need it most.

Today is both Teachers' Day and Mid-Autumn Festival. I wish the faculty a happy Teachers' Day, and I wish all faculty members, students, and your family a happy Mid-Autumn Festival. Thank you all!

第六章实战训练篇参考译文

部落航空研讨会开幕式和闭幕式致辞[①]

美国联邦航空局前任局长　斯蒂芬·M·迪克森

2022年3月14日

开幕词

谢谢阿兰多。我很高兴来到这里。

对航空业来说,这是激动人心的时代。在这个时代,我们有比以往更多的机会来开创事业,发展业务,并成为航空界的一员。在联邦航空局,我们有责任确保每个人都知道这些机会,并了解如何参与其中。这就是今天的意义所在。我们希望使航空业安全、高效、环保,并给予每个人参与其中的权利。

本周的研讨会为我们与部落联系提供了机会。我们希望直接听到你们对航空业的关注和需求。我们想让你们知道联邦航空局正在做什么,以及你们如何从中受益。

今天,我们将讨论联邦航空局的几个拨款项目,如机场改善计划,通过该计划,我们向全美的机场拨款。我们还将讨论两党基础设施建设法案——这是一个历史性的、千载难逢的机会,将投资200亿美元用于机场和空中交通管制基础设施建设。

我们希望你们知道能获得哪些资助,以及如何把握申请程序。

在未来,我们还将促进部落青年和部落居民的航空航天教育和就业机会。

我们还将讨论无人机和先进空中交通如何使原住民受益。例如,联邦航空局已与俄克拉荷马州乔克托族合作,测试无人机如何在低空运送货物,包括包裹。正如我开始所说,对航空业而言,这是激动人心的时代。

---

① 编者译。

我们的目的是在联邦航空局和部落之间建立可信赖的社区，这样我们就可以在所有的航空领域公开和定期地进行沟通，包括融入其中的机会。

要做到这一点，我们必须承认，原住民一直受到联邦政府不公的对待。我们不能忘却这段历史是如何带来挑战，至今还影响各部落的发展。

我们坚决支持拜登总统的行政命令，呼吁联邦政府为所有人促进公平，包括那些在历史上没有得到充分关照的人。

总统还重申，各部落是有主权的民族。我们必须给予你们同样的尊重。我们必须在影响部落的航空政策领域与你们协商，就像我们对待其他国家一样。

美洲印第安人有句名言："只说不听与聋人无异。"

重要的是，我们要倾听你们的意见——真正的倾听——并尊重你们的需求和关切。在倾听的过程中，我们不会带有任何预设的偏见，我们也会认识到每个部落都有其独特观点。

归根结底，我们必须赢得并保持你们的信任。

我们彼此接触越多，就越能更好地制定解决方案，并让每一个人都能受惠于航空业。我给大家举几个例子，说明我们在航空领域实现的积极沟通。

美国联邦航空局已经开始与130多个美国原住民部落开展协商，确保他们拥有真正的机会来表达在美国国家公园上空开展空中游览的关切。其中许多国家公园包含圣地和对部落很重要的区域。

我们还与大平原地区的11个部落协商，在北达科他州威利斯顿的一个新机场项目施工期间保护圣地，包括让41名传统文化专家在施工期间担任监督员。

Kwinhagak 原住民村向我们提到需要修复阿拉斯加西南部机场的跑道，我们努力为这个正在执行的项目提供资金。

这些只是联邦航空局和部落之间合作的几个例子。我们将继续从中学习，以便在未来更好地与部落联系，帮助你们利用现有的机会。

本周的研讨会将帮助我们做更多的事情。谢谢大家！预祝今天的会议取得圆满成功。

闭幕词

谢谢阿兰多。

正如我在之前的发言中指出的，我们希望在联邦航空局和部落之间建立一个可信赖的社区，以解决所有与航空有关的问题。

我们今天谈到了一些重要领域。

通过两党基础设施建设法案，以及联邦航空局的机场改善计划，我们在未来几年有一个巨大的机会来加强整个美国的航空基础设施。这些项目可以帮助改进服务部落的机场。

我们希望原住民当局有信心为部落拥有的机场申请拨款。我们很期待帮你们做到这一点。

除了我们的机场拨款程序，美国本土企业还可以联系联邦航空局的小企业办公室，并查看其网站——*sbo.faa.gov*——了解与 FAA 签约的机会。

此外，我们已经在联邦航空局的 YouTube 网站上发布了一个播放列表，其中有关于两党基础设施建设法的视频，包括有关小企业签约机会的网络研讨会。我鼓励你们在 Facebook 账户和其他社交媒体平台上分享这些资源。

明天，联邦航空局副局长和其他发言人将讨论航空航天领域的教育和就业机会，以及这些努力如何惠及部落的年轻人。

我们还将讨论新兴创新，如无人机、先进空中交通和商业空间运输，以及这些活动对部落的意义。

而且，我们将更详细地探讨在圣地上空飞行的问题，以及我们如何能够更好地相互协作以解决这些问题。

你们中有些人可能知道，我将在本月底离职。做了 43 年多的航空专业人士，我该回家了。届时，本机构将派一位代理局长来指导工作，直到正式任命常任局长。

很多时候，领导层的变动会让人怀疑机构的优先事项会发生什么变化。请记住，联邦航空局将坚定不移地致力于与部落开展合作，这项工作是整个机构的优先事项。

当务之急是，我们必须建立并保持彼此的信任，这样，即使领导层发生变化，我们也能始终保持紧密联系，继续前进。我相信这种牢固的基础已经存在。

部落的后人应该知道，他们在联邦航空局永远有合作伙伴。

我想起了美洲原住民陶斯普韦布洛-克里克的霍华德·雷纳所说的另一句名言。他说到："孩子们从他们看到的东西中学习。我们需要树立一个真理和行动的榜样。"

树立积极的合作榜样，并与部落未来几代领导层保持良好的合作关系。

感谢你们今天与我们沟通。期待明天会议上再次交流。

East Meets West—A New Chapter of Sino-Italian Friendship①

H. E. Xi Jinping

President of the People's Republic of China

20 March 2019

It is a great pleasure for me to pay a state visit to the Italian Republic at the invitation of President Sergio Mattarella in this blossoming season of spring. In 2011, I visited Rome oncelebrations of the 150th anniversary of Italian unification and, in 2016, I had a stopover on Sardinia. I was deeply impressed by the way of life and

① Xinhua. Full text of Xi's signed article on Italian newspaper[EB/OL]. (2019-03-20)[2023-08-20]. https://www.chinadaily.com.cn/a/201903/20/WS5c91dccea3104842260b19e0.html.

industrial outlook of Italy that blend together the ancient and the modern, the classic and the novel. Now that I am about to set foot again on this beautiful country, it feels like I am to be among old friends and get immersed in their wonderful hospitality.

China and Italy are both stellar examples of Eastern and Western civilizations, and both have written splendid chapters in the history of human progress. Being the birthplace of ancient Roman civilization and the cradle of the Renaissance, Italy is known to the Chinese people for its imposing relic sites and masterpieces of great names in art and literature. Friendly ties between our two great civilizations go back a long way. As early as over 2,000 years ago, China and ancient Rome, though thousands of miles apart, were already connected by the Silk Road. During the Eastern Han Dynasty (AD 25—220), Chinese emissary Gan Ying was sent to search for "Da Qin", the Chinese name of the Roman Empire at the time. Roman poet Virgil and geographer Pomponius Mela made many references to *Seres*, the land of silk. The famous explorer Marco Polo's *Travels* roused the first wave of "China fever" among European countries. That pioneer of cultural exchanges between East and West was followed by a long list of personages in search of friendship over the centuries.

In our own era, China-Italy relations, tracing the footsteps of our ancestors, are brimming with dynamism. The People's Republic of China and the Italian Republic established diplomatic relations in 1970. In 2020, the two countries will celebrate the 50th anniversary of our relations. Through the past decades, our two countries have enjoyed mutual trust and close cooperation regardless of changes in the international landscape. Together, we have set a fine example of mutually beneficial relations between two countries that have different social systems, cultural backgrounds and stages of development. The traditional friendship between us, stronger than ever, has become a strong pillar supporting the rapid and steady growth of our bilateral ties.

Sino-Italian friendship is rooted in our long history of exchanges. In the course of over two millennia, our two countries have embraced the principles of mutual respect, mutual learning, mutual trust and mutual understanding in our interactions, principles that underpin our long-lasting, ever-strong friendship. Confronted by the transformations and challenges of today's world and informed by our deep appreciation of history, China and Italy both envision a new type of international relations that are built on mutual respect, fairness, justice and win-win cooperation, and a community with a shared future for all mankind.

Sino-Italian friendship is embedded in our deep strategic trust. Both countries' leaderships approach our relations from a strategic and long-term perspective. Since the establishment of a comprehensive strategic partnership in 2004, our two countries, guided and driven by high-level exchanges, have given each other understanding and firm support on issues concerning our respective core interests and major concerns. Our strategic trust provides a firm underpinning for the long-term and steady growth of China-Italy relations.

Sino-Italian friendship is reflected in our multi-faceted cooperation. As key trading and investment partners for each other, China and Italy have deeply entwined interests. Two-way trade exceeded US $ 50 billion in 2018 and investment surpassed US $ 20 billion in cumulative terms. "Made in Italy" is a guarantee of quality, Italian fashion and furniture are immensely popular with Chinese consumers, and pizza and tiramisu are the love of many young Chinese. Every now and then, we hear stories about the success of Sino-Italian cooperation in satellite R&D and manned space exploration. Initiatives such as the China-Italy science, technology and innovation week, joint police patrols and football training, to name just a few, are applauded by people in both countries.

Sino-Italian friendship is carried forward through our intensive cultural exchanges. Chinese and Italians have a deep interest in each other's cultures. A Chinese professor in his 70s spent 18 years translating Dante's *Divine Comedy*, and after revising several drafts, completed this mammoth task before his final days. From Martino Martini, author of the first Chinese grammar book in Europe, to Giuliano Bertuccioli and Federico Masini who wrote *Italy and China*, many Italian Sinologists have built bridges between Europe and China and contributed to a long-running boom of China studies on the Apennine Peninsula.

The well-known Italian writer Alberto Moravia once wrote, "Friendships are not chosen by chance, but according to the passions that dominate us." In a world that faces profound changes of a kind unseen in a century, the onus is on us to bring China-Italy relations to a higher level and to jointly safeguard world peace, stability, development and prosperity. Through my upcoming visit, I hope to work with Italian leaders to map out the future of our relationship and move it into a new era.

China hopes to work with Italy to strengthen our comprehensive strategic partnership. Our two countries may plan more high-level exchanges and cooperation between our governments, parliaments, political parties and subnational entities, strengthen policy communication, enhance strategic trust and synergy, and continue to give understanding and support to each other on issues of core interests and major concerns, so as to consolidate the political foundation of our relations.

China hopes to work with Italy to advance Belt and Road cooperation. Our two countries may harness our historical and cultural bonds forged through the ancient Silk Road as well as our geographical locations to align connectivity cooperation under the Belt and Road Initiative with Italy's plan to develop its northern ports and the InvestItalia program, and jointly build the Belt and Road of the new era on sea, on land, in the air, in space and in the cultural domain.

China hopes to work with Italy to expand cooperation into new areas. China will open up further to the rest of the world, and share its market opportunities with Italy and other countries through the annual China International Import Expo and other avenues. Our two countries may fully tap our cooperation potential in ports, logistics, ship-building, transportation, energy, telecommunications, medicine and other fields, and encourage our companies to partner with each other in third markets for win-win cooperation.

China hopes to work with Italy to promote closer people-to-people ties. As countries with the largest number of UNESCO world heritage sites, China and Italy have plenty of cultural and tourism resources. We may encourage our world heritage sites to forge twinning relationships and our cultural institutions and individuals to organize premium relic and art exhibitions. We may also encourage joint production of films and TV programs, the teaching of each other's languages, as well as more mutual travel and visits. Through these exchanges, we will make new contributions to the diversity of civilizations and mutual learning between different cultures.

China hopes to strengthen coordination with Italy in international affairs and multilateral organizations. China is ready to enhance communication and collaboration with Italy in the United Nations (UN), the G20, Asia-Europe Meeting and the World Trade Organization (WTO) on global governance, climate change, UN reform, WTO reform and other major issues. Working together, we will promote our shared interests, uphold multilateralism and free trade, and safeguard world peace, stability, development and prosperity.

Looking back at the last five decades, China-Italy relations have struck deep roots and borne rich fruits. Looking ahead, China-Italy cooperation will continue to flourish and prosper. The Chinese people look forward to working hand in hand with our friends in Italy to carry forward our blossoming relationship and imbue our friendship with more vitality and dynamism.

Deepening Exchanges and Mutual Learning Among Civilizations For an Asian Community with a Shared Future —Keynote Speech at the Opening Ceremony of The Conference on Dialogue of Asian Civilizations①

H.E. Xi Jinping

President of the People's Republic of China

Beijing, 15 May 2019

Your Excellencies Heads of State and Government,

Your Excellencies Heads of International Organizations,

Distinguished Guests,

Ladies and Gentlemen, Friends,

In this lovely season of thriving green, I am delighted to join you, distinguished guests from 47 Asian countries and five continents, in a dialogue on deepening exchanges and mutual learning among civilizations. On behalf of the Chinese government and people and in my own name, I extend sincere congratulations on the opening of the Conference on Dialogue of Asian Civilizations and a very warm welcome to all of you!

The world today is moving toward greater multi-polarity, economic globalization and cultural diversity, and is becoming increasingly information-oriented. All this points to promising prospects for the future. Meanwhile, instability and uncertainties are mounting and the global challenges faced by humanity are becoming ever more daunting, calling for joint responses from countries around the world.

To meet our common challenges and create a better future for all, we look to culture and civilization to play their role, which is as important as the role played by economy, science and technology. The Conference on Dialogue of Asian Civilizations is convened just for this purpose, as it creates a new platform for civilizations in Asia and beyond to engage in dialogue and exchanges on an equal footing to facilitate mutual learning.

Ladies and Gentlemen, Friends,

Asia is home to one of the earliest human settlements and an important cradle of human civilizations. This vast and beautiful continent covers a third of the earth's land mass and has two-thirds of the world's population. It has more than 1,000 ethnic

① 中国日报网.习近平在亚洲文明对话大会开幕式上的主旨演讲(双语全文)[EB/OL].(2019-05-16)[2023-08-20]. https://language.chinadaily.com.cn/a/201905/16/WS5cdcb785a3104842260bbe59.html.

groups living in 47 countries. For several thousand years before the Common Era, our forefathers living along the Tigris and the Euphrates, the Indus and the Ganges, the Yellow River and the Yangtze, tilled and irrigated the land, made tools and utensils, and built homes to live in. Generation after generation, our ancestors in Asia, with their tireless endeavors, created a time-honored history and profound and rich civilizations. Our vast and fertile plains, beautiful river basins, large steppes, immense deserts, mighty rivers and oceans, and lofty mountains have nourished and enriched diverse and colorful civilizations across Asia.

In building our civilizations over the course of several millennia, we the people of Asia have made splendid achievements. I think of literary classics such as *The Book of Songs*, *The Analects of Confucius*, *The Talmud*, *One Thousand and One Nights*, *The Rigveda* and *Genji Monogatari*; of inventions such as the cuneiform script, maps, glass, Arabic numerals, paper making and printing techniques; and of majestic structures like the Great Wall, the Great Mosque of Mecca, Taj Mahal and Angkor Wat. They are all invaluable assets of human civilization. Through interactions on this continent, Asian civilizations have enriched each other and written an epic of development.

Our forefathers in Asia have long engaged in inter-civilizational exchanges and mutual learning. The ancient trade routes, notably the Silk Road, the Tea Road and the Spice Road, brought silk, tea, porcelain, spices, paintings and sculpture to all corners of Asia, and witnessed inter-civilizational dialogue in the form of trade and cultural interflow. Today, the Belt and Road Initiative, together with the Two Corridors and One Belt, the Eurasian Economic Union and other initiatives, have greatly expanded inter-civilizational exchanges and mutual learning. Cooperation among nations in science, technology, education, culture, health and people-to-people exchanges are thriving like never before. Thanks to exchanges and mutual learning among themselves and with other civilizations in the world, Asian civilizations have grown from strength to strength.

The great Asian civilizations have a special place in the annals of world civilizations, and they have added to the diversity of human civilizations. Think of what Asia stands to offer in terms of religion, philosophy, ethic code, law, literature, painting, drama, music, and even the building of towns and villages. They speak volumes for Asia's proud achievements: extensive systems of social customs, immortal classics that have endured for millennia, the fine pool of exquisite art, and diverse institutions, among others. All these offer rich choices for civilizations the world over to draw on.

As we review our past and look beyond Asia, we should have greater confidence in our civilizations. We may build on the rich heritage of our forefathers, stay engaged with other civilizations and increase mutual learning. By doing so, we will add new glory to Asian civilizations.

Ladies and Gentlemen, Friends,

We Asian countries are closely connected and share a natural bond of affinity. We went through similar historical trials and hold the same dream for the future. Going forward, we need to see where the world is going, ride on the trend of the times and turn our people's longing for a better life into reality.

—**We Asian people hope to see peace and stability across Asia.** Upholding peace is the responsibility of every country. When peace is interrupted by conflict or war, economic growth, decent lives, social stability and people-to-people exchanges will all be out of the question. We the people of Asian countries wish to live and work in content and security free from fear. We hope that all countries will respect and trust each other, live in harmony, and interact with each other in a manner that transcends national boundaries, time and space, as well as the difference between civilizations. We should work together and jointly safeguard peace, something that is even more precious than gold.

—**We Asian people hope to see common prosperity in Asia.** Economic growth sustains a civilization, and prosperity underpins the progress of a nation. In some parts of Asia, people, women and children in particular, are still suffering from poverty, hunger and disease. This must change. We Asian people long for decent lives free of poverty. We hope that countries will work together to promote economic globalization and make it more open, inclusive, balanced and beneficial to all. Doing so will enable us to eradicate poverty and backwardness that still plague people in some countries. It will make life carefree for our children and bring happiness to all families.

—**We Asian people hope to see an open and better-connected Asia.** Asia's rapid development over the past decades shows that it is important to open one's door to the outside world and ride on the trend of global economic development. If countries choose to close their doors and hide behind them, human civilizations would be cut off from each other and lose all vitality. We Asian people hope that all countries will reject self-exclusion, embrace integration, uphold openness and promote policy, infrastructure, trade, financial and people-to-people connectivity. This way, we can jointly foster a community with a shared future for both us Asians and all humanity.

Ladies and Gentlemen, Friends,

Diversity spurs interaction among civilizations, which in turn promotes mutual learning and their further development. We need to promote exchanges and mutual learning among countries, nations and cultures around the world, and strengthen popular support for jointly building a community with a shared future for both Asia and humanity as a whole. To that end, I believe it is imperative that we act in the following ways:

First, we need to respect each other and treat each other as equals. All civilizations are rooted in their unique cultural environment. Each embodies the wisdom and vision of a country or nation, and each is valuable for being uniquely its own. Civilizations only vary from each other, just as human beings are different only in terms of skin color and the language used. No civilization is superior over others. The thought that one's own race and civilization are superior and the inclination to remold or replace other civilizations are just stupid. To act them out will only bring catastrophic consequences. If human civilizations are reduced to only one single color or one single model, the world would become a stereotype and too dull a place to live in. What we need is to respect each other as equals and say no to hubris and prejudice. We need to deepen understanding of the difference between one's own civilization and others', and work to promote interaction, dialogue and harmony among civilizations.

In the many places I have visited around the world, what fascinates me the most is civilizations in their rich diversity. I cannot but think of the Central Asian city of Samarkand, the Luxor Temple in Egypt, Sentosa in Singapore, Wat Phra Kaew in Bangkok, and the Acropolis in Athens, to mention just a few. China is ready to work with other countries to protect Asian cultural heritage and better preserve and sustain our civilizations.

Second, we need to uphold the beauty of each civilization and the diversity of civilizations in the world. Each civilization is the crystallization of human creation, and each is beautiful in its own way. The aspiration for all that is beautiful is a common pursuit of humanity that nothing can hold back. Civilizations don't have to clash with each other; what is needed are eyes to see the beauty in all civilizations. We should keep our own civilizations dynamic and create conditions for other civilizations to flourish. Together we can make the garden of world civilizations colorful and vibrant.

The beauty of a civilization finds concrete expression in the classic works of philosophy and social sciences and works of literature, music, film and TV drama. Now, a large number of outstanding cultural works from other countries are brought into China, and a lot of fine Chinese cultural products are introduced to other countries. China is happy to launch initiatives with relevant countries to translate Asian classics both from and into Chinese and promote film and TV exchanges and cooperation in Asia. This will help people in Asia better understand and appreciate each other's cultures and build a platform of exchange and mutual learning for the best of Asian civilizations to spread and be known to more in the world.

Third, we need to stay open and inclusive and draw on each other's strengths. All living organisms in the human body must renew themselves through metabolism; otherwise, life would come to an end. The same is true for civilizations. Long-term self-isolation will cause a civilization to decline, while exchanges and mutual learning will sustain its development. A civilization can flourish only through exchanges and mutual learning with other civilizations. Such exchanges and mutual learning should be reciprocal, equal-footed, diversified and multi-dimensional; they should not be coercive, imposed, one-dimensional or one-way. We need to be broad-minded and strive to remove all barriers to cultural exchanges. We need to be inclusive and always seek nourishment from other civilizations to promote the common development of Asian civilizations through exchanges and mutual learning.

People are the best bridge for exchanges and mutual learning among civilizations. Closer people-to-people exchanges and mutual learning, for that matter, is a sure way to eliminate estrangement and misunderstanding and promote mutual understanding among nations. Over the years, China has, in collaboration with other countries, established many platforms and channels for cooperation in education, culture, sports, health and other fields. China will work with other countries to step up exchanges among the youth, non-governmental organizations, subnational entities and media organizations, create a network of exchanges and cooperation between think tanks, explore new models of cooperation, and deliver more solid outcomes in diverse forms. Such efforts will boost exchanges and mutual learning among civilizations.

Fourth, we need to advance with the times and explore new ground in development. For a civilization to endure, efforts must be made to keep it alive and build on its heritage from one generation to the next. More importantly, a civilization needs to adapt itself to the changing times and break new ground. The history of world civilizations tells us that every civilization needs to advance with the times and take in the best of its age in order to develop itself. We need to come up with new ideas to add impetus and inspiration to the development of our civilizations. With these efforts, we will deliver achievements for our civilizations to transcend time and space and have a lasting appeal.

To spur people's innovation and creativity, the best way is to come into contact with different civilizations, see the strengths of others and draw upon them. Last year, Chinese tourists made over 160 million overseas trips and more than 140 million foreign tourists visited China. These visits played an important role in promoting exchanges and mutual learning between China and the rest of the world. In this connection, China will work with other countries to implement a plan to promote tourism in Asia. This will further boost economic development in Asia and deepen friendship among the Asian people.

Ladies and Gentlemen, Friends,

Being an inseparable part of Asian civilizations, Chinese civilization has, since its early days, evolved and grown by drawing on its past achievement, exploring new ground and adapting to changes. It represents the profound pursuit of the Chinese nation and provides a rich source of strength for its lasting development. Chinese inventions such as paper making, gunpowder, printing and the compass as well as China's astronomical knowledge, calendar system, philosophy and the people-centered doctrine have all had a global impact and propelled the development of human civilizations.

Chinese civilization, as an inclusive and integrated whole, has become what it is today through constant interactions with other civilizations. It has been enriched by the introduction of Buddhism and the confluence of Islam and Confucianism in the old days, and by the introduction of Western learning, the launch of the New Culture Movement and the introduction of Marxism and socialism in modern times. All-round opening-up of the country, starting with the reform and opening-up program, has added to its vitality today. For Chinese civilization, amity and good neighborliness is the principle guiding our interactions with other countries; and to deliver prosperity and security to the people is the overarching goal, to keep pace with the times through reform and innovation the abiding commitment, and to achieve harmony between man and nature the underlying philosophy.

China today is more than the country itself; it is very much a part of Asia and the world. In the time to come, China will open its arms wider to embrace the world and contribute the dynamic achievements of Chinese civilization to a better world in the future.

Ladies and Gentlemen, Friends,

The Conference on Dialogue of Asian Civilizations has a wide-ranging agenda, and I look forward to your keen perspectives and insights. By putting our heads together, we will create an even better tomorrow for civilizations in Asia and beyond!

To conclude, I wish this conference every success!

Thank you.

第七章实战训练篇参考译文

2021 年波音公司首席执行官在虚拟年会上向股东们的致辞①

戴维·L·卡尔霍恩　波音公司总裁兼首席执行官

2021 年 4 月 20 日

拉里，谢谢你的介绍，也谢谢你在董事会发挥的领导作用。同时，我要向所有与会嘉宾表示热烈欢迎。

新冠疫情仍在全球肆虐，我衷心祝愿各位平安健康。去年这个时候，我们才刚刚开始了解疫情对人们生活和工作的影响。过去的一年是极不寻常的一年。面对挑战，我们团队始终致力于向客户、相关社区、客户的同事及家人提供支持，对此我倍感自豪。

疫情期间，我们第一时间向疫区伸出援手，展现了人间大爱，这一切深深感动着我。疫情初期，尽管我们许多工厂停产，但我们的团队积极动员，生产 3D 打印面罩，调度自己的机队，向防疫前线运送大量个人防护用品。

事实上，此次全球危机规模之大，让我们始料未及。这场危机对民众生命造成的损失难以估量，造成了破坏性影响。在此，我向因疫情而失去家人、朋友或同事的每个人表示深切慰问。

新冠病毒给人们带来痛苦的同时，也导致需求暴跌，所有行业均受到重创，全球旅游业尤甚。对于航空公司客户、机场、酒店、出租车、餐厅、会议、赛事活动和所有其他旅游相关行业来说，2020 年是具有毁灭性的一年。

虽然困难重重，但我们看到所有行业从未像今天这样同心协力，相互支持。值得庆幸的是，2021 年是一个关键的转折点。尽管全面复苏可能还需数年时间，但我们已经看到了好转的迹象，深感鼓舞。

我们由衷地感谢那些杰出的科学家，他们不知疲倦，以惊人的速度成功研制出新冠疫苗。在 2020 年 5 月举行的第一次疫情影响评估中，我们从未想到，一年后高达 50%的美国成年人接种了疫苗。随着世界各国政府加快疫苗接种速度，我们开始逐渐回到疫情暴发前的正常生活。

过去一年，我们中许多人采用了新的工作方式。事实上，我们正在为大多数员工延长远程办公时间，实施线上线下相结合的工作安排。

尽管这种新形势为改变未来工作模式创造了机遇，但我知道，航空旅行使面对面交流成为可能，而面对面交流的价值是任何东西都无法取代的。面对面时，我们会建立更牢固、更持久的联系，会激发创造力和创新，会增强彼此之间的信任。因此，航空

① 编者译。

旅行一次又一次地证明了其坚韧的能力。我们相信,疫情一旦结束,航空需求势必呈历史性长期增长之势。

在此之前,我们与政府、医疗保健行业和学术界的专家共同推出"放心出行倡议",向世人证明,乘客在乘坐飞机时感染新冠肺炎病毒的可能性要比日常生活小得多。波音飞机使用了高效过滤器,可去除空气中99.9%以上的病毒和微粒。这些过滤器与医用过滤器类似,每两到三分钟就会给整个客舱彻底换气。

机舱设计和气流系统作为全方位的安全措施之一,再加上乘客采取的佩戴口罩及严格的清洁程序等安全措施,确保了整个航空旅行过程中乘客和机组人员的健康。

我们同时还与政府部门和行业协会合作,确保乘客知晓旅行过程中的细节。我们鼓励使用数据驱动和基于风险的新方案,最大限度降低不同国家之间的疾病传播风险。

若要扩大国际旅行,采用标准、稳妥的方式验证旅行者信息,无论是新冠病毒阴性检测结果还是疫苗接种证明,都是必要的。我们鼓励各国政府和行业利益相关者就验证方案达成一致,提升航空旅行安全的同时,保持全球互联互通,推动经济复苏。

在优先保障公共卫生和安全,推动国内和国际旅行复苏的同时,我们同样关注重大国防、太空和安全使命,不仅为我们国家,同时对世界各地的盟友提供保护。

为了满足行业和客户的需求,我们启动内部转型,以适应疫情带来的市场影响,让我们的业务更精简、更敏捷、更可持续。

作为转型计划的一部分:

- 我们正在全方位评估基础设施,尽可能高效利用我们的工厂、仓库、实验室和办公室,减少对环境的影响。
- 我们正在审视我们组织结构,思考如何减少官僚主义的滋长,精简层级。
- 我们正在重新评估我们的资产和投资,确保优先投资核心市场机会、可持续发展和下一代关键技术。
- 我们会向供应商提供支持,帮助他们稳定生产,应对需求降低。
- 我们正在全公司范围内提升运营水平,确保稳定性、提高产品质量、确保工作场所安全并减少返工。最近787客机暂停交付就是明证:在保障生产线稳定的同时,我们对787客机进行了全面检查,为市场复苏做好准备。

我们根据目前的市场状况,作出了裁员的艰难决定。在这个充满挑战的过程中,我们仍然对未来充满憧憬。我们尽量确保我们有必要的团队、资源和投资,以履行对客户的承诺,推动改进计划,面向未来进行创新。

在转型过程中,我们始终脚踏实地,始终遵循核心价值观。

我们不会忘记在印尼狮航610航班和埃塞俄比亚航空302航班空难中逝去的生命,不会忘记我们的不足之处。这无时无刻不在激励我们成为更好的波音。

我们的首要任务是保障每一位乘客的安全。波音领导团队向你们郑重承诺,无论现在还是将来,没有什么比安全更重要。要实现这一承诺,首先要确保所有团队成员以完全透明和高度诚信的方式开展工作。

企业文化始于高层,全球领导团队正在尽可能将更多的时间和注意力转移到我们的日常工作上去。我们正共同培养一种信任文化,去鼓励和认可透明、问责和诚信。

波音是一家工程公司,现在是,将来也是。我们设计、搭建并持续发展世界上最先进的技术,从海洋深处到浩瀚太空都有我们的身影。为了加强自身的基础能力,提升所有工程设计和决策的创新性、透明度、协作性和问责制,我们集聚五万名工程人员,打造一体化的全球社区。

我们还填补了重要的领导岗位,聘任了新的软件工程副总裁、新的首席技术官以及首席航空安全官。

与此同时,我们正在整个公司推进企业安全管理系统,基于及时的数据、分析和洞察,加强日常管理,提高安全性能。该系统将完全嵌入到所有产品和服务的设计、构建和支持流程,向监管机构和客户报告每一步进展。

从 2021 年开始,产品安全、员工安全和质量指标将纳入主要高管、管理层和员工年度激励绩效方案。这一举措进一步推动了公司各层级对安全和质量的关注。

2020 年 11 月,经美国联邦航空管理局批准,737 MAX 飞机安全恢复运营,这是一个重要里程碑。

我们将继续与全球监管机构和客户合作,推动 737 MAX 飞机在全球范围内恢复运营。自恢复运营以来,波音已向全球客户交付了超过 85 架 737 MAX 飞机。21 家航空公司已恢复运营,并安全飞行 23000 多个商业航班,总飞行时间超过 52000 小时。

正如我们之前所讨论和报告的那样,波音发现 737 MAX 驾驶舱特定组件存在潜在的电气问题。波音正在与美国联邦航空管理局合作,确定解决方案,并与美国联邦航空局和客户紧密合作,共同采取行动。

我们对每架飞机进行全面检查确保其符合波音最高标准之后,恢复了 787 飞机的交付。虽然这项工作影响了公司近期的财务状况,但这是正确的做法,也再次证明了我们持续关注质量和业务的长期健康发展。

在国防业务方面,我们在许多项目上取得了重要进展。例如,我们提前为美国空军交付了第一架 F-15EX“鹰Ⅱ”战斗机,并已开始生产美国空军新型 T-7A“红鹰”高级教练机。当然,我们的 X-37B 无人太空飞机还获得了 2020 年航空航天领域科利尔奖,对此我们深感荣幸。

在实现国家太空计划重要里程碑过程中,我们还努力确保关键太空系统的安全性和完整性。今年早些时候,我们成功完成了 NASA 太空发射系统火箭的热火测试,我们十分期待下一次波音星际客机的发射,正在为其首次前往国际空间站做准备。NASA 和我们的团队基于安全第一的原则,决定推迟首次载人飞行任务。

在服务方面,我们的数字产品和政府服务产品发挥了重要作用,帮助我们迅速适应了疫情带来的影响。例如,我们与边疆航空签署了一份为期 10 年的机队数字服务协议,我们还与澳大利亚皇家空军签订了一份为期六年的 P-8A 反潜巡逻机支持合同。此外，随着电子商务推动货运需求不断上升，我们迅速与客户合作改装货机，满足客户对货运机队不断增长的需求。这些及其相关成绩强化了我们在服务业务的长期战略，满足客户不断变化的生命周期需求。

今天，我们向客户安全交付产品的同时，也在推进颠覆性的技术创新。我们预计，研发投资将使下一代飞机性能更高、效率更高、更易于维护和重新装配。

与上一代飞机相比，如今交付客户的飞机燃油效率提高了15%到25%，不仅有利于帮助客户提高经济效益，同时还能帮助我们应对气候变化。此外，我们承诺在2030年前交付可使用100%可持续燃料的商用飞机。

预计未来十年，碳减排需求将大幅增加，我们计划加速推动并引领航空业实现雄心勃勃的气候目标。

为了凸显这项工作的重要性，我们在2020年任命了首席可持续发展官，进一步推进可持续发展，继续关注环境、社会和治理方面的重点工作。

作为对可持续航空承诺的一部分，波音一直在探索、开发先进飞机的概念，参与行业合作，满足2030年及以后具体的能效、环境和运营目标。

波音和基蒂霍克的合资企业Wisk是我们的一个合作伙伴。Wisk是美国首家开发并成功试飞自主、全电动垂直起降飞机的公司，自2017年以来已完成1500多次全面试飞，实现零事故。2020年，Wisk和新西兰政府宣布合作试验世界上第一辆自动驾驶空中出租车Cora，波音子公司Insitu Pacific也将参与其中。

除了商务旅行、可持续发展和先进飞机研发之外，我们还在进行关键投资，为美国军队提供技术优势，确保他们无论面临什么威胁总能领先一步，完成关键安全任务，每次都能平安返航。

我们希望波音的客户未来这样评价我们：在今天这样的历史时刻，我们认真倾听，调整适应，保持专注，孜孜不倦，不断精进。

我们还在整个公司推行反种族主义措施，推进种族平等和包容发展。

我们扩大并重组了执行委员会，形成更加多元化的高管团队，全方位构建公司的战略、运营和文化体系。我们还制定了一项种族平等和包容行动计划，提高了公平和包容措施的衡量标准，并严格遵照执行。

今年，我们将发布公司首个全球公平、多元化和包容性报告。但我们认识到，要实现我们的目标，仍有很多工作要做。我们始终把全方位对抗种族主义和歧视作为我们价值观的核心，并为之付出长久的努力。

方便大家出行，让不同国家、不同文化的人们建立真正的联系，这让我们引以为傲。同事、朋友和家人之间面对面的交流和对话会让我们认识到，我们之间的分歧远没有我们想象的那么多。

从空中旅行到太空旅行，从全球安全到全球贸易，波音在全世界发挥着至关重要的作用。几乎每一次客人登门拜访，每一次包裹送达您家或公司门口，很可能背后就有波音的功劳。同样，美国及盟国的现役军人依赖我们的技术、产品和服务来保护国家和个人安全，完成海陆空甚至太空任务。

波音的宗旨是保护、连接和探索我们的以及我们之外的世界，我们始终致力于了解、满足并超越利益相关方的期望。

过去几年，我们曾经面临破产清算，但我们的公司和员工在艰难时期展现出了最好的精神面貌。在这个充满挑战的时期，我们的客户、供应商和股东都给了我们巨大的支持，让我们深受鼓舞。

我想借此机会感谢格雷格·史密斯。在过去一两年里，他一直是波音应对前所未有挑战不可或缺的一部分。格雷格已决定于七月退休。在30年的职业生涯中，他为波音作出了巨大贡献。近十年，他有效管理公司财务状况，在行业因疫情面临严峻挑战时展现出非凡的领导能力，对于波音走向光明的未来至关重要。

他曾领导波音历史上最大规模的债券发行，启动了一项全面的转型计划，使我们的业务进一步壮大，更能适应未来的发展。格雷格还通过管理企业运营、可持续发展和战略，推动波音成为一家更优秀、更具竞争力的公司。他的领导给波音带来深远的影响。我还要感谢格雷格对未来交接工作的支持，并且感谢他在我们选择继任者时提供的建议。请大家和我一起对格雷格的巨大贡献表示感谢，祝愿他未来一切顺利。

最后，感谢我们的波音团队。你们的激情、你们的奉献、你们的坚韧，每天都鼓舞着我。我非常自豪成为你们中的一员，正是因为你们，我对我们共同的未来充满信心。

拉里，我的致辞到此结束，谢谢大家。

Staying Connected with the World and Abreast with the Times And Making Big Strides on the Path of Sustainable Development①

—Keynote Speech by H. E. Xi Jinping

President of the People's Republic of China

at the Opening Ceremony of the Second United Nations

Global Sustainable Transport Conference

14 October 2021

Your Excellency Secretary-General António Guterres,

Dear Colleagues,

Ladies and Gentlemen, Friends,

It gives me great pleasure to attend the second United Nations Global Sustainable

① China Daily. Staying Connected with the World and Abreast with the Times And Making Big Strides on the Path of Sustainable Development[EB/OL]. (2021-10-15)[2023-08-20]. https://www.chinadaily.com.cn/a/202110/15/WS6168ec27a310cdd39bc6f301.html.

Transport Conference and discuss what is important for global transport and development. Let me begin by extending, on behalf of the Chinese government and the Chinese people and also in my own name, warm congratulations on the convening of the Conference and a hearty welcome to our distinguished guests.

Transport is the artery of the economy and a bond between civilizations. A review of history shows that transport, in the forms of camel caravans and sailing boats on the ancient Silk Road, wave-breaking vessels in the Age of Exploration, and criss-cross transport networks in this modern age, has facilitated economic integration and people-to-people exchanges and turned the world into a close-knit global village.

As we speak, major changes unseen in a century, compounded by a once-in-a-century pandemic, are posing serious challenges to the global efforts of growing the economy and bettering people's lives. It is imperative that we follow the prevailing trend of world development, advance global transport cooperation, and write a new chapter featuring connectivity of infrastructure, unfettered flows of trade and investment, and interactions between civilizations.

First, we need to uphold open interplay and enhance connectivity. A big river is full when its tributaries are filled with water; and tributaries must be dry when there is no water in the big river. Only with openness, inclusiveness and connectivity can countries reinforce each other's efforts and achieve win-win results. It is important that we pursue an open world economy, reject discriminatory or exclusive rules and systems, and make economic globalization more open, inclusive, balanced and beneficial for all. We should strengthen both hard connectivity of infrastructure and soft connectivity of institutions and rules, and develop four-dimensional connectivity of land, sea, air and the Internet.

Second, we need to uphold common development and promote fairness and inclusiveness. Only when countries develop together can there be true development; only when countries prosper together can there be true prosperity. The COVID-19 pandemic has aggravated the wealth gap and widened the North-South divide. Uneven development must be addressed before broader prospects for humanity's common development could be brought about. We should leverage the enabling role of transport and increase related input in poor regions, so that local economies and people's lives could improve as a result of better roads. We should enhance North-South and South-South cooperation and scale up support for developing transport infrastructure in the least developed countries and landlocked developing countries, in an effort to achieve common prosperity.

Third, we need to uphold an innovation-driven approach and create more drivers for development. Our world is going through a new round of scientific and technological revolution and industrial transformation. Digital economy, artificial intelligence and other new technologies and new business forms have become strong technological underpinnings for economic and social development. More should be done to develop smart transport and smart logistics and promote deep integration of new technologies like big data, the Internet, artificial intelligence and blockchain with the transport sector, to ensure easier movement of people and smoother flow of goods.

Fourth, we need to uphold ecological conservation as a priority and pursue green and low-carbon development. The only durable way to achieve sustainable development is to establish an economic system for green and low-carbon development and promote green transition in all respects of economic and social development. More efforts are needed to foster a green and low-carbon way of transport, step up green infrastructural development, promote new energy, smart, digital and light-weight transport equipment, and encourage and advocate green travel, to make transport and travel more environment-friendly and low-carbon.

Fifth, we need to uphold multilateralism and improve global governance. We live in a world where the future and destiny of countries are closely linked and their interests entwined like never before. We need to follow the vision of global governance featuring extensive consultation, joint contribution and shared benefits, pool the wisdom and strength of all, and mobilize resources from across the globe to meet global challenges and promote global development. We should uphold the authority and status of the United Nations and, in the context of implementing the 2030 Agenda for Sustainable Development, advance cooperation in such areas as poverty reduction, health, transport and logistics, and infrastructural development.

Not long ago, I proposed a Global Development Initiative, which is designed to speed up the implementation of the 2030 Agenda for Sustainable Development, promote more robust, greener and more balanced global development, and build a global community of development with a shared future. I hope more will join the Initiative.

Ladies and Gentlemen, Friends,

Since the founding of New China, generation after generation of the Chinese people have worked in the spirit of opening roads through mountains and putting bridges over rivers, and turned China into a country with vast transport infrastructure.

Today, we are redoubling our efforts to build a country with great transport strength. Convinced that transport should come first, we have built the world's largest high-speed railway network, expressway network and world-class port clusters. We have opened air and sea routes that reach all parts of the world. We have set up an integrated transport network exceeding six million kilometers. Convinced of the need for innovation, we have achieved major breakthroughs in equipment manufacturing like high-speed trains and large aircraft. We have more than half of the world's new energy vehicles. Mega transport projects like the Hong Kong-Zhuhai-Macao Bridge and Beijing Daxing International Airport have been completed and put into operation. Transport has become a frontier in China's modernization drive. Convinced of the importance of global connectivity, we have become the economy best connected to the global shipping network and with the highest volume of trade in goods. During the COVID-19 pandemic, the China-Europe Railway Express and ocean-going cargo vessels have been running day and night to keep global industrial and supply chains stable, showcasing China's sense of responsibility in the global community.

Ladies and Gentlemen, Friends,

China will continue to hold high the banner of true multilateralism, and stay connected with the world and abreast with the times. This way, we will contribute more to global development while pursuing our own development.

I wish to reiterate that China will not change course in its pursuit of a new system of open economy of higher standards, and China will not waver in its resolve to promote trade and investment liberalization and facilitation. China's door of opening-up will only open wider, and will never be closed.

China will continue to advance high-quality Belt and Road cooperation, strengthen infrastructure connectivity with other countries, and develop a green Silk Road and a digital Silk Road at a faster pace. Here I announce that China will set up a Global Innovation and Knowledge Center for Sustainable Transport, as a contribution to global transport development.

Ladies and Gentlemen, Friends,

Let us stick together on the promising path of connectivity and mutual benefit, jointly build an open, inclusive, clean and beautiful world that enjoys lasting peace, universal security and common prosperity, and promote the building of a community with a shared future for mankind.

Let me wish the conference a full success.

Toast at the Welcome Dinner Held by The Government of The Hong Kong Special Administrative Region①

H. E. Xi Jinping

President of the People's Republic of China

Hong Kong, 30 June 2017

Chief Executive Leung Chun-ying,

Chief Executive-Elect Carrie Lam Cheng Yuet-ngor,

Fellow Compatriots,

Dear Friends,

Good evening! It gives me great pleasure to come back to this beautiful city after an interval of nine years and celebrate with you the 20th anniversary of Hong Kong's return to the motherland. On behalf of the Central Government and people of all ethnic groups across China, I would like to extend our warmest greetings and best wishes to you and, through you, to all Hong Kong residents.

—Time keeps rolling on. In the blink of an eye, the Hong Kong Special Administrative Region is 20 years old. The moving occasion of Hong Kong's return to the motherland in 1997, like a long-separated child coming back to the warm embrace of his mother, is still vivid in our memory. We still recall the solemn ceremony of the transfer of government in Hong Kong, the playing of the stirring national anthem of the People's Republic of China and the raising of the national flag of the People's Republic of China and the regional flag of the Hong Kong Special Administrative Region. We still recall the joy and excitement of Hong Kong people who cheered the processions of the People's Liberation Army Hong Kong Garrison despite downpours. And we still recall the festive celebrations across China where people were singing and dancing to hail Hong Kong's return. These historical scenes have become a part of the collective memory of all the Chinese people.

—All crops grow through cultivation. Over the last two decades, the practice of "One Country, Two Systems" in Hong Kong, like the growth of a seedling, has become strong and robust despite wind and rain and yielded many fruits. The SAR system established under *the Constitution of the People's Republic of China* and *the Basic Law of the Hong Kong Special Administrative Region* has operated effectively, and democracy in Hong Kong has been advanced in accordance with law. The multiple

① xinhua. Full text of speech by President Xi Jinping at welcome dinner in HK[EB/OL]. (2017-07-01)[2023-08-20]. http://www.chinadaily.com.cn/china/hk20threturn/2017-07/01/content_29958522.htm. (文字有校改)

indicators of governance of the HKSAR such as government effectiveness and the rule of law are much higher than those before Hong Kong's return to the motherland. Hong Kong enjoys steady economic growth and remains one of the most competitive and free economies in the world. Hong Kong has maintained social stability and scored great progress in various fields of endeavor. Its average life expectancy is one of the highest in the world. While such achievements would not have been possible without the strong support of the Central Government and the mainland, they are primarily attributed to the unity and hard work of the HKSAR Government and people of Hong Kong. Here, I wish to salute you, the people of Hong Kong and all those in Hong Kong who have made such achievements possible.

—Now is the time to build on past achievements and make new progress. "One Country, Two Systems" is a great pioneering initiative by China. Its practice, featuring socialism in the main body of the country and capitalism in certain regions, is unprecedented in human political history. It is a breakthrough those before us made through exploration and with extraordinary courage. We the succeeding generation should practice and develop "One Country, Two Systems" with firm resolve. The road ahead may not be smooth, but our commitment to "One Country, Two Systems" remains unchanged, and our resolve remains firm and strong. We are fully confident that we are able to "ride the wind and cleave the waves", and achieve even greater success in the practice of "One Country, Two Systems" in Hong Kong through persistent and unrelenting efforts like sifting through sand thousands of times to find gold.

On this occasion, I wish to share with Hong Kong compatriots the importance of having confidence in the following three areas:

First, we should have confidence in ourselves. We Chinese are a great people. Our time-honored 5,000-year civilization is the only ancient civilization that has survived with no interruption. For a great part of recorded history, the Chinese nation led the world in economic, scientific, cultural, art and other fields, and contributed much to the progress of human civilization. China lagged behind other countries in modern times, but that has changed since the founding of New China in 1949. The Chinese nation, under the leadership of the Communist Party of China and thanks to dedicated efforts of the Chinese people of several generations, has proudly taken its place among the nations of the world.

Thanks to the hard work of several generations of its people, Hong Kong has transformed itself from an unknown fishing village into a world-renowned modern metropolis. The key to Hong Kong's success lies in its people's love of the country and

Hong Kong, and their perseverance, pursuit of excellence and adaptability. I want to particularly commend Hong Kong people's participation in and significant contribution to China's reform, opening-up and modernization drive. The Central Government and people across the country have never forgotten what you have done. The people of Hong Kong have the ability and wisdom to administer Hong Kong well and achieve greater development and progress in Hong Kong. I am sure you will continue to do all you can for the development of the country and distinguish yourselves on the world stage.

Second, we should have confidence in Hong Kong. Hong Kong is blessed with many favorable conditions and unique strengths for development. It has a highly free and open economy featuring free flow of factors of production such as personnel, goods and capital, and this is a major factor in both attracting international capital and retaining local capital. With its internationally recognized legal, accounting and regulatory systems, a full-fledged service sector, clean and efficient government and business-friendly environment, Hong Kong has the full confidence of outside investors. Hong Kong is an important international financial, shipping and trade center, a major conduit connecting the mainland and international markets and a two-way service platform for China both to attract foreign investment and enter the international market. Hong Kong is by far the mainland's biggest source of external direct investment and non-local financing platform, and it has grown into the mainland's largest non-local investment destination and the biggest offshore RMB business center in the world. More importantly, the practice of "One Country, Two Systems" has given Hong Kong an institutional advantage, enabling it both to share in the mainland's vast market and development opportunities and serve as a testing ground for the country's new opening-up initiatives, and this gives Hong Kong an edge in pursuing development. Pilot programs of the Shanghai-Hong Kong Stock Connect and the Shenzhen-Hong Kong Stock Connect have both been launched in Hong Kong, so will be the "Bond Connect". By building on and leveraging these strengths, Hong Kong will surely be able to keep and attract investment and talent, seize opportunities presented by economic globalization and regional cooperation to promote innovative local business start-ups, and develop new growth drivers. The legendary city of Hong Kong by the Lion Rock will surely open a new chapter of development and prosperity.

Third, we should have confidence in our country. The motherland has given and will always give a strong backing to Hong Kong. Thanks to close to 40 years of reform and opening-up, China has made big strides forward: from first managing to stand on its feet to becoming prosperous and strong. China is now the world's second largest economy and its leading manufacturer and trader in goods. China has the world's largest foreign exchange reserves, and it has contributed more to global economic growth than any other country. Its scientific and technological strength is rising, with advanced achievements made that include the high-performance computer, manned spaceflight, lunar exploration program, quantum communications, BeiDou Navigation Satellite System and manned deep-sea submersible. China's high-speed railway has entered the world market. The independently-developed C919 airliner made a successful maiden flight. Our circle of friends is growing: The China-initiated Asian Infrastructure Investment Bank now has 80 approved members, and over 100 countries, regions and international organizations have actively participated in the Belt and Road Initiative launched by China. The Belt and Road Forum for International Cooperation was successfully held in Beijing six weeks ago, and China's call to boost global development has added strong impetus to the endeavor of building a community of shared future for mankind. An increasingly prosperous motherland serves as a source of strength for Hong Kong to overcome difficulties and challenges; it also presents a reservoir of opportunities for Hong Kong to break new ground, foster new driving force and create new space for development. When our country does well, Hong Kong will do even better.

Fellow Compatriots,

Dear Friends,

A popular song in Hong Kong has this line: "Self-confidence is so important. Open up your mind and your dream will come true." We should have full confidence in ourselves, in Hong Kong and in our country, fully and faithfully implement the policies of "One Country, Two Systems", "Hong Kong people administering Hong Kong" and a high degree of autonomy, abide by *the Basic Law of the HKSAR*, and make focused and concerted efforts to pursue development. By doing so, we can certainly deliver an even brighter future for Hong Kong.

To conclude, I wish to propose a toast:

To a prosperous and strong China and the great renewal of the Chinese nation;

To the long-term prosperity and stability of Hong Kong; and

To the health of all the friends present and your families.

Cheers!

第八章实战训练篇参考译文

“我们一起飞得更高”

——在旋翼机安全会议上的讲话①

美国联邦航空管理局局长　斯蒂芬·M·迪克森

2020 年 10 月 27 日

大家好，感谢各位参加美国联邦航空管理局(FAA)旋翼机安全会议。我想感谢 FAA 合规和适航部主管兰斯·甘特和整个团队，让我们能够在线下见面之前，持续在线上开展安全对话。

我还要感谢史蒂夫、兰斯和韦恩在会议开幕时就 FAA 旋翼机认证和飞行标准优先事项进行了精彩的讨论。

认证和飞行标准是支持 FAA 广泛安全要求的众多组成部分中的两个，旨在提供世界上最安全高效的航空航天系统。

旋翼机对该航空航天系统的效率和生产率来说至关重要。

这一点在今年春季体现得淋漓尽致，当时新冠疫情导致航空运输网络大规模瘫痪，但旋翼机作业并未受到影响。而直升机作业在五月中旬就恢复到了正常水平。

无论是服务于警方、快递、公用事业、企业通勤，还是实现其他数百种用途，旋翼机都是必不可少的。这些事情是其他飞行器无法做到的。在座所有人都明白这一点，而且各位都想促进行业发展，这也是大家今天汇聚在此的原因。

我之所以来到这里，以及 FAA 之所以召开这次会议，是为肯定这一独特行业的发展，并集思广益，汇聚来自政府、行业和学术界的各方智慧，不断提升行业的安全水平。

为了充分发挥航空业各部门的潜力，保障安全是重中之重。众所周知，航空业的安全水平堪称“黄金标准”。

实现成功的关键因素之一是成员间的合作、伙伴关系和信息与数据的共享，参与者包括在系统中发挥作用的每个机构和人员，如 FAA、制造商、飞行员、机械师、调度员、空乘人员，等等。我把幸存者也包括在内了，因为他们承受了由于我们失误而造成的后果。

我们越来越多地使用安全管理系统，即 SMS，来规范和简化组织内的信息和数据流。大家知道，SMS 是航空规章 121 部的规定内容，我们正在整个航空航天产业逐步推广 SMS 实践。

SMS 的重要组成部分是飞行数据监控和安全报告。这些都是将安全置于首位、

① 编者译。

主动的、数据驱动的监督方法。为了取得成功,这些方法遵循公正文化,高度重视一线员工提出或报告的安全问题。

有了公正文化,飞行员和航空工作人员就会感觉自己有如实上报错误和问题的权利,而且不用担心受到责罚。这种氛围使工作人员可以自由报告,也为管理层提供了数据,从而预先察觉可能发生的事故。

我们鼓励航空公司采用并将飞行数据监控反馈到培训计划中,最好是将其作为SMS流程的一部分。

我们将安全管理原则整合到设计和制造流程中,通过协调风险管控流程和设计、制造、运行与维护之间的反馈回路来确保系统安全。

大家可以看到,我们坚信SMS的力量。事实上,现如今FAA的目标是在2022年春季发布一项SMS规则提议,该规则将适用于空中出租车、航空旅行社、维修站和PMA零部件供应商。我们还在为机场开发SMS。

当然,大家不需要等待规则的出台。通过自愿实施SMS,落实安全风险管控流程,从而识别危险,防止事件、事故的发生。关键是要能够识别和理解运营风险,而这正是SMS的作用。

具体到旋翼机部分,有各种渠道可以分享信息,让我们所有人都更加安全。

我们开展"面向基层"的研讨会,把FAA安全小组,或者FAAST小组,以及行业安全专家直接带到当地飞行员身边,一起讨论某些事故场景,作为飞行员决策培训的起点。

由于新冠疫情,我们不得不暂时中止这些线下会议,但好消息是,我们即将推出线上研讨会,与会者将就飞行员在危险的直升机飞行中应当如何应对挑战进行实时投票。

分享经验、取长补短的一个好方法就是参加我们新设立的直升机InfoShare计划。InfoShare是我们与航空公司合作开展的一个项目,这一项目的成功正在引导其他部门采用同样的模式,包括商务航空,以及现在的旋翼机。

另一个途径则是利用直升机安全咨询委员会,该委员会为石油和天然气行业的旋翼运营企业制定了推荐做法,这些做法也可以灵活适用于其他直升机部门。

老话说,我们要"跳出条条框框思考问题"。对于旋翼机行业来说,现在就需要大家这样做。十五年来,直升机死亡事故率基本保持不变。正如我刚才所说,我们必须努力推动"零死亡事故"这一安全目标的实现。

任何事故,尤其是死亡事故,都是不可接受的。这就是我们支持美国直升机安全小组的原因,他们的主要任务就是实现零死亡事故。政府和行业团体,包括FAA的一些安全专家,正在采取科学方法,敦促(有关部门)采用基于数据的安全建议。

我们还强烈提倡运营企业进行自愿安全升级,包括直升机乘员保护功能。

为什么这一点如此重要?因为90%以上的直升机死亡事故都与钝器外伤有关。

对于新的直升机设计,认证规则要求通过防撞座椅和周围结构提供潜在的救生保护。但我们并未要求传统机队中的数千架直升机具备这些功能。为什么不考虑加装这些升级设备呢?

我们还希望看到防撞燃料系统等安全改装项目。根据我们2018年重新授权的要求,FAA要求2020年4月5日之后制造出的直升机必须拥有这些即开即用的系统。我们真心希望把这些相同的系统提供给客户,同时运营企业也自愿在传统直升机队中安装这些系统。

我想大家可以看到,我们已经有许多方法来提升旋翼机行业的安全记录,而且我们也一直在寻找新的想法。

这就是在座各位的用武之地。请利用这次会议重新评估,重视直升机的安全,还请您告诉那些还没有加入我们的朋友。现在是时候加入我们了。我们有能力做出真正的改变。

安全必须是头等大事,是指引我们前进方向的北极星。

各位不必只相信我的一面之词,接下来大家将会听到三个人的发言,他们会展示安全的重要性。像其他许多在航空事故中失去家人、亲人的人一样,戴夫、阿曼达·雷普舍尔以及凯伦·马哈尼已然促成了行业的变革。

听了他们的故事,大家就会深刻地理解我们为何需要全力以赴,以提高安全水平。

我们要尽可能保证航空行业的安全,拯救生命,努力争取实现零死亡事故。也许达成这一目标的进展缓慢,但却稳定而坚决。

谢谢大家,祝愿会议圆满成功!

Speech at the First International Summit on BDS Applications (Excerpt)①

Changsha, 16 September 2021

Leaders, Distinguished guests, Ladies and Gentlemen, Friends,

Good morning!

On the occasion of the first anniversary of President Xi's inspection of Hunan Province and the completion and global operation of the BeiDou-3 Global Navigation Satellite System, the first International Summit on Scale Applications of BeiDou is held in Changsha. First of all, on behalf of Hunan Provincial Party Committee and the government, I wish to extend congratulations on the opening of the summit. On behalf of the Organizing Committee, I would like to extend my sincere welcome to all the guests attending the summit both online and in person. I also wish to thank all those who have been caring and supporting Hunan's development.

---

① 编者译。

The BeiDou Navigation Satellite System (BDS) is a major national science and technology project and a critical spatiotemporal infrastructure implemented under the decision of the CPC Central Committee. With Comrade Xi Jinping at its core, the CPC Central Committee attaches great importance to the construction and application of the BDS. Last year, on July 31, President Xi announced the official launch of the BDS-3, which marked the completion of an independent, open and compatible global navigation satellite system in China. It also marked an important milestone in our strive for sci-tech breakthrough in aerospace industry. Today, President Xi also sent a congratulatory letter to the Summit, speaking highly of the fruitful achievements China has made in satellite navigation, elaborating on the positive impact of the BDS applications on the development of China and the world at large, and clarifying the vision of extending BeiDou services worldwide for the betterment of humanity. With President Xi's backing, I am very optimistic about the success of the summit and the BDS applications. We will work hard to deliver on his initiative.

Looking up at the starry sky, the Big Dipper shines bright. Since its official launch in 1994, from BDS 1, to BDS 2 and BDS 3, China's BDS has gone through a three-step evolution, marking a historical progression from inception to optimization and from regional to global coverage. Since its official operation, the BDS-3 has demonstrated stable performance, high-caliber services, and extensive coverage, showcasing a thriving scenario where the industry is flourishing, covered regions are diverse, and various sectors are in vigorous competition. With the deepening of the new scientific and technological revolution and industrial transformation, BeiDou is leading us to an era of the Internet of Everything with unprecedented speed, breadth and depth. As a veteran who has worked in the space industry for 32 years, I had the honor to participate in the organization of the launch of the first BDS-1 satellite in 2000, which successfully sent the BDS-1 satellite A and the BDS-1 satellite B into the scheduled orbit. When I worked at the National Space Administration in 2013, I focused on promoting the integrated development of space technology, space science and space applications, the last of which included the promotion of BDS applications. Over the years, wherever I was, I could never forget those passionate years dedicated to space exploration, like an enduring flame that fuels my unwavering commitment to propel the satellite application industry.

"Move the stool to stand by the window, teach the kid to tell the big Dipper." Hunan Province and BeiDou have a deep historical bond. More than 2,000 years ago, Qu Yuan proposed "Where do the Sun and Moon hang? Where do all the stars lie?" in *Heavenly Questions* alongside the Miluo River. *Prognostics of the Five Stars* of the

Western Han Dynasty unearthed in Mawangdui, Changsha, contained the world's earliest atlas describing the shape of a comet. Chairman Mao further expressed the lofty aspirations of "reaching for the moon in the ninth heaven and catching turtles in the five oceans" and the ambition of "all trying to challenge the heights of the heavens" in his poems. With the deeply-rooted "questioning the heavens" sentiment, generations of Chinese people devoted themselves to the country's aerospace industry, and made positive contributions to the development of the BDS. Among these heroes are a large number of leading figures from Hunan Province. The National University of Defense Technology, as part of the navigation technology expert group, has been involved in the whole process of the construction of the BDS-1, 2 and 3 systems, becoming an important force in the construction of the BeiDou System and a leader in technological innovation.

Remarks at the Launch of the *Global Development Report*①

Foreign Minister Wang Yi

20 June 2022

Colleagues, Ladies and Gentlemen,

It is a great pleasure to join you at today's event. Congratulations to the Center for International Knowledge on Development (CIKD) on launching the first *Global Development Report*.

Development is an eternal theme of humanity and a major yardstick of the progress of times. At the United Nations Sustainable Development Summit in 2015, President Xi Jinping pointed out that to the people of all countries, development bears on their survival and hope and symbolizes their dignity and rights. At the Summit, leaders around the world unanimously adopted the 2030 Agenda for Sustainable Development, drawing a blueprint of actions for global development and charting a new course for international development cooperation.

Seven years have passed fairly quickly, and we are approaching the halftime in the implementation of the 2030 Agenda. Thanks to our joint efforts, the cause of global development has made encouraging progress. But it also faces unprecedented challenges brought about by the changes unseen in a century. The global pandemic, regional

① Ministry of Foreign Affairs of the People's Republic of China. Remarks by State Councilor and Foreign Minister Wang Yi at the Launch of the *Global Development Report* [EB/OL]. (2022-06-20) [2023-08-20]. https://www.mfa.gov.cn/eng/zy/jj/GDI_140002/xw/202206/t20220620_10706286.html.

conflicts, major-power rivalry, backlash against globalization and other complicated factors are intertwining with each other. Global poverty reduction is gravely frustrated. And over 800 million people still live in extreme poverty. What is more worrisome is that development is getting more and more marginalized on the international agenda. Assistance obligations by developed countries are far from being fulfilled. And there is a huge shortfall in global development resources. The outlook of attaining the SDGs as scheduled is not that optimistic.

At this historical juncture, President Xi Jinping, bearing in mind the well-being of the entire humanity, proposed the Global Development Initiative (GDI) at the United Nations. Aiming at building a global community of development, the GDI gives priority to development, puts people at the center, and seeks to expedite the implementation of the 2030 Agenda and to promote stronger, greener and healthier global development. The Initiative, upon its introduction, has been warmly received and strongly supported by more than 100 countries.

The GDI is a rallying call to galvanize greater attention on development and bring it back to the center of the international agenda. It offers a "fast track" to promote development, as well as an effective platform for all parties to coordinate development policies and deepen practical cooperation. It provides strong boost to the implementation of the 2030 Agenda, as well as an impetus to pool resources, address difficulties holding back development and promote greater coordination and efficiency. This week, President Xi Jinping will chair the High-level Dialogue on Global Development. He will be joined by leaders of emerging markets and developing countries in a discussion on global development to reinvigorate international cooperation on development.

Ladies and Gentlemen,

China places high importance on common development. We are among the first to adopt a national plan on the implementation of the 2030 Agenda. We have released multiple progress reports, sharing unreservedly with the world China's philosophies, initiatives and wisdom.

The *Global Development Report* issued by the CIKD today reviews the progress on the implementation of the 2030 Agenda and existing challenges. It underscores that in the face of the major changes and the pandemic both unseen in a century, the international community should be more focused on development cooperation and on building global partnerships, embrace the digital and green transformation, and strive for common, sustainable development for humanity.

Drawing on the useful experience in China and other countries, the Report also lays out policy recommendations in eight aspects for the implementation of the 2030 Agenda. This is an important step to deliver on the GDI, which will provide useful reference for countries to pursue development as well as intellectual support for global development.

One can only travel far by keeping in mind the journey of the past. As the largest developing country, China is a consistent advocate for global development cooperation. We will stay committed to true multilateralism and an open and inclusive spirit of partnership, actively share development expertise and experience, and work with all parties to implement the GDI, step up efforts to advance the 2030 Agenda, and build a global community of development.

Thank you.

附录 航空航天词汇

北斗全球卫星导航系统	the BeiDou Navigation Satellite System
卫星通信广播	satellite communications and broadcasting
探月工程“三步走”	the three-step lunar exploration program (“orbit，land，and return”)
地月系	the earth-moon system
“天问一号”火星探测器	Tianwen-1 Mars probe
空间科学、空间技术、空间应用	space science，technology and applications
国家发展大局	China’s growth as a whole
航天强国	a Strong Space Presence
创新引领、协同高效、和平发展、合作共享	innovation-driven，coordinated，efficient， and peaceful progress based on cooperation
航天领域国家战略科技力量	state strategic scientific and technological strength in the space industry
航天重大科技工程	major space programs
外空武器化、战场化	turn outer space into a weapon or battlefield
外空军备竞赛	an arms race in outer space
航天运输系统	space transport system
发射任务	launch missions
长征系列运载火箭	Long March carrier rocket series
“长征五号”“长征五号乙”运载火箭	Long March-5 and Long March-5B carrier rockets
“长征八号”	Long March-8
“长征七号甲”	Long March-7A
“长征十一号”	Long March-11 carrier rocket
“捷龙一号”	Smart Dragon-1
“快舟一号甲”	Kuaizhou-1A
“双曲线一号”	Hyperbola-1
“谷神星一号”	CERES-1
可重复使用运载器	reusable launch vehicles
飞行演示验证试验	demonstration flight tests
运载火箭型谱	launch vehicle family

续 表

载人运载火箭	manned carrier rockets
大推力固体运载火箭	high-thrust solid-fuel carrier rockets
重型运载火箭	heavy-lift launch vehicles
重复使用航天运输系统	reusable space transport systems
航班化发射需求	regular launches
火箭发动机	rocket engines
组合动力	combined cycle propulsion
上面级	upper stage
空间基础设施	Space Infrastructure
卫星遥感系统	satellite remote-sensing system
高分辨率对地观测系统天基部分	space-based section of the China High-resolution Earth Observation System
对地观测	earth observation
高空间分辨率	high-spatial-resolution
高时间分辨率	high-temporal-resolution
高光谱分辨率	high-spectrum-resolution
陆地观测业务	land observation services
"资源三号"03 星	Ziyuan-3 03 earth resources satellite
"环境减灾二号"A/B 星	Huanjing Jianzai-2A/2B satellites for environmental disaster management
高分多模综合成像卫星	high-resolution multi-mode imaging satellite
高光谱观测卫星	hyper-spectral observation satellite
商业遥感卫星	commercial remote-sensing satellite
海洋观测	ocean observation
多要素	multiple indexes
"海洋一号"C/D 星	Haiyang-1C/1D satellites
"海洋二号"B/C/D 星	Haiyang-2B/2C/2D satellites
新一代静止轨道气象卫星"风云四号"A/B 星	new-generation Fengyun-4A/4B meteorological satellites in the geostationary orbit
全天候	all-weather
"风云三号"D/E 星	Fengyun-3D/3E satellites
"风云二号"H 星	Fengyun-2H satellite
"一带一路"	Belt and Road Initiative
遥感卫星地面系统	ground system of remote-sensing satellites

续表

卫星遥感数据全球接收、快速处理与业务化服务能力	remote-sensing satellite data receiving and quick processing services across the world
卫星通信广播系统	satellite communications and broadcasting system
固定通信广播卫星系统	fixed communications and broadcasting satellite network
广播电视业务	broadcasting and television services
"亚太"6D卫星	APSTAR-6D satellites
通信容量	capacity
"高通量"	high-capacity service
移动通信广播卫星系统	mobile communications and broadcasting satellite network
"天通一号"02/03星	Tiantong-1 02/03 satellites
"天通一号"01星	Tiantong-1 01 satellite
手持终端用户	hand-held terminal users
中继卫星系统	relay satellite system
"天链二号"01星	Tianlian-2 01 satellites
卫星通信广播地面系统	satellite communications and broadcasting ground system
卫星通信广播、互联网、物联网及信息服务	satellite communications and broadcasting, Internet, Internet of Things, and information service
卫星导航系统	satellite navigation system
遥感、通信、导航卫星融合技术	remote-sensing, communications, navigation, and positioning satellite technologies
泛在通联	extensive connection
精准时空	precise timing and positioning
全维感知	all dimension sensoring
静止轨道微波探测	satellites for geostationary microwave monitoring
新一代海洋水色、陆地生态系统碳监测	new-type ocean color observation, carbon monitoring of the territorial ecosystem
大气环境监测	atmospheric environmental monitoring
双天线X波段干涉合成孔径雷达	dual-antenna X-band interferometric synthetic aperture radar (InSAR)
高低轨协同的卫星通信系统	a satellite communications network with high and low orbit coordination
新型通信卫星	new communications satellites
第二代数据中继卫星系统	a second-generation data relay satellite system
低轨增强	low-orbit augmentation

续 表

通信融合	navigation-communications integration
国家综合定位导航授时(PNT)体系	national positioning, navigation and timing (PNT) system
卫星遥感、通信、导航地面系统	ground systems for remote-sensing, communications and navigation satellites
载人航天	Manned Spaceflight
“天舟一号”货运飞船	Tianzhou-1 cargo spacecraft
“天宫二号”空间实验室	earth-orbiting Tiangong-2 space laboratory
货物运输	cargo transport
推进剂在轨补加	in-orbit propellant replenishment
“天和”核心舱	Tianhe core module
“天舟二号”“天舟三号”货运飞船	Tianzhou-2 and Tianzhou-3 cargo spacecraft
“神舟十二号”“神舟十三号”载人飞船	Shenzhou-12 and Shenzhou-13 manned spacecraft
出舱活动、舱外操作	extravehicular activities
在轨维护	in-orbit maintenance
“问天”实验舱	Wentian experimental modules
“梦天”实验舱	Mengtian experimental modules
“巡天”空间望远镜	Xuntian space telescope
“神舟”载人飞船	Shenzhou manned spacecraft
“天舟”货运飞船	Tianzhou cargo spacecraft
地月空间	cislunar space
深空探测	deep space exploration
“鹊桥”卫星	Queqiao satellite
软着陆	soft landing
“嫦娥五号”探测器	Chang'e-5 lunar probe
探月工程“绕、落、回”三步走	three-step lunar exploration program of orbiting, landing and return
行星探测工程	planetary exploration
“天问一号”火星探测器	Tianwen-1 Mars probe
“祝融号”火星车	Zhurong Mars rover
月球极区高精度着陆	a precise landing in the moon's polar regions
阴影坑飞跃探测	a hopping detection in lunar shadowed area
国际月球科研站	an international research station on the moon
小行星探测器	asteroid probes
近地小行星采样	near-earth asteroids

续表

主带彗星	main-belt comets
木星系	Jupiter system
太阳系边际探测	boundary exploration of the solar system
发射场	space launch sites
测控	Telemetry，Tracking and Command（TT&C）
适应性改造	adaptive improvements
液体火箭商业发射工位	commercial launch of liquid fuel rockets
沿海内陆相结合、高低纬度相结合、各种射向范围相结合的航天发射格局	a launch site network covering both coastal and inland areas，high and low altitudes，and various trajectories
空间站舱段	space station modules
深空探测器	deep space probes
天地一体化航天测控体系	an integrated space-ground TT&C network
商业卫星测控站网	TT&C station networks for commercial satellites
航天产品统一技术体制	unified technical standard-setting for its space products
商业发射工位和商业航天发射场	commercial launch pads and launch sites
天地基测控资源融合运用能力	capacity to utilize and integrate space- and ground-based TT&C resources
国家太空系统	national space system
深空测控通信网	the deep-space TT&C communications network
新技术试验卫星	new technological test satellites
新一代通信卫星公用平台	common platforms of new-generation communications satellites
甚高通量通信载荷	very high throughput satellites' telecommunication payload
Ka 频段宽带通信	Ka-band communications
星地高速激光通信	satellite-ground high-speed laser communications
新型电推进	new electric propulsion
航天器智能自主管理	smart self-management of spacecraft
空间扩展飞行器	space mission extension vehicle
新型空间动力	innovative space propulsion
航天器在轨服务与维护	in-orbit service and maintenance of spacecraft
空间碎片清除	space debris cleaning
空间环境治理	space environment governance
空间碎片监测网络	space debris monitoring system
碰撞预警	collision warning

续 表

空间事件感知应对能力	space event perception and response
在轨航天器	in-orbit spacecraft
国际空间碎片减缓准则	Space Debris Mitigation Guidelines
外空活动长期可持续准则	Guidelines for the Long-term Sustainability of Outer Space Activities
运载火箭末级钝化	upper stage passivation to all its carrier rockets
主动离轨	active deorbit
任务末期	end of life
近地小天体	near-earth objects
空间碎片减缓	mitigating space debris
空间天气保障业务体系	space climate service system
编目数据库	cataloguing database
近地小天体防御系统	near-earth object defense system
天地结合的空间天气监测系统	integrated space-ground space climate monitoring system
灾害性空间天气事件	catastrophic space climate events
空间应用产业	space application industry
卫星公益服务	public services with satellites
资源环境与生态保护	protection of resources and the eco-environment
防灾减灾	disaster prevention and mitigation
气象预报与气候变化应对	weather forecasting and climate change response
区域协调发展	coordinated regional development
脱贫攻坚	poverty eradication
应急监测	emergency monitoring
重特大自然灾害	major and catastrophic natural disasters
数据回传	returned data
手机通信基站	mobile phone base stations
定位和短报文通信服务	positioning and short message communication services
海洋渔船	seagoing fishing vessels
精准位置服务	precise positioning services
物资运输	freighting of supplies
人员流动管理	tracking of individual movement
平安中国、健康中国、美丽中国、数字中国	a safe，healthy，beautiful and digital China
陆地、海洋、气象遥感卫星数据	remote-sensing satellite data on land，ocean and meteorology
地面通信	ground communications

续表

卫星遥感高精地图	high-accuracy maps using remote-sensing data
全维影像	full dimensional images
灾害损失评估	assessment of disaster losses
保险理赔	insurance claims
不动产登记	registration of real estate
4K超高清频道上星	4K Ultra HD television channels
远洋船舶	ocean vessels
民航客机	passenger aircraft
“天通一号”卫星移动通信系统	Tiantong-1，a satellite mobile communication system
北斗兼容型芯片模块	chips compatible with the BeiDou system
无人驾驶	unmanned driving
生物制药	biomedicine
宇宙起源和演化	origin and evolution of the universe
空间天文	space astronomy
“悟空”号暗物质粒子探测卫星	The Dark Matter Particle Explorer (DAMPE) Satellite
精细结构	precise measurements
能谱	energy spectrums
宇宙射线	cosmic ray
电子	electrons
质子	protons
氦核	GCR helium
“慧眼”硬X射线调制望远镜卫星	Huiyan (Insight) Hard X-ray Modulation Telescope
宇宙磁场	magnetic field
黑洞双星爆发过程	black hole binary explosion process
“羲和号”太阳探测科学技术试验卫星	Xihe observation satellite
太阳光谱图像	multiple solar spectroscopic images
波长	wavelengths
Hα波段	Hα waveband
月球与行星科学	lunar and planetary science
月球探测工程	lunar exploration program
月球地质	moon's geology
浅层结构	subsurface structure
月球岩浆活动	lunar magmatic activity
矿物学特征	mineralogical features

续 表

火星地质演化	geological evolution of Mars
空间地球科学	space earth sciences
“张衡一号”电磁监测试验卫星	Zhangheng-1, also known as the China Seismo-Electromagnetic Satellite
全球地磁场	global geomagnetic field
电离层	ionosphere parameters
原位数据	in situ data
全球高精度二氧化碳分布图	a high-precision global carbon flux map
全球二氧化碳监测科学实验卫星	Chinese Global Carbon Dioxide Monitoring Scientific Experimental Satellite
空间基础物理	space physics
“墨子”号量子科学实验卫星	Mozi, the world's first quantum communication satellite
千公里级星地量子纠缠分发和隐形传态实验	experiments on satellite-based quantum teleportation and entanglement distribution over thousand kilometers
引力诱导量子纠缠退相干实验	experiments on gravitational induced decoherence of quantum entanglement
基于纠缠的无中继千公里量子密钥分发实验	experiments on entanglement-based secure quantum cryptography over thousand kilometers with no trusted relay
“太极一号”和“天琴一号”卫星	Taiji-1 and Tianqin-1 satellites
空间引力波探测	space gravitational wave detection
空间引力波探测卫星	satellite for space gravitational wave detection
爱因斯坦探针	Einstein Probe
先进天基太阳天文台	advanced space-based solar observatory
太阳风-磁层相互作用全景成像卫星	panoramic imaging satellite for solar wind and magnetosphere interaction
高精度地磁场测量卫星	high precision magnetic field measurement satellite
极端宇宙	extreme universe
时空涟漪	ripples in time and space
日地全景	the panoramic view of the sun and the earth
宜居行星	habitable planets
日球物理	heliospheric physics
“实践十号”卫星	Shijian-10 satellite
哺乳动物细胞胚胎发育	mammalian embryonic development
世界首台空间冷原子钟	the world's first space cold atom clock

续表

在轨验证	in-orbit verification
微重力颗粒分聚	particle segregation in microgravity
煤粉燃烧	pulverized coal combustion
"嫦娥"系列探测器	Chang'e lunar probe series
上中下游协同	the upper, middle and lower industrial chains are coordinated
大中小企业融通	large, small and medium-sized enterprises advance in an integrated way
产学研用深度融合	bringing together enterprises, universities, research institutes and end-users
航天科技跨越发展	the leapfrog development of space science and technology
航天技术二次开发	the secondary development of space technologies
工业基础能力	basic industrial capabilities
系统集成商	system integrators
专业承包商	specialized contractors
产业链供应链	industrial and supply chains
发射运营	launch operations
智能化脉动生产线、智能车间、智慧院所	intelligent production lines, workshops and institutes
卫星应用产业	satellite application industry
共享共用渠道	the channel for sharing and utilization
卫星应用服务体系	satellite application services
产品标准化	unified standards
服务个性化	customized choices
增值产品	value-added products
"航天+"产业生态	a "space plus" industrial ecosystem
采购商业航天产品	procurement of space products
重大科研设施设备	major scientific research facilities and equipment
航天活动市场准入负面清单	a negative list for market access to space activities
航天技术转移转化	the transfer and transformation of space technologies
法治航天建设	law-based governance
卫星导航条例	regulations on satellite navigation
航天法	a national space law
卫星导航活动管理	management of satellite navigation activities
空间物体登记管理办法	measures for the registration of space objects

续表

民用航天发射许可管理	the licensing of civil space launches
卫星频率轨道资源管理条例	regulations on the management of satellite frequency and orbit resources
申报、协调和登记	declaration, coordination and registration
国际空间法	international space law
国际电联规则	International Telecommunication Union standards
外空国际规则	international rules regarding outer space
世界重要人才中心和创新高地	a world center for talent and innovation
人才发展沃土	favorable conditions for the development of professionals
人才培养机制	personnel training mechanism
战略科学家、科技领军人才、青年科技人才和创新团队建设	strategic scientists, leading and young scientists, and teams with strong innovation capacity
卓越工程师	outstanding engineers
高素质技术技能人才和大国工匠	top technicians championing fine craftsmanship
具有国际视野和社会责任感的优秀企业家	visionary entrepreneurs with a sense of social responsibility
航天后备人才队伍	a reserve force of aerospace personnel
航天科普教育和文化建设	space education and culture
"中国航天日"系列活动	Space Day
"世界空间周"	World Space Week
"全国科技活动周"	National Science and Technology Week
"天宫课堂"	Tiangong Classroom
"两弹一星"精神和载人航天精神、探月精神、新时代北斗精神	spirit embodied in the development of the atomic and hydrogen bombs, missiles, man-made satellites, manned spaceflight, lunar probes and the BeiDou Navigation Satellite System
体验园	experience parks
构建人类命运共同体	build a global community of shared future
平等互利、和平利用、包容发展	equality, mutual benefit, peaceful utilization, and inclusive development
《关于各国探索和利用包括月球和其他天体在内外层空间活动的原则条约》	*Treaty on Principles Governing the Activities of States in the Exploration and Use of Outer Space*
外空活动长期可持续发展	greater sustainability of space activities
亚太空间合作组织	Asia-Pacific Space Cooperation Organization (APSCO)

续表

上海合作组织框架	the framework of the Shanghai Cooperation Organization
二十国集团合作机制	Group 20 mechanisms
空间合作协定或谅解备忘录	space cooperation agreements or memoranda
外空活动长期可持续性	long-term sustainability of outer space activities
空间资源开发利用	development and utilization of space resources
防止外空军备竞赛	prevention of arms race in outer space
联合国空间2030议程	Space 2030 Agenda of the UN
联合国灾害管理与应急反应天基信息平台北京办公室	the Beijing office of the United Nations Platform for Space-based Information for Disaster Management and Emergency Response
联合国全球卫星导航系统国际委员会	the International Committee on Global Navigation Satellite Systems
空间任务规划咨询组	Space Missions Planning Advisory Group
国际小行星预警网	International Asteroid Warning Network
中俄航天合作分委会空间碎片工作组	Space Debris Work Group of China-Russia Space Cooperation Sub-committee
中美空间碎片与空间飞行安全专家研讨会	Sino-US Expert Workshop on Space Debris and Space Flight Safety
国际电信联盟	International Telecommunication Union
地球观测组织	Group on Earth Observations
机构间空间碎片协调委员会	Inter-Agency Space Debris Coordination Committee
国际空间数据系统咨询委员会	Consultative Committee for Space Data Systems
国际空间探索协调组	International Space Exploration Coordination Group
机构间互操作顾问委员会	Interagency Operations Advisory Group
伽马暴偏振探测研究	gamma-ray burst polarization monitoring research
人体医学研究	human body medical research
微重力环境	micro-gravitational environment
“神舟十一号”载人飞行任务	Shenzhou-11 manned spaceflight mission
洞穴训练	CAVES training
海上救生训练	maritime rescue drills
欧洲航天员中心	European Astronaut Centre
美国全球定位系统	the United States' Global Positioning System
俄罗斯格罗纳斯系统	Russia's GLONASS system

续 表

欧洲伽利略系统	Europe's Galileo system
兼容与互操作、监测评估、联合应用	compatibility, interoperability, monitoring and assessment, and joint application
国际电工委员会	International Electrotechnical Commission
民航、海事、国际搜救、移动通信	civil aviation, maritime affairs, international search and rescue, and mobile communications
北斗合作论坛机制	BeiDou cooperation forum mechanisms
阿盟	the League of Arab States
非盟	the African Union
海外北斗中心	overseas BeiDou center
国际月球科研站计划	the international lunar research station project
中俄月球与深空探测联合数据中心	Sino-Russian Joint Data Center for Lunar and Deep-space Exploration
"嫦娥七号"月球极区探测任务	Chang'e-7's lunar polar exploration mission
俄罗斯月球-资源-1轨道器任务联合实施	Russia's LUNA-Resource-1 orbiter mission
月球探测工程"嫦娥四号"任务	Chang'e-4 lunar exploration mission
欧洲空间局	the European Space Agency
载荷	payloads
国际载荷搭载合作	international onboard payload cooperation
月球探测工程"嫦娥六号"任务	Chang'e-6 lunar exploration mission
火星探测器轨道数据交换机制	a Mars probe orbit data exchange mechanism
小行星探测任务	asteroid exploration mission
月球与深空探测	lunar and deep-space exploration
中法海洋卫星	China-France Oceanography Satellite
中巴(西)地球资源04A星	China-Brazil Earth Resources Satellite 04A
埃塞俄比亚遥感微小卫星	Ethiopian Remote-Sensing Satellite
大学生小卫星	Student Small Satellites (SSS)
埃及二号遥感卫星	MisrSat-2 Remote-sensing Satellite
在轨交付	the in-orbit delivery
巴基斯坦遥感卫星一号	Pakistan Remote-sensing Satellite (PRSS-1)
委内瑞拉遥感卫星二号	Venezuelan Remote-sensing Satellite (VRSS-2)
苏丹一号遥感卫星	Sudan Remote-Sensing Satellite (SRSS-1)
阿尔及利亚一号通信卫星	Algerian Communications Satellite (Alcomsat-1)

续表

卫星搭载发射服务	satellite carrying or launching services
"一带一路"空间信息走廊	the Belt and Road Initiative Space Information Corridor
风云气象卫星国际用户防灾减灾应急保障机制	an emergency support mechanism for disaster prevention and mitigation for international users of the Fengyun meteorological satellites
气象卫星	meteorological satellites
金砖国家遥感卫星星座合作协定	BRICS Remote-sensing Satellite Constellation
对地观测卫星数	earth observation satellite
中国-东盟卫星信息(海上)服务平台	China-ASEAN Satellite Information Offshore Service Platform
遥感卫星数据共享服务平台	Remote-sensing Satellite Data-sharing Service Platform
澜沧江-湄公河空间信息交流中心	Lancang-Mekong Space Information Exchange Center
卫星数据接收站	satellite data receiving stations
空间与重大灾害国际宪章机制	the mechanism of the International Charter on Space and Major Disasters
值班卫星	on-duty satellites
星座	constellations
2018 年阿富汗大旱	the severe drought in Afghanistan in 2018
2018 年老挝溃坝事故	the dam collapse in Laos in 2018
2019 年莫桑比克台风	the cyclone that struck Mozambique in 2019
《中国面向全球的综合地球观测系统十年执行计划(2016—2025 年)》	*GEO Strategic Plan 2016—2025: Implementing GEOSS*
轮值主席国	rotating chair
国际空间气候观测(SCO)平台机制	the international Space Climate Observatory platform
"悟空"号、"墨子"号、"实践十号"和"慧眼"	Wukong, Mozi, Shijian-10, and Insight
中意电磁监测试验卫星	China-Italy Electromagnetic Monitoring Experiment Satellite
中欧太阳风-磁层相互作用全景成像卫星	Sino-European Panoramic Imaging Satellite for Solar Wind and Magnetosphere Interaction
中法天文卫星	Sino-French Astronomic Satellite (Space-based multiband astronomical Variable Objects Monitor)
中意电磁监测卫星 02 星	China-Italy Electromagnetic Monitoring Experiment Satellite 02

续 表

增强型X射线时变与偏振空间天文台	enhanced X-ray timing and polarimetry observatory
中国-巴西空间天气联合实验室	China-Brazil Joint Laboratory for Space Weather
南美地区空间环境综合监测研究平台	the space environment monitoring and research platform for South America
国际宇航联合会	International Astronautical Federation
国际空间研究委员会	International Committee on Space Research
国际宇航科学院	International Academy of Astronautics
国际空间法学会	International Institute of Space Law
全球空间探索大会	2017 Global Space Exploration Conference
全球卫星导航系统国际委员会第十三届大会	the 13th Meeting of the International Committee on Global Navigation Satellite Systems
中国/联合国航天助力可持续发展大会	United Nations/China Forum on Space Solutions: Realizing the Sustainable Development Goals
文昌国际航空航天论坛	Wenchang International Aviation and Aerospace Forum
珠海论坛	Zhuhai Forum
北斗规模应用国际峰会	International Summit on BDS Applications
风云气象卫星国际用户大会	Fengyun Satellite User Conference
联合国空间科技教育亚太区域中心(中国)	the Regional Centre for Space Science and Technology Education in Asia and the Pacific (China) (Affiliated to the United Nations)
“一带一路”航天创新联盟	“Belt and Road” Aerospace Innovation Alliance
中俄工科大学联盟	Association of Sino-Russian Technical Universities
中欧空间科学研讨会	China-Europe Space Science Bilateral Meeting
中欧空间科技合作对话	China-EU-ESA Dialogue on Space Technology Cooperation
中欧“龙计划”	Dragon Programme—a joint undertaking between ESA and the Ministry of Science and Technology of China
行星保护	planet protection
太空交通管理	space traffic management
太空危机管控和综合治理效能	efficiency of space crisis management and comprehensive governance
空间科学观测台	space science observatory
空间天文观测	space-based astronomical observations

续表

微重力环境下	under conditions of microgravity
星基增强系统	satellite-based augmentation systems
行星际探测	interplanetary exploration
埃及二号卫星	MisrSat-2
中巴(西)资源系列后续卫星	China-Brazil Earth Resources Satellites program
地面站网	ground station networks
卫星整星	whole satellites
分系统	sub-systems
零部件	spare parts
电子元器件	electronic components
巴基斯坦航天中心	Pakistan Space Center
埃及航天城	Egypt's Space City
世界气象卫星组织	World Meteorological Organization
数据共享服务平台	data-sharing service platform
空间气候观测平台	Space Climate Observatory
地外样品	extraterrestrial samples
行星起源演化	planetary origin and evolution
暗物质粒子	dark matter particles
太阳爆发活动	solar burst activities
空间引力波	spatial gravitational wave